Surface Engineering for Enhanced Performance against Wear

Manish Roy
Editor

Surface Engineering for Enhanced Performance against Wear

Springer

Editor
Manish Roy
Defence Metallurgical Research Laboratory
DRM Laboratory
Hyderabad, India

ISBN 978-3-7091-0100-1 ISBN 978-3-7091-0101-8 (eBook)
DOI 10.1007/978-3-7091-0101-8
Springer Wien Heidelberg New York Dordrecht London

Library of Congress Control Number: 2013935951

Printed on acid-free paper

Springer is part of Springer Science+Business Media (www.springer.com)

My wife Papri, my daughter Mayuri and my son Moinak, a small boy with big heart, are the stimuli for my enhanced protection against wear.

Preface

One of the most effective ways of enhancing the performance of industrial components is to engineer the surface so as to make them suitable for functions that are different from the primary function of the bulk substrate. Although surface engineering has been practised from the ancient times, the science and technology of surface modification technology has been explored only in last several decades. Major progress in this direction has been achieved only when a significant gain in performance along with financial benefit is realised by practising surface modification techniques. Numerous choices for modification of the surfaces are available today; one needs to consider the thickness and adhesion of the modified layer, its interaction with environment, mechanical process and substrate. Selection of coating will be governed by component specification, functional and non-functional requirements, coating and process characteristics and economic implications. Thus the aim of this book is to guide one to select the most suitable process and the layer for a given application. This attempt is by no means the last word rather it is just initiation.

This book consists of eight chapters written by experts in various fields in surface engineering. The book begins with tribology of thermal-sprayed coatings. In this chapter, various thermal-sprayed coatings, their features, advantages, disadvantages and tribological performances are discussed. Typical applications of thermal-sprayed coatings with few examples are also highlighted. This is followed by a chapter on nanocomposite films for wear resistance applications. This chapter contains deformation behaviour of nanostructured films in addition to their microstructural features, mechanical properties and tribological performances. Chapter 3 deals with diamond films and their tribological performances. Various deposition techniques and microstructural characterisation of diamond films have been provided in this chapter along with the influence of various parameters on tribological and nanotribological performances of these films. Diffusion-treated surfaces and their tribological properties are the subject matters for Chap. 4. Chapter 5 concerns with the details of hardfacing for wear erosion and abrasion. Various hardfacing processes, mechanical properties, sliding wear, abrasive wear and erosive wear of various hardfaced surfaces are described in this chapter.

Electroplating for tribological applications is outlined in Chap. 6. After introducing electrolytic and electroless plating technologies, a description of sliding wear performances of plated surfaces is made. Laser surface modification is the subject matter of Chap. 7 in which tribology of laser-modified surface has been discussed. Finally, surface engineering of bio-tribological application is made the concluding chapter of this book. Tribology of various bioactive, bioinert and biotolerant coatings for human body applications forms the main focus of this chapter. Each of the above chapters contains a summary and perspective for future study.

Although this book can be utilised both as a text book and as a reference book, one of the major objectives of this compilation has been to highlight the many recent developments in the field of tribology of coatings. The book targets undergraduate and post-graduate students, practising engineers, researchers in various academic institutions, research establishments and production units. It is expected that this book will be a valuable guide and a useful reference for those who are working in the area of surface engineering for enhanced wear performances.

While compiling this book, I am indebted to several individuals. Many thanks to all my co-authors who have spent considerable time for preparation of various chapters. I am also thankful to D.K. Das, Scientist, Defence Metallurgical Research Laboratory, India; Rahul Mitra, Professor, Indian Institute of Technology, Kharagpur; Andreas Pauschitz, CEO, Austrian Center of Competence for Tribology; K, Balasubramanian, Director, NonFerrous Materials Technology Development Center, Hyderabad and Subir Kr. Roy, Scientist, Defence Metallurgical Research Laboratory, India, who have helped me in terms of constant encouragement, stimulating discussion, providing valuable information. I am also thankful to M. Ramakrishna, Senior Technical Assistant of Defence Metallurgical Research Laboratory, for his help in many technical plots and sketches.

Hyderabad, India
December, 2012

Manish Roy, Ph.D

Contents

Chapter 1
Tribology of Thermal-Sprayed Coatings

R.J.K. Wood and Manish Roy

1.1 Introduction

Thermal spraying refers to a group of processes that apply consumable powders or wires in the form of finely divided molten and semi-molten droplets to produce coatings. Thermal spraying dates back to 1911 when Dr. Schoop was successful in atomising molten metal by high pressure gas and propelled them on to a surface. In 1912 he produced a device to spray metal wires and this process is known as flame spraying. Subsequently he introduced new technique for thermal spraying by using electricity to melt feedstock materials. This method in present times is referred as arc spraying. The idea of using powder for the flame spraying process was developed for the first time by F. Schori in 1930.

The above-mentioned processes can use only low melting metallic powders. Slowly demand for powders with high melting temperature and oxidation resistance aroused. The demand for spraying cermet or ceramic powder saw development of detonation spraying in 1955 and atmospheric plasma spraying in 1960. This was followed by development of vacuum plasma spraying (VPS) and low-pressure plasma spraying (LPPS) in late 1970 and 1980. The major development of thermal spraying occurred in 1980 when a novel technique for spraying powder employing high oxygen, known as high velocity oxy-fuel (HVOF) technique was introduced. The year 1990 experienced the development of most important technology cold spraying. This process has the ability to produce unique coating that is not possible by other coating method.

R.J.K. Wood (✉)
National Center for Advanced Tribology, University of Southampton, Southampton, UK
e-mail: rjw3@soton.ac.uk

M. Roy
Defence Metallurgical Research Laboratory, Kanchanbagh, Hyderabad 500058, Andhra Pradesh, India

M. Roy (ed.), *Surface Engineering for Enhanced Performance against Wear*,
DOI 10.1007/978-3-7091-0101-8_1, © Springer-Verlag Wien 2013

Important advantages of thermal spraying process are ease of processing, ease of selection of substrate and no heating of the substrate. Several extensive reviews on this process can be found in [1–5]. This chapter tries to give recent development and the data base of thermal-sprayed coatings for tribological applications in various industrial sectors.

1.2 Process Classifications and Fundamentals

Thermal spray processes can be classified into three different categories. The first group uses combustion as heat sources. The second variety uses electrical energy either in the form of plasma or as an arc. The third variety is cold spraying. This is a process by itself. Low velocity combustion also known as flame spraying (FS) and high velocity combustion are two different combustion-spraying methods. The maximum temperature attained in both cases are, in general, comparable as similar fuel gases such as propylene, propane, acetylene or hydrogen or liquid fuel such as kerosene, etc., are commonly used. In all cases, oxygen is used as oxidant. In case of high velocity process, it is called high velocity oxy-fuel (HVOF) as against flame spraying for low velocity combustion; sometimes atmospheric air is used as oxidant and the process is termed as high velocity air fuel (HVAF). In case of low velocity technique, the dwell time of particles in the gas is more than the high velocity technique and consequently oxidation and degradation of the particles are more. This results in formation of coating with higher porosity and lower or moderate (around 50–75 MPa) bond strength. However, flame spraying process permits higher deposition rate as course powder or wire is used. This process is portable, less noisy and economical.

In case of high velocity technique, the dwell time of the powders is much less of the order of microsecond resulting in less oxidation and degradation of coating powders. This permits use of fine powders, which can be propelled to the substrate with high velocity. Consequently, a coating with minimum porosities less than 0.5 % and good bond strength is possible. Further deposition with high velocity permits spraying powders with higher melting point and thus widening the range of application to include high temperature coatings, thermal- and shock-resistant coatings. Deposition of coating at high velocity using wire permits continuous deposition. However, there are far less varieties of wire than powders. Most of these HVOF processes are continuous combustion process. In contrast, detonation gun process which uses spark discharge to propel powder is a repeated cyclic process. However, D-gun process is extremely noisy.

HVOF processes can broadly be divided under two different heads depending on their burners. These are throat combustion burners and chamber combustion burners. In throat combustion process, powder is introduced axially. Detonation gun, as shown schematically in Fig. 1.1, can be considered as an example of throat combustion process. Many system based on throat combustion process suffered from the disadvantage of short barrel, lower velocity and reduced particle heating.

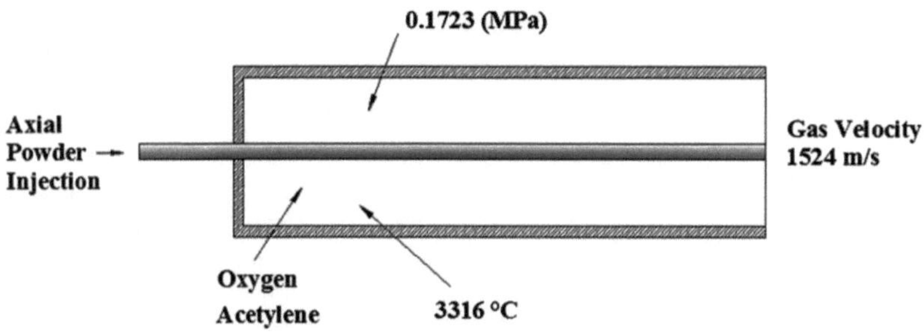

Fig. 1.1 Schematic representation of detonation-sprayed coatings

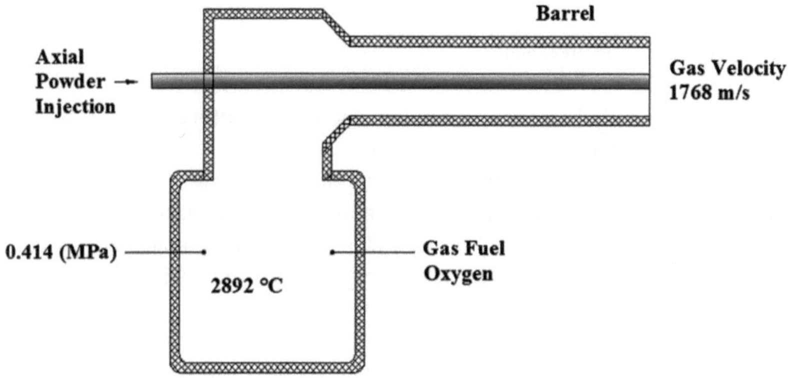

Fig. 1.2 Schematic diagram of jet kote process

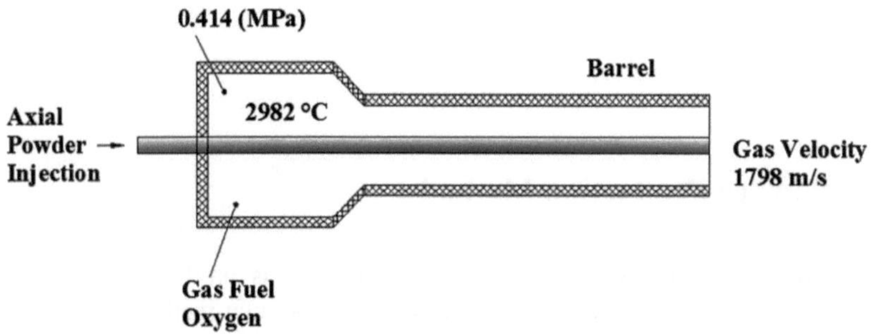

Fig. 1.3 Schematic diagram of top gun or continuous detonation process

In chamber combustion process, powder is fed both radially or axially. Jet Kote shown in Fig. 1.2 pertains to an axially injected water cooled process having combustion chamber at right angle. In contrast, Top gun or continuous detonation gun presented schematically in Fig. 1.3 is an example of throat combustion process with coaxial combustion chamber for axial injection. Finally JP 5000 represents a

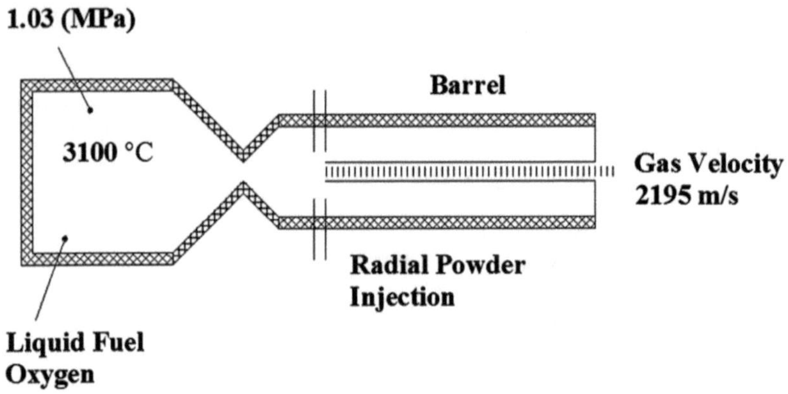

Fig. 1.4 Schematic of JP 5000 gun

double chamber, radial injection chamber combustion process which uses liquid fuel. JP 5000 is shown schematically in Fig. 1.4.

Plasma and electric arc are two different processes of thermal spraying techniques driven by electrical energy. In the wire arc spraying process, high temperature generated in the arc is used to melt feedstock materials. In contrast, in plasma spraying process, electrical energy is utilised to ionise gas medium. Plasma spraying is extremely versatile as it can be used to spray metallic, cermet as well as ceramic powders. A large variety of powder with coarser dimension can be used. This equipment is quite sophisticated, emits ultraviolet rays and needs skilled operator. There are three different types of plasma spraying processes, namely, atmospheric plasma spraying, (APS), low pressure plasma spraying (LPPS) and vacuum plasma spraying (VPS). Choice of the process depends on the powders to be sprayed and their applications.

1.3 Powder Production Methods

Depending on the nature of powder required, various production methods are adopted. In general, atomisation is one of the most popular methods of production of powders of metals and alloys. In atomisation method, molten metal or alloy is poured in a heated funnel connected to a nozzle where it is finely dispersed either by water or by gas. Sometimes dispersion is carried out by centrifugal method which is far more energy efficient than gas or water atomisation methods. Low melting solder powders are produced by spinning disc atomisation techniques which has yield as high as 70 %. Higher melting point powders such as zinc (for batteries), aluminium (for chemicals), etc., are produced by spinning cup atomisation technique with very high yield, even higher than 70 % yield. Highly corrosive powders such as Ti powders are produced by rotating electrode method. This method provides powder with exact standard of cleanliness and smaller range of size distribution than gas atomisation method. Ultrasonic gas atomisation or vibrational ultrasonic atomisation also finds applications when perfectly spherical powder with narrower range of composition is required. Roller atomisation, vibrating electrode atomisation and melt drop atomisation are other atomisation techniques which are still in experimental stages. Powders produced by atomisation technique are spherical. The composition and the phases of the powders depend on the methods of atomisation or dispersion.

Most of the ceramic powders are produced by fusion and crushing methods. In this method the required material is melted in a furnace and then solidified as an ingot. The ingot is subsequently crushed in industrial crushing mills. Powders manufactured in this method are dense, blocky and are of irregular shapes.

Metal powders are also produced by chemical or electrolytic method. These methods allow wide variety of powder production with close control of composition, shape and size of powders. Oxide reduction for powder production is a well-established commercial process. Powders of metals with high melting point such as tungsten and molybdenum can be produced by this method for economic reason. Further, these powders are stabilised against tarnishing.

Production of metal powders from hydrometallurgical processing is based on leaching an ore concentrate followed by precipitation of metal from the leach solution by electrolysis, cementation or by chemical reduction. It is possible to produce a large variety of powders with regard to size, shape, density and specific surface area employing this process with additives or control of particle growth, particle nucleation and particle agglomeration. By using co-precipitation or successive precipitation, it is possible to produce powders of various alloys and composites.

Thermal decomposition is another important chemical method for production of metal powders. Generally this method produces highly pure fine powders. Among other chemical powder production method, precipitation from salt solution, precipitation from gas, hydride decomposition and thermal reactions are worth mentioning.

It is possible to produce a large variety of metal powders by electro deposition. By direct electro deposition, loosely adhering powdery and spongy deposit can be obtained and these deposits can be disintegrated mechanically in fine powders. Dense, smooth and brittle layers of refined metal can be deposited and crushed into fine powders. Metals with high electrolytic polarisation can form highly brittle coherent deposit by brittle cathode process. This deposit can subsequently be granulated into fine powders.

Spray drying is a versatile method, which allows production of any kind of agglomerated powders using an organic binder phase. In this method, a mixture of organic binder, water and the materials to be agglomerated is sprayed in a chamber where hot dry gases are flowing. The water in the mixture evaporates and the organic binder covers the material particles, producing agglomerated powder. Powders produced by this method are porous and sometimes subjected to densification.

1.4 Types of Thermal Spray Coatings

Thermal spray coatings can be of different types. Many coatings are deposited as pure metals such as molybdenum, nickel, copper, tungsten, etc. Pure metals are deposited for both tribological and corrosion resistance applications. Various types of stainless steel, Hastalloys, Monels, etc., constitute alloy coatings. These coatings find applications primarily as corrosion resistance layer. WC–Co and Cr_3C_2–NiCr are most popular variety of cermet powders used in general for wear resistance applications. Finally, a large variety of ceramic powders such as ZrO_2, Al_2O_3, yttria-stabilized zirconia, etc., are employed for high temperature applications. Chemical inertness, excellent sealing properties due to high density and low porosity and outstanding friction characteristics have made polymer a viable protective coating. The variety of polyethylene, polyamide, poly-ether ether ketone, poly methylmethacrylate and other thermoplastic are sprayed successfully and found several engineering applications. Low friction materials dispersed in soft matrix are often termed abradable coatings. These coatings are used for clearance control between two moving parts. (Al–Si)–graphite, Ni–Graphite, (Al–Si) polyester are some examples of abradable coating. Self-fluxing powders constitute another variety of coatings. These are the powders that are designed to remelt on spraying. These powders contain elements like boron and silicon which reacts with oxygen and form a slag over the molten coating producing an oxygen-free metallic coating containing finely dispersed boride and nitride particles. These coatings are excellent for high temperature tribological applications.

1.5 Properties of Thermal Spray Coatings

The properties of thermal spray coatings are governed by chemical composition, porosities and oxide content of the coatings. In thermal spray coatings, many properties are determined by splat morphology, interfacial strength of splat–splat interface and splat–substrate interface.

1.5.1 Microstructural Features

WC–17 % Co with a range of particle size '08–64 μm' has been characterised. Typical SEM images of commercial spherical WC–17 % Co powders, spherical Cr_3C_2–25(Ni20Cr) powder, angular WC–12 % Co powder and spherical WC-12% powder are shown in Fig. 1.5 [6, 7]. All granulated powders are of spherical shapes with presence of comparatively large initial porosities. Presence of surface oxide on the spray granules were detected by SEM analysis. There are many applications where angular powders with different sizes and shapes are used. Angular powders are

Fig. 1.5 SEM images of various powders. (**a**) Spherical WC–17 % Co powders, (**b**) spherical Cr_3C_2–25(Ni20Cr) powder, (**c**) angular WC–12 % Co powder and (**d**) spherical WC–12 % Co powder

easy to produce and if they are not sufficiently fine, resultant coatings have porosities. Very fine powders are difficult to feed at constant rate and sometimes they are responsible for high degree of oxidation or undesirable phase transformation.

The low magnification images of the sectioned surfaces of HVOF-sprayed Cr_3C_2–25(Ni20Cr) coating, detonation-sprayed WC–12 % Co coating and plasma-sprayed WC–12 % Co coating are shown in Fig. 1.6. The coatings exhibit reasonably good interface with the substrate as evident from the figure. The thickness of the coatings is around 200–300 μm. No cracks can be seen on the sectioned surfaces of the coatings. The coating is built by depositing lenticular splat one after the other. This type of stratified structure is common in HVOF spraying [8].

The high magnification SEM image of the detonation-sprayed WC–12 % Co coating on MS is illustrated in Fig. 1.7. In detonation-sprayed coating, there are evidences of insufficient melting. Splats of lenticular shapes are deposited one after the other. Figure 1.7 represents one of the cross-sectional micro-structure of the WC–17 % Co coating as observed in the plasma spray process (Fig. 1.7b) and HVOF process (Fig. 1.7c). The porosities on WC–17 % Co-coated surface as determined by image analysis were observed to be less than 0.5 %. The related percentage of unmelted particle was less than 0.2 % with exception to WC particles which are difficult to melt. WC is one of the hardest carbides with melting point as around 3,043 K. However the hot flattening behaviour of WC is significant in plasma spray process compared to that behaviour in HVOF process. This phenomenon leads to production of a dense and adherent coating by the plasma process. In such a process, the as-sprayed surface roughness (R_a) was about 1.5–2.1 μm.

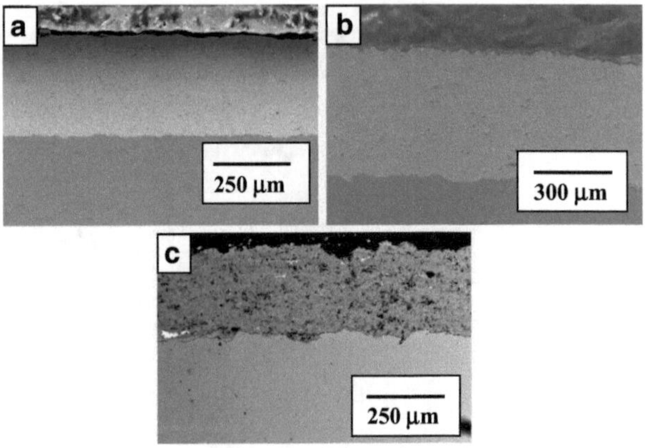

Fig. 1.6 The low magnification images of the sectioned surfaces of (**a**) HVOF-sprayed Cr_3C_2–25 (Ni20Cr) coating, (**b**) detonation-sprayed WC–12 % Co coatings and (**c**) plasma-sprayed WC–12 % Co coating

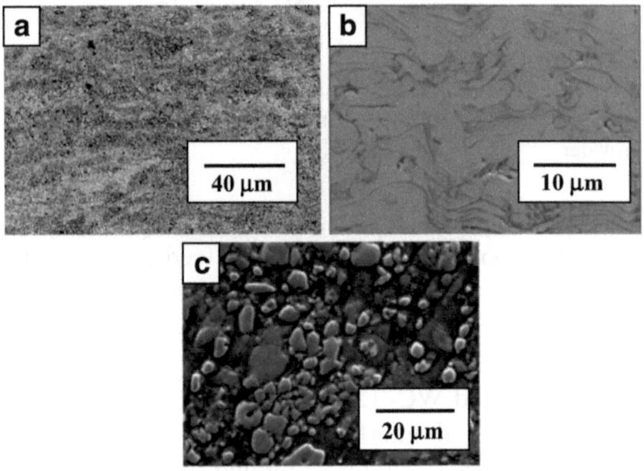

Fig. 1.7 The high magnification back-scattered images of the sectioned portion of (**a**) detonation-sprayed WC-12 % coating, (**b**) plasma-sprayed WC–12 % Co coating and (**c**) SEM image of HVOF-sprayed WC–17 % Co coating

The high magnification back-scattered images of the sectioned portion of detonation-sprayed WC-12 % coating and HVOF-sprayed Cr_3C_2–25(Ni20Cr) coating are given in Fig. 1.8. Pores (black areas) and oxides (circular black areas) are evident in coating. Image analysis of HVOF-sprayed Cr_3C_2–25(Ni20Cr) coating indicates 1.5 porosity and 0.75 % oxides in the coatings respectively. Examination of the coated microstructure also reveals the presence of three different types of zones. The first zone appears dark. The EDS spectrum from the dark area indicates that this area contains primarily Cr and C indicating orthorhombic Cr_3C_2 phase. The second zone is of grey colour. The EDS spectrum of this area reveals that this area contains all the three important elements, Ni, Cr and C. Quantitative analysis indicates this zone contains both Cr_3C_2 and NiCr. The third zone is white and the EDS spectrum pertaining to this zone reveals NiCr as the important phase. The quantitative analysis of these zones is given in Table 1.1.

The transmission electron micrograph from an arbitrary zone of the Cr_3C_2–25 (Ni20Cr) coating is illustrated in Fig. 1.9a along with the electron diffraction pattern and EDS spectrum [9]. Very fine grain size of the order of 40–80 nm is

Fig. 1.8 The high magnification back-scattered images of the sectioned portion of detonation-sprayed WC-12 % coating and HVOF-sprayed Cr_3C_2–25 (Ni20Cr) coating

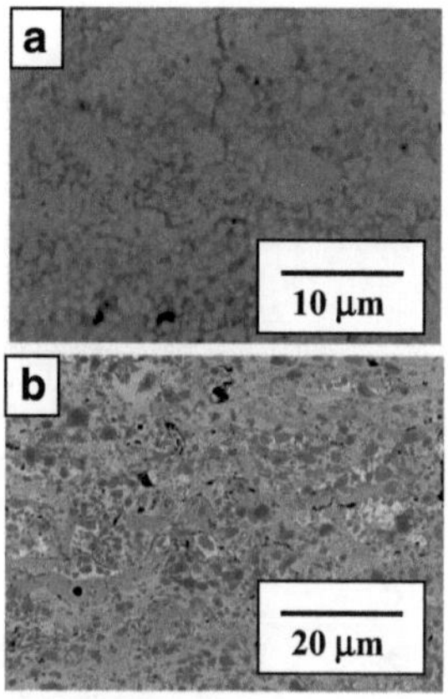

Table 1.1 The composition of various elements as measured by EDAX

	Elements		
	C	Cr	Ni
Dark zone			
Wt%	16.05	83.04	0.91
At%	45.40	54.07	0.53
Grey zone			
Wt%	13.31	65.86	20.83
At%	40.68	46.30	13.03
White zone			
Wt%	12.72	58.66	28.62
At%	39.67	42.08	18.25

evident. The selected area diffraction (SAD) pattern also suggests fine grains. The EDS spectrum and the indexed diffraction pattern confirm the fact that the micrograph belongs to the dark zone containing Cr_3C_2 phase. There is some tendency for amorphisation as can be noted from the SAD pattern. This is consistent with the observation from X-ray diffraction pattern. The bright field transmission electron micrographs along with electron diffraction pattern and EDS spectrum from another area are presented in Fig. 1.9b. The indexed electron diffraction pattern and EDS spectrum confirm a mixture of FCC NiCr and orthorhombic Cr_3C_2 phases. Representative TEM bright field images, corresponding electron diffraction pattern of another area, are given in Fig. 1.9c. The grains in this area are very fine, around 25 nm of size. The crystallinity of the area is evident. Based on Fig. 1.9c, it can be concluded that the area contains primarily FCC NiCr structure. This observation is consistent with that noted by He and Lavernia [10], who found crystalline matrix of NiCr for Cr_3C_2–NiCr coating. In contrast, Guilemany and Calero [11] observed amorphous matrix phase in HVOF sprayed conventional Cr_3C_2–NiCr coating. There are certain areas in the grey zone, which shows presence of an amorphous zone along with nanocrystalline region. This region is possibly the one which was melted during spraying. A rapid solidification is responsible for this amorphisation.

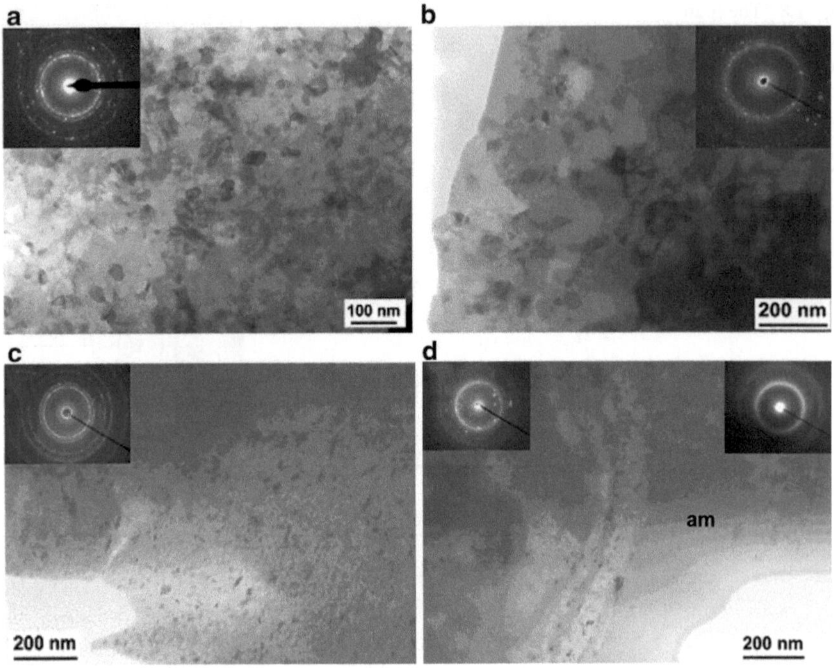

Fig. 1.9 Transmission electron microscopy image of Cr_3C_2–25(Ni20Cr) coating from an (**a**) area containing Cr_3C_2 phase, (**b**) area containing a mixture of FCC NiCr and orthorhombic Cr_3C_2 phases, (**c**) area containing primarily FCC NiCr structure and (**d**) area containing an amorphous zone

A typical TEM micrograph of this region is shown in Fig. 1.9d. A transition zone separates the purely amorphous area from an area containing a mixture of small crystals and remaining amorphous phase. Thus the coating obtained contains three different zones containing different proportion of orthorhombic Cr_3C_2 and FCC NiCr phases.

XRD pattern of WC–12 % Co powder, detonation-sprayed WC–12 % Co coating, WC–17 % Co powder and detonation-sprayed WC–17 % Co coating is shown in Fig. 1.10. Powders in both composition exhibit WC and Co phases. However, on spraying pure Co ceases to exist. Detonation spraying has resulted in partial amorphisation. While the powder contains primarily WC and Co, the coating contains W_2C and Co_6W_6C in addition to WC and phases. Small amount of amorphous Co can be seen in detonation-sprayed WC–17 % Co coating. Comparison of detonation-sprayed coatings and plasma-sprayed coating can be seen in Fig. 1.11 for WC–12 % Co coating and WC–17 % Co coating. In plasma-sprayed coating, unreacted Co can be seen. Further extent of decarburisation resulting in formation of W_2C phase is more in plasma-sprayed coating as compared to detonation-sprayed coatings.

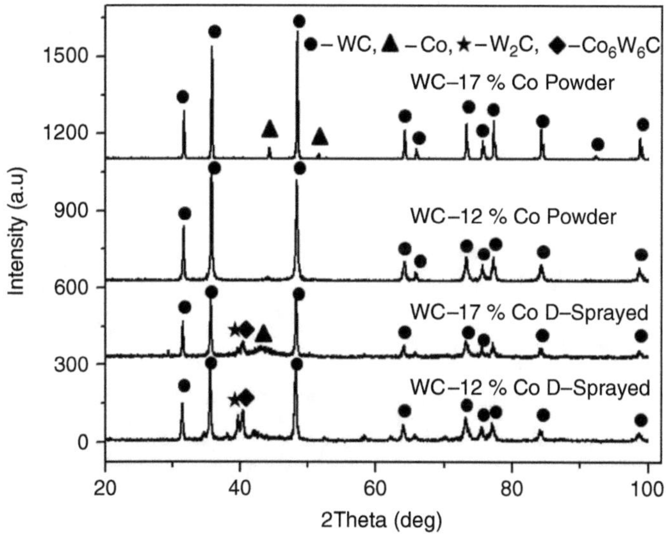

Fig. 1.10 XRD pattern of WC–12 % Co powder and detonation-sprayed coating and WC–17 % Co powder and detonation-sprayed coating

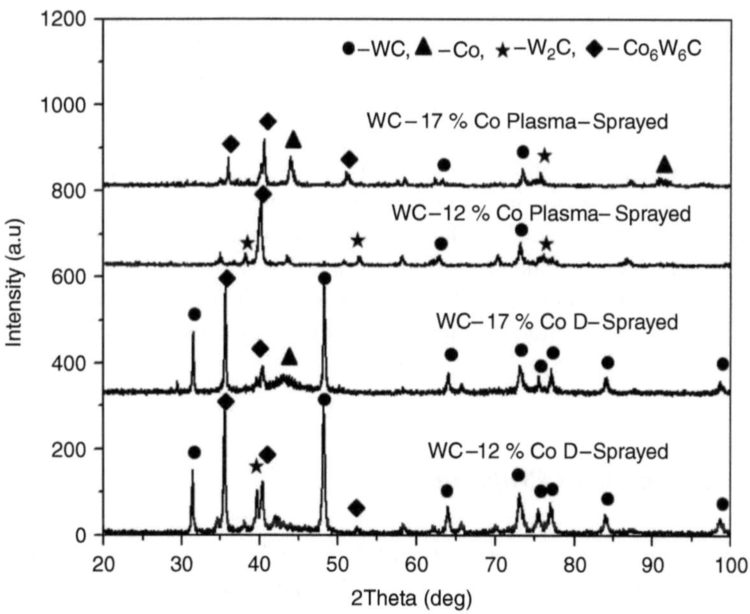

Fig. 1.11 XRD pattern of WC–12 % Co plasma-sprayed and detonation-sprayed coating and WC–17 % Co plasma-sprayed and detonation-sprayed coating

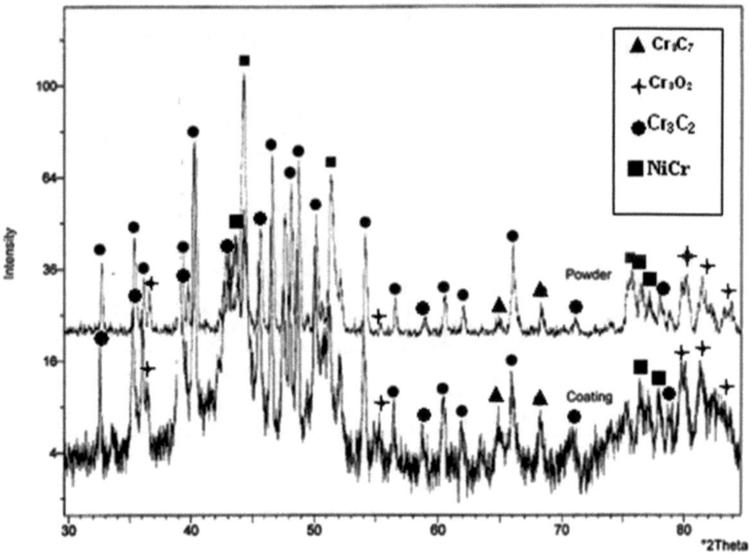

Fig. 1.12 XRD pattern of Cr_3C_2–25(Ni20Cr) powder and coating obtained by JP 5000 gun

The XRD pattern of the Cr_3C_2–25(Ni20Cr) powder and coating is shown in Fig. 1.12. Most important point to be noticed is that, there is no change of phases present in the powder. This is contrary to what is commonly observed for thermal-sprayed WC–Co coatings [12]. Interestingly, Mohanty et al. [13] observed a transformation of Cr_3C_2 phases to $Cr_{23}C_6$ phases while spraying Cr_3C_2–NiCr powder with conventional grain size by Jet Kote technique. Such dissociation of Cr_3C_2 phases is not noted in the present study even in the case of nanocrystalline grains. One possible reason for this phenomenon is that the residence time of the particles in the flame is much less for the spraying system with liquid fuel. Presence of Cr_3O_2 and Cr_3C_7 can be traced in the powder and in the coated surface. There is significant amount of amorphisation of the powder as a result of spraying.

1.5.2 Mechanical Properties

Most important mechanical property of thermal spray coatings is hardness. Mostly Knoop indenter is used to measure the hardness of thermal-sprayed coatings although Vickers indenter is used for measuring both hardness and indentation toughness. Details of Knoop indentation testing is given by Marshall et al. [14] and Leigh et al. [15]. Hardness using Knoop indenter can be given by

$$\text{KHN} = 14,229\frac{P}{a^2} \tag{1.1}$$

where KHN is Knoop hardness number expressed in kg/mm^2, P is the applied load in gf and a is the length of the major diagonal of the indenter in μm. Elastic modulus can be determined using Knoop indenter following the equation given below:

$$\frac{b}{a} \approx \frac{b}{a'} = \frac{b'}{a'} - \frac{\alpha \mathrm{KHN}}{E} \tag{1.2}$$

where b is the minor diagonal of the indentation, a', b' are the major and minor diagonals of the ideal Knoop indentation, α is a constant and generally assumed to be 0.45 and E is the elastic modulus having same dimension as that of KHN. In general, b'/a' is equal to 0.14.

The hardness values so obtained follow mostly Normal, Lognormal and Weibull distributions [16]. However, Weibull distribution appears to be most suitable for thermal-sprayed coatings as properties distributions are highly screwed and broadly distributed [17]. The Weibull distribution in two parameter form may be given as

$$F(x) = 1 - \mathrm{e}^{-(x/x_o)^m} \tag{1.3}$$

where $F(x)$ is the cumulative density function of probability, x_o is scale parameter below which 63.2 % of the data lie and m is the Weibull modulus.

Indentation toughness of thermal-sprayed coatings is evaluated by using Vicker's indentor. The indentation can be carried out on the traverse section of the coating in the mid-plane region. The indentor should be loaded so that one of the horizontal diagonals is parallel to the coating substrate interface. The diagonals of the indentation ($2a$) and lengths of the cracks (l) at the corners of the indentation are measured and used to estimate characteristics crack length ($c = l + a$) and fracture toughness (K_{i}). For median/radial cracks ($c/a \geq 2.5$), expression proposed by Antis et al. [18], Evans and Charles [19] and Niihara et al. [20] are used. The equations are

$$\text{Anstis et al.} \quad K_{\mathrm{i}} = 0.016\left(\frac{E}{H}\right)^{1/2} x \frac{P}{c^{3/2}} \tag{1.4}$$

$$\text{Evans and Charles} \quad K_{\mathrm{i}} = 0.16H\sqrt{a}\left(\frac{c}{a}\right)^{-3/2} \tag{1.5}$$

$$\text{Niihara et al.} \quad K_{\mathrm{i}} = 0.0309\left(\frac{E}{H}\right)^{0.4}\left(\frac{P}{c^{3/2}}\right) \tag{1.6}$$

where E is the young modulus, H is the hardness and P is the applied load. For Palmqivst cracks ($c/a \leq 2.5$) expression proposed by Niihara et al. [21] as given below is used

$$K_i = 0.0123E^{0.4}H^{0.1}\left(\frac{P}{l}\right)^{1/2} \tag{1.7}$$

1.6 Tribological Properties

In this section, sliding wear, defined as degradation of coatings due to rubbing of two surfaces in relative motion, erosive wear which is degradation of coatings due to impact of particles travelling with significant velocity and abrasive wear known as material loss when a hard particle is made to slide against a relatively soft-coated surface are addressed. Both forms of sliding wear such as unidirectional sliding wear and reciprocating sliding wear are covered. Erosive wear as a result of liquid drop impact, cavitation erosion, liquid metal, etc., are not discussed. Description of abrasive wear is primarily confined to three body abrasion estimated by rubber wheel abrasive wear test, although high stress abrasion and low stress abrasion are taken into consideration.

1.6.1 Sliding Wear

The most important governing parameters for sliding wear are load and sliding velocity. The influence of load and sliding velocity on coefficient of friction of Cr_3C_2–25(Ni20Cr) coating is presented in Fig. 1.13 [22]. The coefficient of friction increases with increase in applied load. According to Greenwood and Williamson [23], the real contact area is proportional to the normal load. Based on the classical theory of adhesion, the frictional force is defined as [23]:

$$F = \tau_a A_r \quad \text{and} \quad \frac{F}{(L)} = \frac{\tau_a A_r}{(L)} = \mu_a \tag{1.8}$$

where τ_a is the average shear strength during sliding and (L) is the applied load. μ_a is adhesion induced friction coefficient.

For elastic contact of a spherical indenter and a homogenous half-space, the contact area A_r can be estimated as:

$$A_r = \pi\left(\frac{3(L)R}{4E}\right)^{\frac{2}{3}} \tag{1.9}$$

where E is the effective elastic modulus and R is the effective radius of curvature. Combining Eqs. (1.8) and (1.9), the friction coefficient corresponding to the pure elastic adhesion is given by

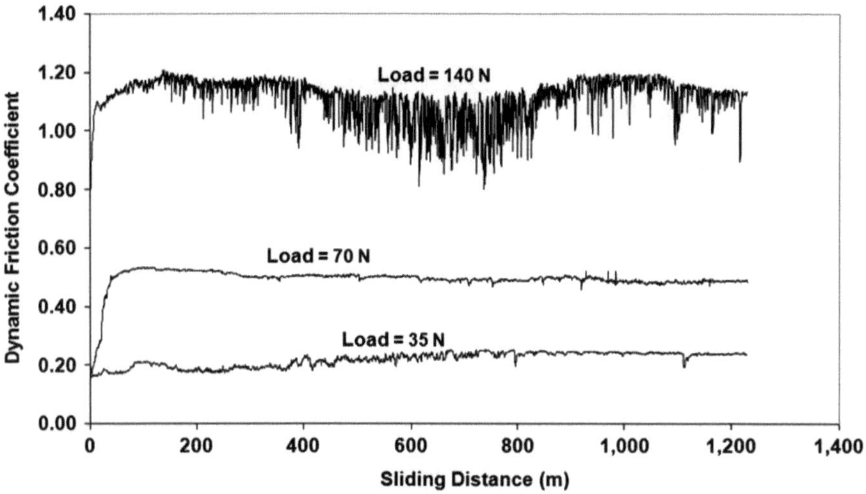

Fig. 1.13 Influence of applied load on the coefficient of friction of HVOF-sprayed Cr_3C_2–25 (Ni20Cr) coating [22]

$$\mu_a = \pi\tau\left(\frac{3R}{4E}\right)^{\frac{2}{3}} L^{-\frac{1}{3}} \tag{1.10}$$

where μ_a is the coefficient of friction, τ is the average shear strength during sliding and L is the applied load. E is the effective elastic modulus and R is the effective radius of curvature. Thus the coefficient of friction is inversely related to the cube root of the applied load and the observation of Fig. 1.13 contradicts this relation. If the adhesion is the dominant friction mechanism, then the coefficient of friction would decrease with the applied load. Thus coefficient of friction in the above observation is not governed by adhesive component.

However, coefficient of friction initially increases with increase in sliding velocity but then remains independent of velocity. Such behaviour is in contradiction of what normally observed and described by the equation given below [24].

$$\mu = K_1 - K_2\text{Ln}(v) \tag{1.11}$$

where μ is coefficient of friction and v is sliding velocity. K_1 and K_2 are constants. This contrasting behaviour can be attributed to small range of sliding velocity over which the measurement is made. For a wider range of sliding velocity, appropriate relation can be established.

Nanostructured thermal-sprayed coatings are known for their improved wear and friction properties. Conventional and nanostructured WC–12 % Co coatings were deposited on 1Cr18Ni9Ti stainless steel substrate using air plasma spraying by Zhao et al. [25]. The hardness of the coatings was measured, while their friction and wear behaviour sliding against Si_3N_4 at room temperature and elevated

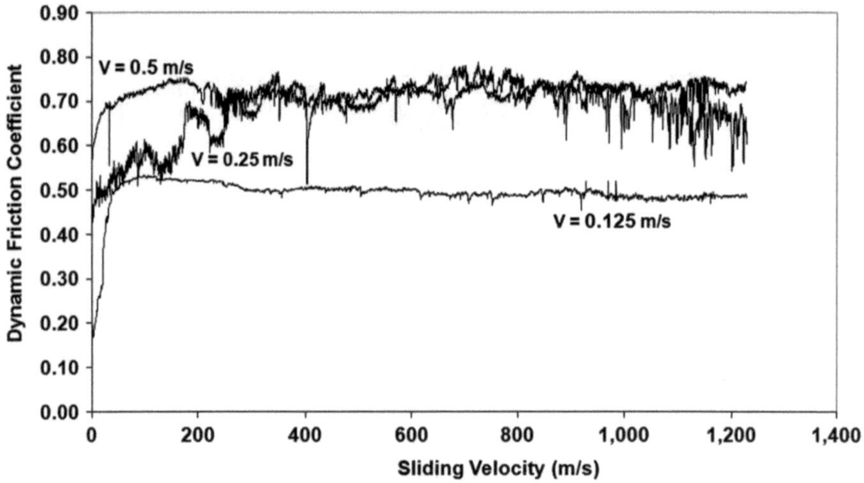

Fig. 1.14 Influence of sliding velocity on the coefficient of friction of HVOF-sprayed Cr_3C_2–25 (Ni20Cr) coating [22]

temperatures up to 673 K were comparatively studied. It was found that the as-sprayed WC–12 % Co coatings were composed of WC as the major phase and W_2C, WC_{1-x}, and W_3Co_3C as the minor phases. The plasma-sprayed nanostructured WC–12 % Co coating had much higher hardness and refined microstructures than the conventional WC–12 % Co coating. This largely accounted for the better wear resistance of the nanostructured WC–12 % Co coating than the conventional coating. Besides, the two types of WC–12 % Co coatings showed minor differences in coefficient of frictions, though the nanostructured WC–12 % Co coating roughly had slightly smaller coefficient of friction than the conventional coating under the same sliding condition as given in Fig. 1.14. Moreover, both the conventional and nanostructured WC–12 % Co coatings recorded gradually increased wear rate with increasing temperature, and the nanostructured coating was less sensitive to the temperature rise in terms of the wear resistance. The nanostructured WC–12 % Co coating with a wear rate as small as 1.01×10^{-7} mm^3/Nm at 673 K could be promising candidate coating for the surface-modification of some sliding components subject to harsh working conditions involving elevated temperature and corrosive medium. The friction behaviour of HVOF-sprayed Cr_3C_2–25 (Ni20Cr) coating having nanocrystalline grains and microcrystalline grains has been studied by Roy et al. [26]. Their work is presented in Fig. 1.15. The coefficient of friction of nanocrystalline grain coating was found to be lower than that of microcrystalline coating. Similar results have been reported by Zhu and Ding [27] for WC–Co coatings.

Thermal-sprayed coatings are extensively used to improve the sliding wear properties of various engineering component. An improvement of wear resistance of mild steel by 200 times by depositing detonation-sprayed WC–12 % Co coating is reported by Roy and Sundararajan [12]. Their work is presented in Fig. 1.16. The

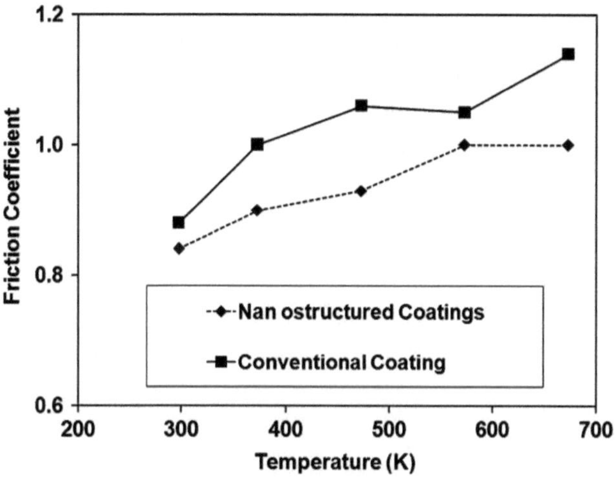

Fig. 1.15 Influence of temperature on the coefficient of friction of plasma-sprayed conventional and nanostructure WC–12 % Co coatings [25]

influence of applied load on wear rate of a NiCrBSi alloy is examined by Rodrıguez et al. [28]. As shown in Fig. 1.17, the load influences the wear rate significantly. However, no specific trend about influence of applied load can be established. Another important thermal-sprayed coating for tribological application is Cr_3C_2–25 (Ni20Cr). The study due to Mohanty et al. [13] indicates that for this variety of coating the wear rate is governed by splats, porosity, forms and distribution of second phase. The coefficient of friction decreases with increase in sliding velocity in accordance with Eq. (1.9). However, wear rate initially decreases with velocity but subsequently increases with sliding velocity. Such behaviour of wear rate can be explained by the fact that at low sliding velocity the wear mechanism is mild oxidational wear and the wear rate in this regime, assuming a flat pin sliding against flat disc, is given by [24, 29]

$$W = \frac{FC^2 A_o}{v Z_c H_o} \exp\left(-\frac{Q}{RT_f}\right) \tag{1.12}$$

where A_o is Arrhenius constant, Q is the activation energy for oxidation, F is applied load, H_o is hardness of the oxide scale, Z_c is the critical thickness of spalling, T_f is the temperature at the place of contact of the asperities, v is the sliding velocities, C is a constant which depends on the composition of the oxide scale and R is the molar gas constant. As the velocity term is in denominator, increase in velocity will decrease the expression of Eq. (1.12) and hence the wear rate will decrease. Subsequently when the sliding velocity increases, the wear takes place by severe oxidational wear mechanism. The wear rate in this regime can be given by [24, 29]

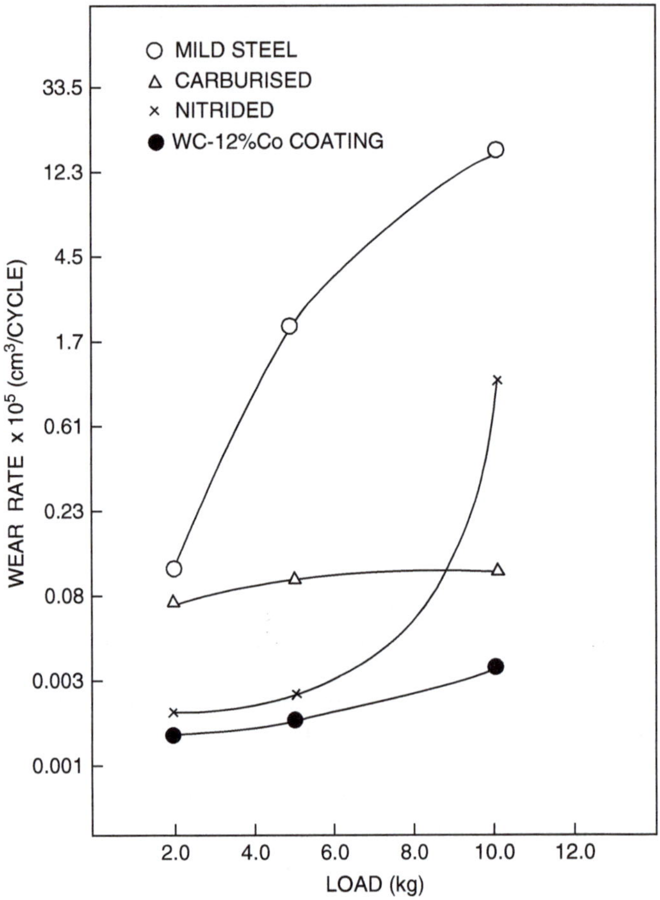

Fig. 1.16 Influence of applied load on the wear rate of detonation-sprayed WC–12 % Co coating, nitrided and carburised mild steel [12]

$$W = \frac{f_m \alpha \mu F}{L_{ox}} - \frac{f_m K_{ox} (T_m^{ox} - T_b) F^{1/2} N^{1/2} A_n^{\frac{1}{2}}}{L_{ox} \beta v r_0 H_o^{1/2}} \tag{1.13}$$

where f_m is volume fraction of molten material during sliding, r_0 is the radius of the nominal contact area, L_{ox} is the latent heat for fusion per volume of the oxide scale, β is a dimensionless parameters for bulk heating, N is the total number of asperities, H_o is the hardness of the oxide scale, T_m^{ox} is the melting temperature of the oxide scale, α is the fraction of generated heat that is conducted by the wearing component, μ is friction coefficient, F is applied load, v is sliding velocity and T_b the bulk temperature of the wearing material. According to this equation, as sliding velocity increases, the second term of Eq. (1.13) decreases and the wear rate increases.

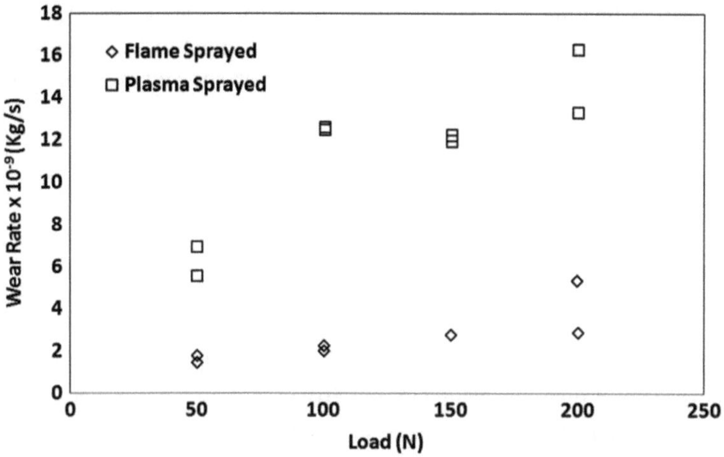

Fig. 1.17 The influence of applied load on wear rate of flame-sprayed and plasma-sprayed NiCrBSi alloy [28]

Wear performance of WC–Co coatings reported to improve significantly by heat treatment [30–32]. Studies by Lenling et al. showed that post-heat treatment of WC base coatings increased the compressive residual stress of these coatings [30]. But research of Stewart et al. [32] showed that heat treatment at all the examined temperatures reduced residual stress of the coating. Increasing of residual stress in thermal spray coatings is very harmful as it may damage the coatings by spalling. On the whole, residual stress is one of the major problems in thermal spray layers especially in thick coatings. Khameneh Asl et al. [33, 34] also found deterioration of tribological properties of WC–Co coatings on heat treatment. Heat treatment also resulted in improvement in wear resistance of Cr_3C_2–25(Ni20Cr) coating as reported by Guilemany et al. [35]. In general, heat treatment improves the wear rate. However, heat treatment at 1,033 K for 1 h in inert atmosphere improves the wear rate dramatically. This happens due to precipitation of Cr_3C_2 in Ni20Cr matrix. During heat treatment in oxidising atmosphere, Cr combines with oxygen and forms Cr_2O_3. This is minimised in inert atmosphere and thus exhibiting best wear performances. Stoica et al. [36] deposited functionally graded WC–NiCrBSi coatings by thermally sprayed using a high velocity oxyfuel (JP 5000) system and heat treated at 1,473 K in argon environment. The relative performance of the as-sprayed and heat-treated coatings was investigated in sliding wear under different tribological conditions. Results indicate that by heat treating the coatings at a temperature of 1,473 K, it is possible to achieve higher wear resistance, both in terms of coating wear, as well as the total wear of the test couples. This was attributed to the improvements in the coating microstructure during the heat treatment and this resulted in an improvement in coating's mechanical properties through the formation of hard phases, elimination of brittle W_2C and W and the establishment of metallurgical bonding within the coating microstructure.

Post-coating treatments other than heat treatment do alter the wear behaviour. Hipping is one such process which results in improvement of wear rate as noted by Stoica et al. [37]. The relative performance of the as-sprayed and hot isostatically pressed functionally graded WC–NiCrBSi coatings was investigated in sliding wear conditions. Coatings were deposited using a high velocity oxy-fuel (HVOF) JP 5000 system and HIPed without any encapsulation at temperatures of 1,123 and 1,473 K. The results indicated significant alteration of the coating microstructure, brought about by the coating post-treatment, particularly when carried out at the higher temperature of 1,473 K. As a consequence, improvements in the coating mechanical properties took place that led to higher wear resistance of the HIPed coatings. They have also examined effect of hipping on WC–Co coating [38]. WC–12Co coatings sprayed by a HVOF system were deposited on SUJ-2 bearing steel substrate and then encapsulated and HIPed at 1,123 K for 1 h. Hot isostatic pressing of thermal spray coatings was seen to improve the microstructural features by reducing porosities and increasing intersplat bond strength, physical properties, and, correspondingly the wear resistance of cermet coatings. As a result, the wear resistance of the post-treated coatings was approximately twice that of the as-sprayed coatings as shown in Fig. 1.18. Hipping resulted in elimination of secondary phase W_2C and metallic tungsten W, alteration of amorphous binder phase through recrystallisation of Co leading to precipitation of the η carbides and development of metallurgical bonding at the interface between the constituent lamellae of the coating, thereby increasing the coatings modulus after HIPing. Wear mechanism of as-sprayed coatings, which lost higher amounts of material, is by microcracking, spallation, ploughing and material transfer. The wear mechanisms involved in the HIPed coating/ball test couples were extrusion of binder followed by the removal of the carbide grains, some levels of abrasion, plastic deformation and material transfer. Similarly Mateos et al. [39] found improvement of wear rate of plasma-sprayed Cr_3C_2–25(Ni20Cr) coating by laser glazing. Laser glazing made the coating more homogeneous, free from porosity, harder and improved adhesion. This in turn resulted in improve wear rate.

Bolelli et al. [40] produced Al_2O_3 coatings by the high-velocity suspension flame spraying (HVSFS) technique using a nanopowder suspension. Their structural and microstructural characteristics, micromechanical behaviour and tribological properties were studied and compared to conventional atmospheric plasma-sprayed and high-velocity oxygen-fuel-sprayed Al_2O_3 coatings manufactured using commercially available feedstock. The HVSFS process enables near full melting of the nanopowder particles, resulting in very small and well-flattened lamellae (thickness range 100 nm to 1 μm), almost free of transverse microcracking, with very few unmelted inclusions. Thus, porosity is much lower and pores are smaller than in conventional coatings. Moreover, few interlamellar or intralamellar cracks exist, resulting in reduced pore interconnectivity (evaluated by electrochemical impedance spectroscopy). Such strong interlamellar cohesion favours much better dry sliding wear resistance at room temperature as described in Fig. 1.19.

The kind of fuel used also governs wear performances. Work due to Sudaprasert et al. [41] as presented in Fig. 1.20 indicates that wear rate of WC–12 % coating

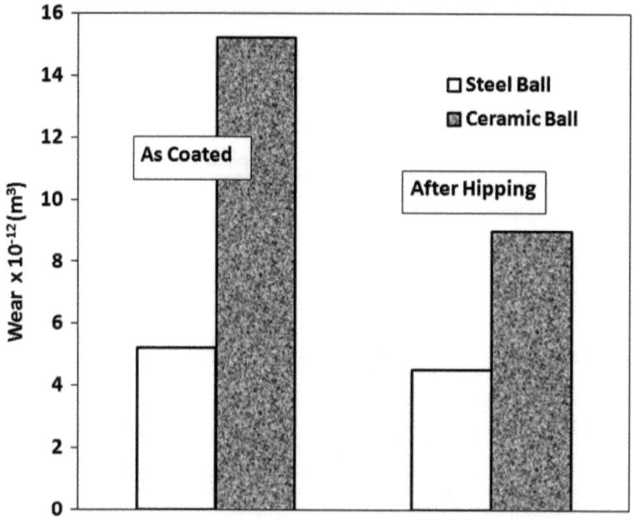

Fig. 1.18 The influence of hipping on the wear of HVOF-sprayed WC–Co coating [38]

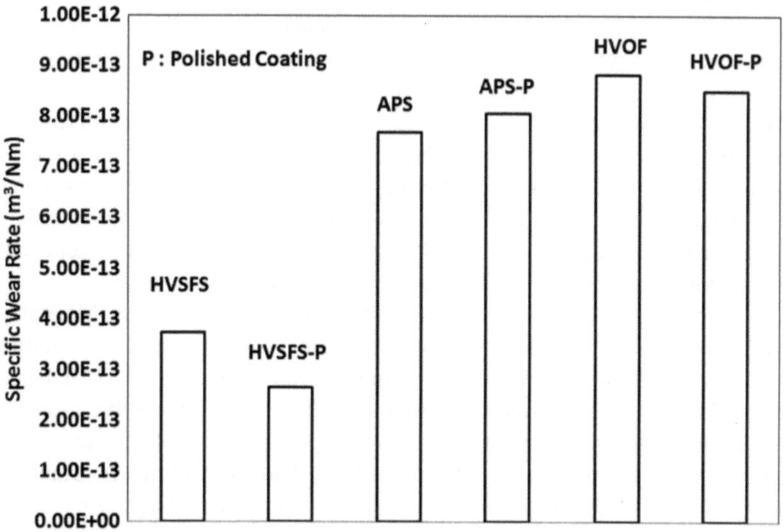

Fig. 1.19 The influence of spraying techniques on the wear rate of Al_2O_3 coatings in as-received and polished condition deposited by high velocity suspension flame (HVSFS), HVOF and air plasma spraying [40]

increases as oxy gaseous fuel is changed to oxy liquid fuel. During spraying with gaseous fuels, the binder material is totally molten and carbide dissolution is extensive forming a highly alloyed matrix. On impacting, binders solidify leaving well-bonded carbides with the matrix phase. In contrast, during liquid fuel spraying,

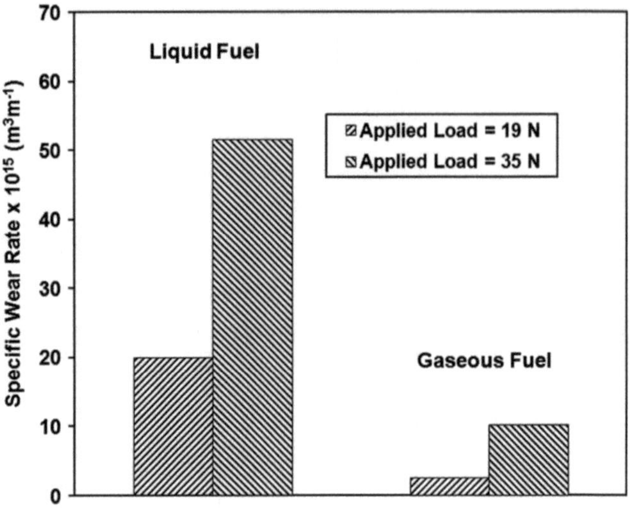

Fig. 1.20 The influence of type of fuel on the wear rate of HVOF-sprayed WC–Co coating [41]

binder is partially molten leaving solid core and liquid shell. On impact the solid core are damaged. This leads to poor wear resistance.

In recent times WC–Co–Cr powders are employed to improve wear and corrosion resistance [42–45]. Addition of Co to Cr results in improvement of adhesion of the WC particles with the matrix [46]. Lee et al. [45] noted that the mixing of fine and coarse powders results in improvement of mechanical and tribological properties. Their observation shown in Fig. 1.21 indicates best wear resistance for mixed powder containing 70 % fine powder and 30 % coarse powders. This observation is attributed to the fact that spaces generated by insufficient melting of coarse powders are filled by large amount of molten fine powders.

Counterbody governs the wear resistance to a significant extent. Mechanically alloyed Al–12Si/TiB_2/h-BN composite powder was deposited on aluminium substrates by atmospheric plasma spraying by Ozdemir et al. [47]. It has been observed that TiB_2 and in situ formed AlN and Al_2O_3 phases in combination with h-BN solid lubricant strongly affect the wear performance of the coating. The presence of solid lubricant, h-BN (7.7 wt.%), had no significant influence in reducing the coefficient of friction of the plasma-sprayed coatings as compared to the that of the substrate. Counter material plays an important role in reducing the volume loss of the coatings after wear test. The volume loss of the coating was measured approximately two times higher when it was wear tested against Al_2O_3 ball instead of 100Cr6 ball as illustrated in Fig. 1.22.

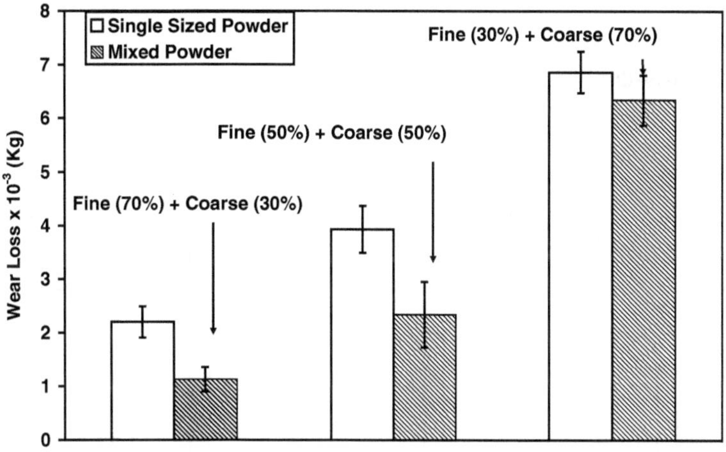

Fig. 1.21 Influence of powder mixing on the wear rate of HVOF-sprayed WC–10Co–4Cr coating [45]

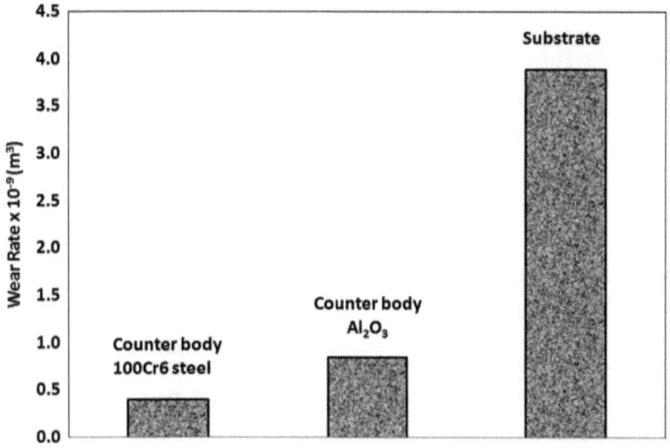

Fig. 1.22 Influence of counterbody material on the wear rate of Al–12Si/TiB_2/h-BN composite coating [47]

1.6.2 Erosive Wear

Thermal-sprayed coating is extensively used for protection against erosive wear as well. In order to examine what is the most erosion resistance thermal-sprayed coating, work of Hawthorne et al. [48] appears to be the most relevant. The performance of ten high velocity oxy-fuel HVOF-sprayed coatings has been evaluated under normal and oblique impingement conditions. Their results are presented in Fig. 1.23. Erosion test was conducted with 50 mm quartz particles at an impact velocity of 84 m/s. It is clear that WC–Co coating has the highest erosion

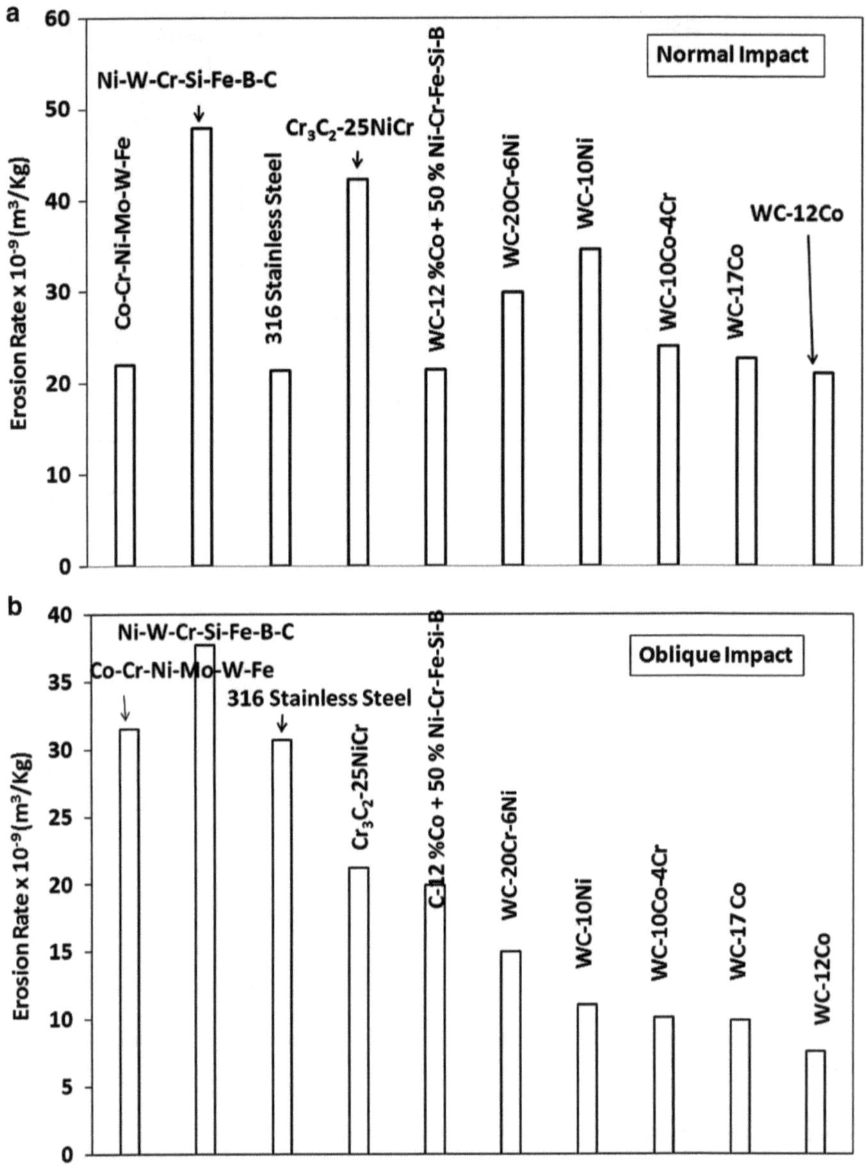

Fig. 1.23 The erosion rate of various thermal-sprayed coatings [48]

resistance. For this coating, material loss was by cutting, platelet removal and occasionally by carbide particle removal. Levy and Wang [49] noted brittle erosion response characterised by lower erosion rate at oblique impact than at normal impact of plasma-sprayed WC–Co coating. Observation of Barbezat et al. [50] indicated CDS-sprayed WC–Co coating exhibits improved erosion resistance.

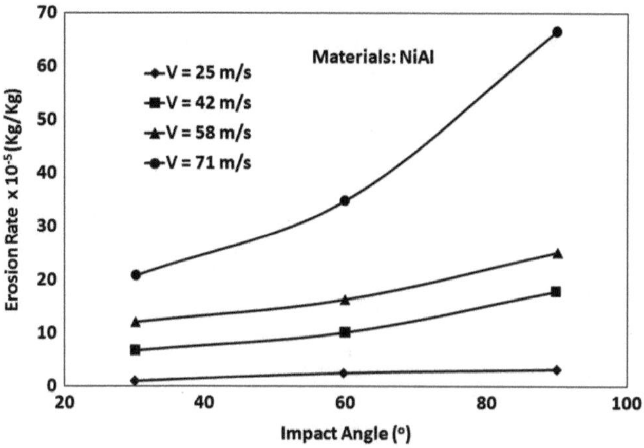

Fig. 1.24 The influence of impact angle on the erosion rate of HVOF-sprayed NiAl coating [54]

Impact angle independent erosion rate at low impact velocity and impact angle dependent erosion rate at high impact velocity for WC–Co coating is noted by Wood et al. [51]. As reported by Karimi et al. [46], addition of Co to Cr results in improvement of adhesion of WC particles with the matrix resulting in improved erosion resistance. According to Kim et al. [52] erosion rate of plasma-sprayed WC–Co coating decreases with increase of cohesive strength. Roy et al. [53] demonstrated that the erosion resistance of detonation-sprayed WC–Co coating is better than plasma-sprayed and HVOF-sprayed coatings at normal impact and comparable at oblique impact.

The influence of erosion conditions on the erosion resistance can be noted from the work of Hearley et al. [54]. The influence of impact angle on the erosion rate of HVOF-sprayed NiAl coating can be seen from Fig. 1.24. Erosion rate is maximum at normal impact indicating classical brittle erosion response. However, examination of eroded surfaces indicates ductile materials removal mechanism. Materials removal is governed by deformation rather than fracture. This apparent contradictory behaviour lies on the nature of erodent, their size and shape. They also showed the effect of impact velocity on erosion response of HVOF-sprayed NiAl coating. Their result, shown in Fig. 1.25, indicates increase of erosion rate with increase of impact velocity. Relation between erosion rate and impact velocity for this type of behaviour can be obtained from the following expression:

$$E \infty V^{3.2} r^{3.7} \rho^{1.3} H^{-1.25} \tag{1.14}$$

where V is impact velocity, r is radius of the erodent and ρ and H are density and hardness of target material.

Influence of test temperature on erosion rate is studied by Wang and Verstak [55] as shown in Fig. 1.26. The target materials used for the erosion test were the HVOF-sprayed Cr_3C_2/TiC–NiCrMo cermet produced with a self-propagating high-

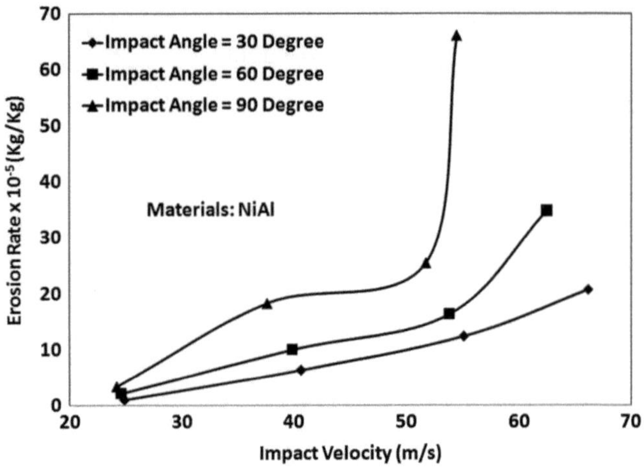

Fig. 1.25 The influence of impact velocity on the erosion rate of HVOF-sprayed NiAl coating [54]

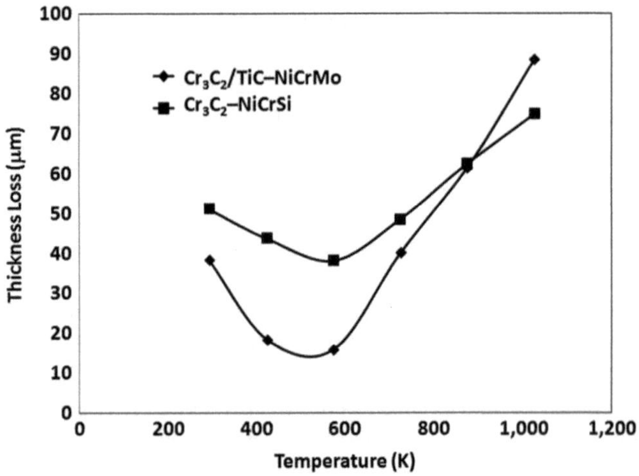

Fig. 1.26 The influence of test temperature on the erosion rate of HVOF-sprayed Cr_3C_2/TiC–NiCrMo cermet coatings [55]

temperature synthesis method [56, 57] and blended Cr_3C_2–NiCrSi cermet coatings on mild steel. The coating was sprayed using the Sulzer-Metco Diamond Jet spray system. Propane was used as a fuel gas. It is observed that the thickness loss decreased in the range from room temperature to 673 K and then increased in the range from 673 to 1,023 K. Similar results for alumina and zirconia plasma-sprayed coatings have been reported in the literature [58]. It is evident that the erosion behaviour of the HVOF Cr_3C_2/TiC–NiCrMo coating is more sensitive to temperature than that of the HVOF Cr_3C_2–NiCrSi coating. Below 873 K the thickness loss

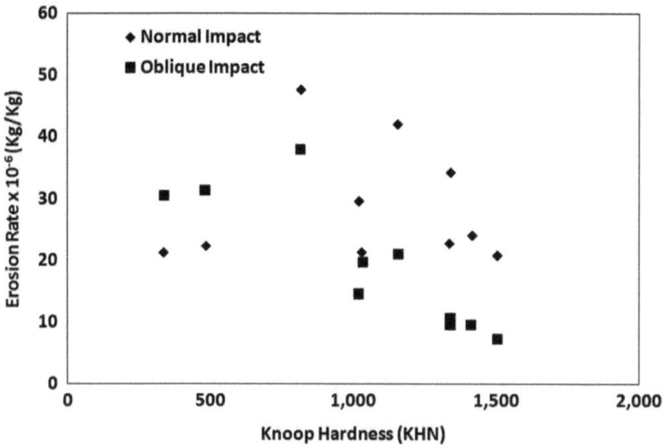

Fig. 1.27 The influence of hardness on the erosion rate of thermal-sprayed cermet coatings [48]

of the HVOF Cr_3C_2/TiC–NiCrMo coating was lower than that of the HVOF Cr_3C_2–NiCrSi coating. Yet with increasing temperature, the thickness loss of the former increased faster than the latter. Over 873 K the erosion wastage of the HVOF Cr_3C_2/TiC–NiCrMo coating was even higher than that of the HVOF Cr_3C_2–NiCrSi coating. The higher sensitivity of the HVOF Cr_3C_2/TiC–NiCrMo coating to temperature may be related to oxidation of titanium carbide TiC. Therefore, the HVOF Cr_3C_2/TiC–NiCrMo coating is not suitable for applications under such high temperatures.

Work of Hawthorne et al. [48] also exhibited the influence of hardness on the erosion rate of thermal-sprayed coatings. As shown in Fig. 1.27, there is no direct correlation between hardness and erosion rate. However, at normal impact there is a tendency for decrease of erosion rate with increase of hardness. The lamellar structure determines the mechanical properties of thermal-sprayed coatings. Li et al. [59] measured the erosion rate of plasma-sprayed Al_2O_3 coating under normal impact condition. They derived a microstructural parameter mean bonding ratio between lamellae and thickness of the lamellae. The result shown in Fig. 1.28 clearly indicates that the erosion resistance is inversely proportional to the mean bonding ratio. They were able to establish the relation as given below:

$$\frac{1}{E} = \frac{2\gamma_c \alpha}{E_{\mathrm{eff}} \delta} \tag{1.15}$$

where γ_c is the effective surface energy of the lamella material, α is the bonding ratio of the interface between lamellae, δ is the thickness of the lamella, $1/E$ is the erosion resistance and E_{eff} is the fraction of kinetic energy per unit mass of impacting particle utilised for cracking of the bonded lamellar interface.

The post-coating treatment also alters the erosion performances of coatings. Tsai et al. [60] investigated the improvement of erosion resistance of plasma-sprayed

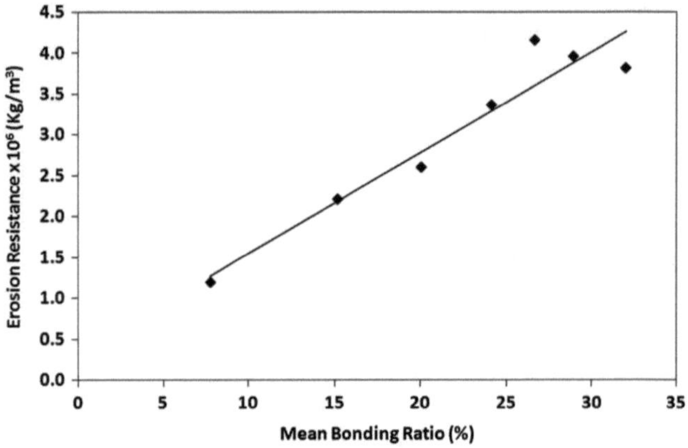

Fig. 1.28 The influence of mean bonding ratio between lamellae and thickness of the lamellae on the erosion rate of plasma-sprayed Al_2O_3 coating [59]

thermal barrier coatings on laser glazing. Several erosion tests were conducted at room temperature using 50 μm silica erodent particles with impact velocity of 50 m/s. The erosion rate increased as the impingement angle increased for both plasma-sprayed and laser-glazed TBC. Laser glazing enhanced the erosion resistance of plasma-sprayed TBC by about 1.5–3 times with the impingement angle ranging between 30 and 75°, while the erosion resistance did not significantly improve when the impingement angle reached 90° as shown in Fig. 1.29. Erosion morphology analysis clearly indicates that the erosion of the plasma-sprayed TBC is deemed to be the erosion of the protrusions and the sprayed splats. The erosion of the laser-glazed TBC is proven to be the spallation of the glazed layer. Spallation occurred in the laser-glazed layer/plasma-sprayed splats interface. Similarly the nature of coating materials also reflects the performance under erosion condition. The investigation due to Kulu et al. [61] presented in Fig. 1.30 indicates that WCCoCr coating has significantly lower erosion rate than self fluxing NICrSiB coating.

A series of hot erosion and erosion–corrosion (E–C) tests was carried out on thermal-sprayed coatings, diffusion coatings and boiler steels using a burner-rig type elevated temperature E–C tester by Uusitalo et al. [62]. Carbide containing HVOF coatings, diffusion coatings and nickel-based high-chromium HVOF coatings performed well. A series of elevated temperature erosion tests was carried out on AlSl 1018 low-carbon steel and four thermal spray-coated mild steel specimens using a nozzle-type elevated temperature erosion tester by Wang and Lee [63]. The thermal-sprayed coatings included a high velocity oxy-fuel (HVOF) $75Cr_3C_2$–20NiCr cermet coating, HVOF Cr_3C_2 low velocity flame-sprayed Cr_3C_2, ceramic coatings and an arc-sprayed Fe–CrSiB coating. The HVOF Cr_3C_2, coating specimen exhibited the lowest erosion wastage due to its favourable composition and morphology. It had a finer structure and smaller splat size than other coatings. Guilemany et al. [64] successfully sprayed the near stoichiometric Fe40Al by

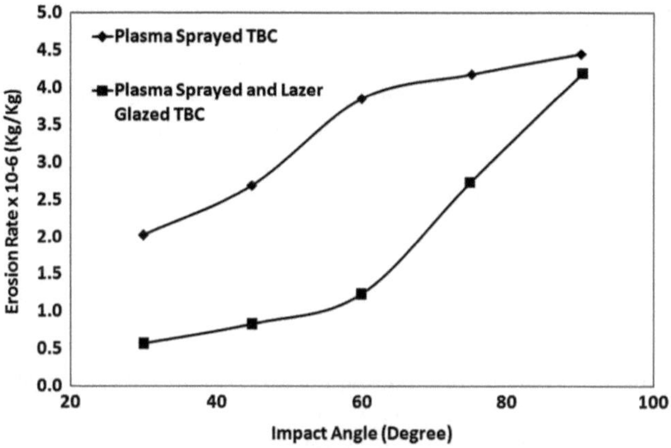

Fig. 1.29 The improvement of erosion resistance of plasma-sprayed thermal barrier coatings on laser glazing [60]

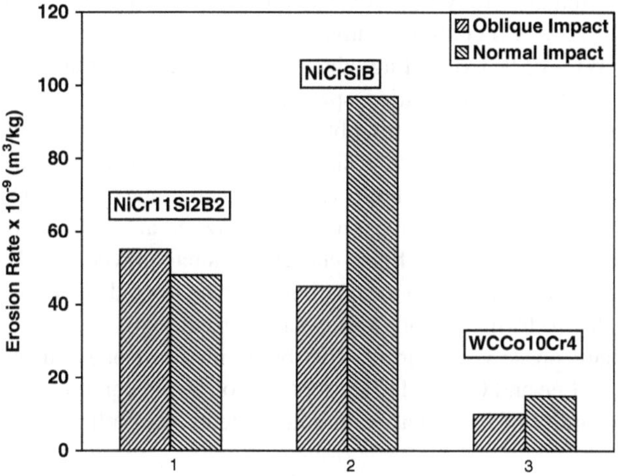

Fig. 1.30 Higher erosion resistance of WCCo10Cr4 coating than self-fluxing NiCr11Si2B2 coating [61]

means of high velocity oxy-fuel. Erosion tests demonstrated that iron aluminides possess a ductile behaviour even when the impingement angle is 90°.

Unlike sliding wear, the work on solid particle erosion of nanocomposite coatings is very less. Only work due to Dent et al. [65] indicates that erosion resistance of nanocomposite coating decreases compared to conventional coatings. However, in nanostructure coating, erosion resistance increases with decrease in Co binder phase.

1.6.3 Abrasive Wear

Most important thermal-sprayed coatings other than WC–Co that are used against abrasion are Al_2O_3, TiO_2, Cr_2O_3, etc. [66, 67]. According to Liu et al. [68], the abrasive wear resistance of thermal spray coating can empirically be related to hardness and indentation toughness of the coating as given below:

$$\mathrm{WR}_{\mathrm{abrasion}} = C\frac{H^{1/2}K^{2/3}}{(1+nP)} \tag{1.16}$$

where H is hardness, K is indentation toughness, P is porosity, n is a parameter to be determined experimentally and C is a constant. The results of Habib et al. [69] which shows the influence of hardness on abrasive wear resistance is given in Fig. 1.31. It can be seen that abrasive wear resistance is directly related to the hardness of the coating and toughness and porosity have minimum influence. Work due to Kim et al. [52] exhibited decrease of abrasive wear rate with increase of bond strength. A systematic study by Barbezat et al. [50] indicated that under three body abrasion WC–12 % Co coating exhibited lower abrasion rate than WC–17 % Co coating. For WC–12 % Co coating, abrasion resistance was maximum for coating obtained by continuous detonation spraying whereas for WC–17 % Co coating abrasion was maximum for coating obtained by cyclic detonation spraying. It is to be stated that minimum abrasion resistance can be seen for coating obtained by vacuum plasma spraying for both variety of coatings. This observation was attributed to inter splat adhesion strength. Nerz et al. [70] demonstrated that abrasive wear resistance of high energy plasma-sprayed or HVOF-sprayed WC–Co coating increases when the coating was heat treated above recrystallisation temperature due to formation of various hard carbides.

When a hard phase is uniformly distributed in a soft phase using quantitative metallography, Lee and Gurland [71] proposed following equation to estimate grain sizes and contiguity of hard phase and binder mean free path:

$$d_{\mathrm{WC}} = \frac{2V_{\mathrm{WC}}}{2N_{\alpha\alpha} + 2N_{\alpha\beta}} \tag{1.17}$$

$$C_g = \frac{2N_{\alpha\alpha}}{2N_{\alpha\alpha} + N_{\alpha\beta}} \tag{1.18}$$

$$\lambda = \frac{d_{\mathrm{WC}}(1 - V_{\mathrm{WC}})}{V_{\mathrm{WC}}(1 - C)} \tag{1.19}$$

where d_{WC} is the tungsten carbide grain size, C_g is the contiguity of tungsten, λ is the binder mean free path, V_{WC} is the volume fraction of tungsten carbide, $N_{\alpha\alpha}$ is the average number of intercepts per unit length of test line for carbide–carbide

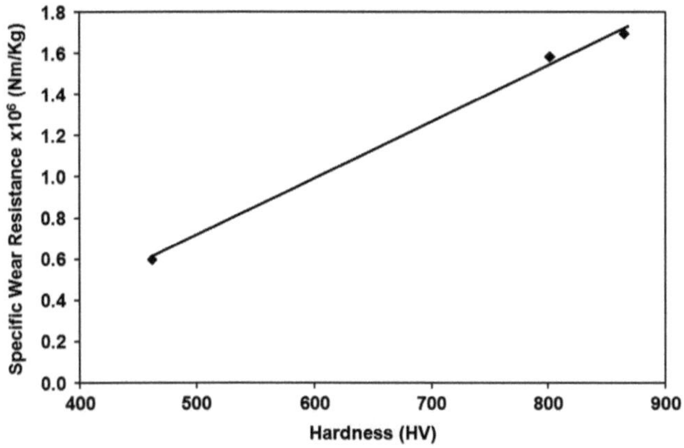

Fig. 1.31 The influence of hardness on abrasive wear resistance of ceramic layers of Al_2O_3/TiO_2 deposited by flame spray [69]

intercepts and $N_{\alpha\beta}$ is the average number of intercepts per unit length of test line for carbide binder intercepts. An interesting study was reported by Kumari et al. [72]. According to their observation shown in Fig. 1.32, the abrasive wear rate of WC–10Co–4Cr coating deposited by JP 5000 gun increased with increase of mean free path of the carbides. As materials loss takes place primarily by abrasion of binder followed by pulling out of carbide particle, binder abrasion becomes difficult due to small mean free path as the abrasive directly interact with hard carbides.

Post-coating heat treatment can alter abrasive wear rate significantly. Stewart et al. [73] sprayed WC–17 wt.% Co powders by the HVOF method to form coatings approximately 200 μm thick on steel substrates. Analysis of the microstructure of the coatings found to consist of WC, W_2C and anamorphous binder phase. Some of the carbide particles are found to decarburise and dissolve in the liquid metal binder during spraying, resulting in a brittle binder phase in the coating. The coatings also exhibit tensile stresses in the as-sprayed condition. Coatings have been heat treated at a range of temperature between 523 and 1,373 K. Heat treatment above 873 K resulted in significant phase changes with in the coating. However, heat treatment at all the temperatures resulted in changes in the integrity and residual stress state of the coating due to thermal expansion coefficient mismatch between the coating and substrate. The abrasive wear behaviour of the as-sprayed and heat-treated coatings demonstrates that heat treatment improves the wear behaviour. Heat treatment at temperature as low as 573 K has resulted in improvement in wear resistance up to 35 %.

The constituents of spraying feedstock play an important role in determining abrasive wear resistance. Wang et al. [74] studied the abrasive wear performance of coatings deposited by high velocity oxy-fuel (HVOF) thermal spraying using of multimodal and conventional feedstocks. WC–12Co coatings were deposited under same conditions using multimodal and conventional WC–12Co powder feedstocks.

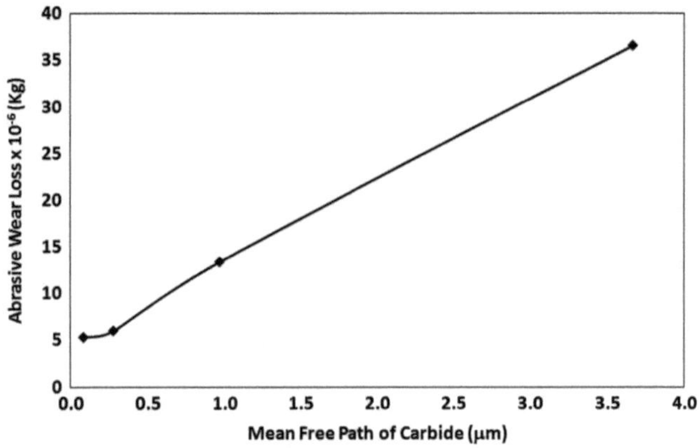

Fig. 1.32 The influence of mean free path of the carbides on the abrasive wear rate of WC–10Co–4Cr coating deposited by JP 5000 gun [72]

Abrasive wear resistances of coatings were carried out on sand rubber wheel abrasion tester. The results shown in Fig. 1.33 indicated the multimodal coating showed better abrasive wear resistance than the conventional counterpart. Also, the thermally sprayed carbide-based coatings have excellent wear resistance with respect to the hard chrome coatings. The nanophase WC–Co component in the multimodal material melts with relative ease and forms a strong and tough matrix that holds the coarse WC particles in place. In turn, the coarse particles provide a hard, 'blocky' surface that is needed for good wear resistance. According to Sudaprasert [41], the dense conventional powder will tend to result in fracture of the carbides on impact, whereas the more porous multimodal powder will cushion impact and result in better wear resistance. Wirojanupatump et al. [75] also studied the influence of powder feedstock on abrasive wear rate of Cr_3C_2–25(Ni20Cr) coating. They deposited the coating using 'Miller Thermal' top gun employing three different types of feedstocks namely sintered and crushed powder, blended powder and composite powder. Sintered and crushed powder was produced by furnace route to give Cr_3C_2 with NiCr metallic binder. Blended powder was prepared by gas atomising NiCr alloy with Cr_3C_2. Composite powder was supplied by 'Praxair' and prepared through a series of thermal and chemical reaction to obtain a fine grain carbide dispersed in a metal matrix. The coating obtained using composite powder feedstock exhibited best abrasion resistance as it has highest hardness due to high degree of carbide retention and small degree of reaction of these particles during spraying.

The influence of amount of reinforcement on the abrasive wear behaviour can be found in the work of Gawne et al. [67]. A ball-milled mixture of glass and alumina powders has been plasma sprayed to produce alumina–glass composite coatings. The coatings have the unique advantage of a melted, ceramic secondary phase parallel to the surface in an aligned platelet-like composite structure. The alumina raises the hardness from 300 HV for pure glass coatings to 900 HV for a 60 wt.%

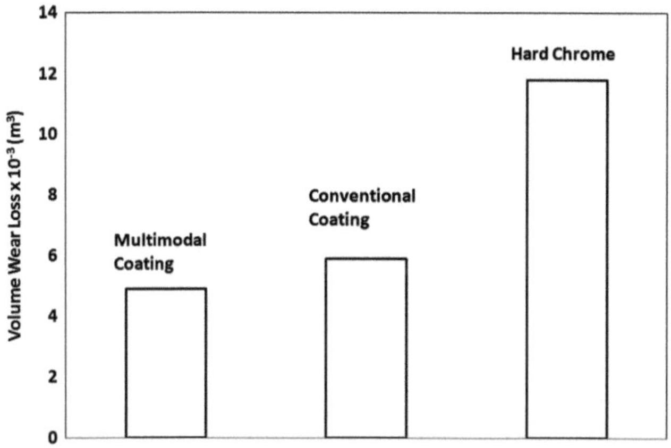

Fig. 1.33 The influence of multimodal and conventional feedstocks on the abrasive wear rate of HVOF-sprayed WC–12 % Co powder [74]

alumina–glass composite coating. The scratch resistance increases by a factor of 3, and the wear resistance increases by a factor of 5. The variation of wear resistance with the amount of reinforcing alumina is shown in Fig. 1.34. Such variation follows the expression provided by Zum Ghar and Eldis [76] as given below:

$$\frac{1}{W} \propto \frac{d^{3/2} v}{\lambda} \tag{1.20}$$

where W is the wear resistance, d is mean diameter of reinforcing particles, v is the volume fraction of reinforcing particles and λ is the mean free path of the particles in the matrix. Thus wear resistance increases with increase in volume fraction of reinforcing particles. The wear resistance reaches a maximum at 40–50 vol.% alumina, above which there is little further improvement. The glass wears by the formation and intersection of cracks, while the alumina wears by fine abrasion and supports most of the sliding load.

Micro-abrasion–corrosion of detonation-sprayed WCCoCr samples in alkaline solutions for different sliding distances as reported by Thakare et al. [77] show generation of a negative synergy that increases with increasing sliding distance. Corrosion takes place at the binder–carbide interface in a region which is likely to have a composition of W_2C. The localised corrosion observed along the carbide grain boundary is due to the corrosion of the decarburised surface of the carbide grain which progresses towards the carbide grain centre and not into the binder phase. Anodic treatment of the WC–Co–Cr samples shows a mild negative synergy. Such treatment has not replicated the wear–corrosion processes seen for the exposed samples. A non-linear relationship between the abrasion–corrosion specific wear rate and E_{corr} is seen and is related to subtle changes in the surfaces such as the formation of passive films during exposure which are not formed during the anodic treatments. Despite the fact that mechanical abrasion losses dominate the

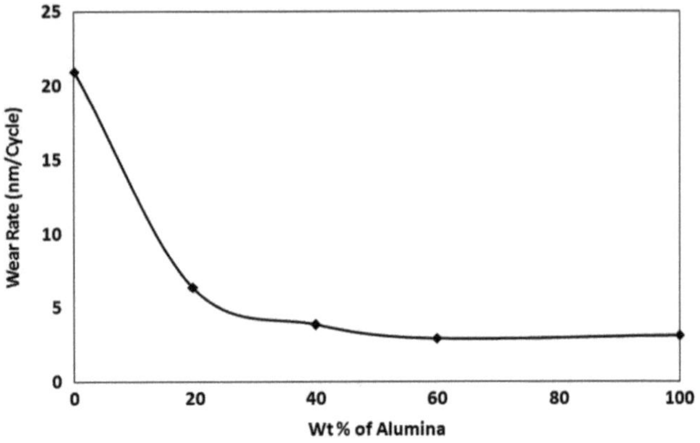

Fig. 1.34 The influence of volume fraction of reinforcing alumina on plasma-sprayed alumina–glass composite coatings [67]

abrasion–corrosion processes, negative synergies are seen due to corrosion which only accounts for 1/500th of the material loss.

Nanocomposite WC–Co coatings are also received attention for abrasive wear related application. However, unlike sliding wear, nanocomposite coating exhibited lower abrasion resistance than conventional coatings [78] due primarily to the decarburisation of wear-resistant WC particles and formation of amorphous phases. Similar behaviour is also noted by Dent et al. [65] although they found that abrasive wear resistance increases with decrease in Co binder content as shown in Fig. 1.35. The abrasive wear rate of nanocomposite coating is also found to be superior to microcrystalline coatings although they are sensitive to the size of abrasives as presented in Fig. 1.36 [79]. In this study, the abrasive wear behaviour of nanocrystalline coatings of the composition (Ti, Mo)(C, N)–45 vol.% NiCo, prepared by vacuum plasma spraying (VPS) of high-energy-milled powder, is characterised and compared to microcrystalline coatings of the same composition. Two-body abrasive wear tests are applied to produce wear traces on the surfaces of the nano- and microcrystalline coatings. Nanocrystalline VPS coatings show superior wear resistance. The wear mechanisms and failure of nano- and microcrystalline coatings are distinctly different.

The potential of post-spraying heat treatment for improving the wear properties of WC–Co nanocomposite coatings has been investigated by Kim et al. [80]. Heat treatment of the coating at 873 K provided an improvement in abrasive wear resistance by 45 %, with an increase in microhardness of the coating as shown in Fig. 1.37. Microstructural examinations showed that heat treatment promoted the formation of additional carbide phases in the splat boundary layer. This increased carbide volume fraction made a major contribution to the improved coating performance. However, higher heat treatment temperatures above 1,073 K reduced the volume fraction of WC and W_2C phases, resulting in detrimental effects on microhardness and abrasive wear resistance of the coatings. Jordan et al. [81] and

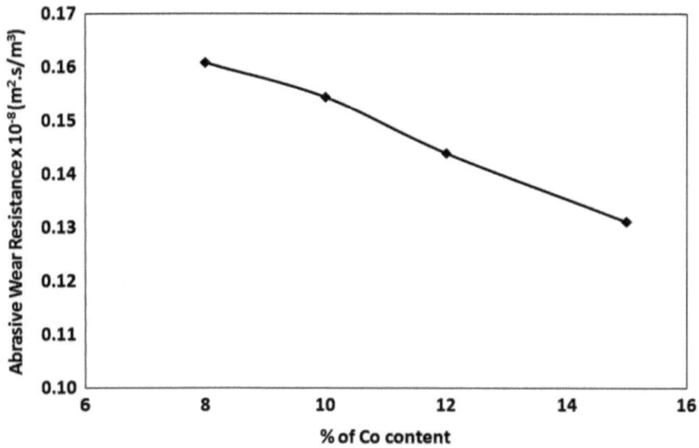

Fig. 1.35 The influence of Co content on abrasive wear resistance of thermal-sprayed WC–Co coatings [65]

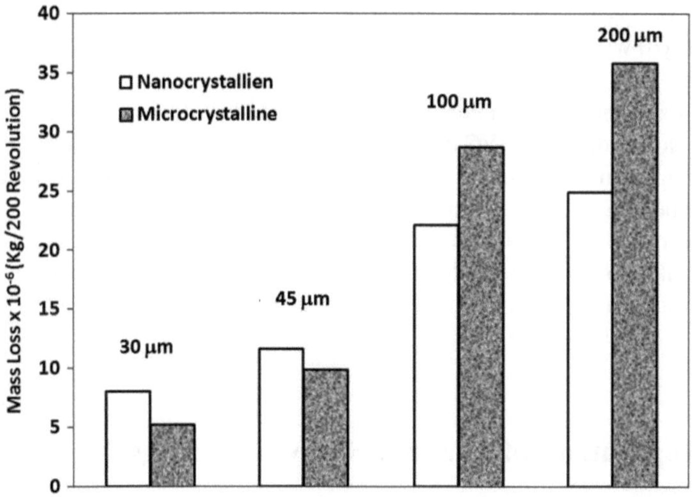

Fig. 1.36 The influence of size of abrasives on the abrasive wear resistance of nanocomposite and microcrystalline vacuum plasma-sprayed (Ti, Mo)(C, N)–45 vol.% NiCo coatings [79]

Gell et al. [82] also found improvement of abrasive wear resistance of nanostructured alumina–titania coatings by modifying the powder with addition of nanostructured oxide powder followed by heat treatment at high temperature (1,073–1,473 K).

Nanocomposite coatings are extremely popular in recent time for tribological application. The mechanical and tribological properties of nanocomposite WC–Co coating can be improved if carbide powders are coated with cobalt. In a study by Baik et al. [83], a porous spray-dried WC–Co nanocomposite powder has been modified by a protective Co surface layer and then sprayed onto a substrate by high

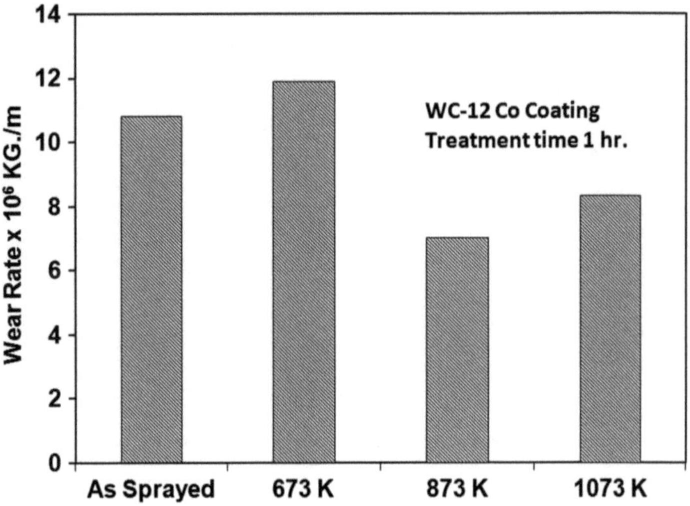

Fig. 1.37 The influence of post-coating heat treatment temperature on the abrasive wear rate of HVOF-sprayed WC–12 % Co coating [80]

velocity oxy-fuel to form a coating. The Co-coated powder promoted coarser size of splats, lower degree of WC decomposition and higher retained carbon content in the coating, compared to the spray-dried powder, mainly because of a limited powder melting. Improvements in hardness and abrasive wear resistance were obtained from the Co-coated powder coating, and the results depicted in Fig. 1.38 are attributed to more retained WC and consequently a lower amount of non-WC phases whose presence is detrimental to wear resistance.

1.7 Applications of Thermal-Sprayed Coatings

Figure 1.39 shows an application of WC–Co coating in aircraft engine. The internal coating of stage III disc of an aircraft engine is illustrated in this figure. With this coatings disc, life is reported to increase substantially. There are of course numerous other applications where WC–Co coatings are applied to protect the components from tribological degradation. Similarly, thermal-sprayed WC–Co coated mixer impeller and electric arc-sprayed stainless steel coated in tapered centre portion of forged lifting lid is shown in Fig. 1.40a and Fig. 1.40b, respectively.

Jet engines of fighter aircraft are equipped with fans and compressors. The vibrations of the blades of fans and compressor blades are controlled by midspan dampers. The dampers in the process are subjected to wear and tear. HVOF-sprayed WC–Co coatings are deposited on dampers to improve the life [84]. HVOF

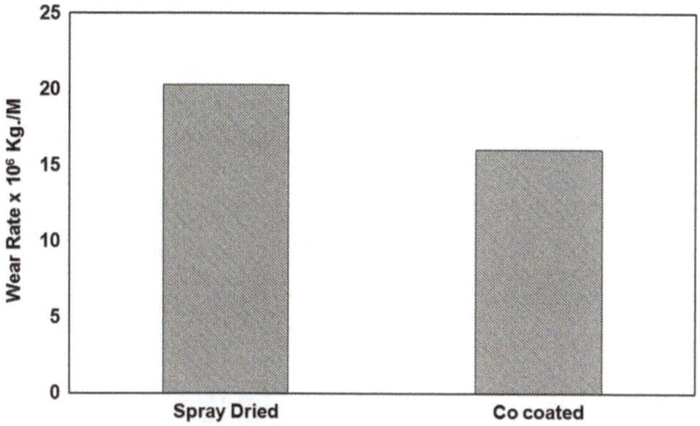

Fig. 1.38 Improvements in abrasive wear resistance of HVOF-sprayed coating obtained by Co-coated porous spray-dried WC–Co nanocomposite powder [83]

Fig. 1.39 WC–Co coating the internal coating of stage III disc of in aircraft engine

WCCoCr used on landing gear for fighter aircraft in US Navy. Also this coating is used on offshore gate valves [85].

Thermal-sprayed WC–Co coating is extensively used in several places of steel industry. Continuous casting moulds are coated with WC–Co coating. The top of the mould is coated with thin WC–Co coating which can withstand high thermal stress in the zone of the molten steel meniscus. At the bottom, thick coating which can withstand high ferrostatic pressure and thereby prevent cracking and spalling of the coating [86] are used.

Briddle rolls are used to control the tension of steel strips as it passes through the continuous pickling, annealing and galvanising lines. These rolls are conventionally coated with Cr plating and are increasingly replaced by HVOF-sprayed WC–Co coatings due to their superior wear resistance [87]. Similar result is obtained from detonation-sprayed WC–Co coating against Cr plated deflector roll [88].

Fig. 1.40 Thermal-sprayed (**a**) WC–Co-coated mixer impeller and (**b**) electric arc-sprayed stainless steel coated in tapered centre portion of forged lifting lid

WC-(10–15)% Co thermal-sprayed coatings are applied to galvanising hardware due to their superior liquid metal erosion resistance relative to Fe and cobalt alloy [89–91]. However, success of this coating under such circumstance depends on reaction between WC and Co phases to improve the poor corrosion properties of Co in molten Zn.

Murakawa and Watanabe [92] demonstrated that an ironing die when coated with plasma-sprayed WC–Co coatings followed by vacuum sintering and hipping, not only resulted improved life of the die and increased output but also resulted in an improved surface finish of the product. Picas et al. [93] demonstrated that thermal-sprayed CrC–NiCr is a very good substitute for hard chromium plating for pistons and valves of automotive engines.

Oxide ceramics offer the benefit of a high melting point, high hardness and wear resistance and chemical stability at high temperature in molten metal environments [94, 95]. Many oxides, particularly Al_2O_3 and yttria-stabilized zirconia are not wetted by liquid zinc and are therefore particularly resistant to corrosion. Such properties make oxide-based coatings particularly attractive for high temperature, high Al-content, and galvalume baths where the more common WC–Co-based coatings do not perform well [96] against liquid metal erosion.

As the solidifying steel exits the mould, it is supported by a series of retaining rolls which redirect it from a vertical orientation to a horizontal run-out bed and cutting operations. These rolls are subjected to high temperatures (steady state of 500–600 °C) and thermal cycling [97] resulting from unavoidable stoppages during casting operations [97]. Large stresses are generated in the rolls from the ferrostatic pressure head of molten steel within the solidified steel shell and the weight of the steel strand itself [97]. To accommodate this, all four faces are supported by closely spaced rolls at the exit of the mould, while only two faces are supported further along the casting line. Cooling water sprays introduce corrosion issues from steam oxidation and contribute to issues of thermal cracking [97]. Abrasive wear of the roll surfaces results from steel oxides, casting slag and mineral deposits which reduce the roll diameters below acceptable limits [97]. In assessing the performance of detonation-sprayed Cr_3C_2–25NiCr coatings, Wang et al. [98, 99] noted the formation of 'alligator cracks' on the surface of uncoated rolls after 3,740 heats, while coated rolls showed no changes after 12,000 heats [99]. Sanz [97] trailed several chemically hardened TS coatings (HVOF WC–17 % Co coated with a dense chemically blended Cr_2O_3 + SiO_2 + Al_2O_3 based oxide layer) which showed superior wear resistance and lower abrasion rates relative to a new generation weld overlay coating (weld composition: Fe, Nb 0.637, Ni 13.78, Mo 1.39, Cr 17.43, Mn 3.58, Si 0.67 and C 0.062) [98].

1.8 Direction of Future Research

Ever increasing demand for technology has resulted in development of new technique, new materials and new architecture for thermal-sprayed coatings. New era should see development of super hard thermal-sprayed coating. Considerable effort will be diverted for further development of composite and functionally graded thermal-sprayed coatings. Development of smart thermal-sprayed coatings which respond in a selective way to external factors such as stress, temperature, etc. and innovations of new processes such as warm spraying, hybrid thermal spraying, plasma-assisted thermal spraying will draw lot of attention in the years to come.

One important recent development in thermal spraying is cold spray process. Cold spray process can best be used by using expensive helium gases. Improvement in the recovery process made this process quite competitive. Further development in nozzle design, powder quality and process optimisation will result in improved coating performances. To date only limited number of feed stock materials is available for cold spraying. This needs to be upgraded.

Another important development for thermal spraying is nearnet shape components. Substantial research is expected to be directed in the development of removable mandrel and development of complex shape. A larger variety of coatings for near net shape tribology-related components and their increased usage is expected to come in near future.

Over the last decades, various sensors relevant for thermal spraying process and capable of operating under harsh environment of spray booth are developed [100]. Today sensors are available to measure trajectories, temperatures, velocities, sizes and shapes of in-flight particles [101–106]. Infrared cameras and pyrometers are employed to understand the temperature profile of substrate and coatings during preheating, spray process and cooling down [107–115]. Sensors are also developed to measure the stresses within the coatings and evolution of thickness of the coating during spraying [116]. Further development towards improved precision on measurement is expected. New techniques such as shadowgraphs and laser allow precise measurement of particle diameter, which was not hitherto possible.

Development of sensors will allow online control of thermal spraying and thus enhance coating quality and reliability.

There should be more comprehensive database on tribological properties of a larger variety of compositions of cermet coatings. Further effort should be directed for generating data on high temperature wear resistance of cermet coatings and wear behaviour of the said coatings in presence of corrosive fluid.

References

1. Tucker RC Jr (1994) Thermal spray coatings. In: Surface engineering, vol 5, ASM handbook. ASM, Materials Park, OH, p 497
2. Kharlamov YA (1987) Mater Sci Eng 93A:1
3. Fauchais P, Vardelle A, Dussoubs B (2001) J Therm Spray Technol 10:44
4. Pawlowski L (1995) The science and engineering of thermal spray coatings. Wiley, Chichester
5. Wood RJK (2010) Int J Refract Met Hard Mater 28:82
6. Roy M, Pauschitz A, Franek F (2002) In: Proceedings of the 15th international corrosion congress, Grenada, Spain, 22–27 Sept 2002
7. Roy M, Narkhede BE, Paul SN (1999) In: Prashad H (ed) Tribology in 2000 and beyond. Kalajyothi, Hyderabad, India, p 205
8. Kear BH, Sadangi RK, Jain M, Yao R, Kalman Z, Skandan G, Mayo WE (2000) J Therm Spray Technol 9:399
9. Roy M, Pauschitz A, Benardi J, Franek F (2006) J Therm Spray Technol 15(3):372
10. He J, Lavernia EJ (2000) Metall Mater Trans 31A:555
11. Guilemany JM, Calero JA (1997) In: Berndt CC (ed) A united forum for scientific and technological advances. ASM, Materials Park, OH, p 15
12. Roy M, Sundararajan G (1971) In: Proceedings of the 12th international colloquim of tribology, Stuttgart/Ostfilderm, Germany, Jan 2000
13. Mohanty M, Smith RW, De Bonte M, Celis JP, Lugscheider E (1996) Wear 198:251
14. Marshall BD, Noma T, Evans AG (1982) J Am Ceram Soc 65(10):C175
15. Leigh SH, Lin CK, Berndt CC (1995) J Am Ceram Soc 80(8):2093
16. Walpole RE, Myers RH (1978) Probability and statistics for engineers and scientists, 2nd edn. Macmillan, New York
17. Lin CK, Berndt CC (1995) J Mater Sci 30:111
18. Antis GR, Chantikul P, Lawn BR, Marshall DB (1981) J Am Ceram Soc 64:533
19. Evans AG, Charles EA (1976) J Am Ceram Soc 59:371
20. Niihara K, Morena R, Hassleman DPH (1983) In: Bradt RC, Evans AG, Hassleman DP, Lange FF (eds) Fracture mechanics of ceramics. Plenum, New York, p 97
21. Niihara K, Morena R, Hassleman DPH (1982) J Mater Sci Lett 1:13
22. Roy M, Pauschitz A, Franek F (2006) Tribol Int 39:29
23. Greenwood JA, Williamson JBP (1966) Proc R Soc Lond A 295:300
24. Lim SC, Ashby MF (1987) Acta Metall 35:11
25. Zhao XQ, Zhou HD, Chen JM (2006) Mater Sci Eng A431:290
26. Roy M, Pauschitz A, Franek F (2004) Wear 257:799–811
27. Zhu YC, Ding CX (1999) Nanostruct Mater 11:319
28. Rodrıguez J, Martın A, Fernandez R, Fernández JE (2003) Wear 255:950
29. Roy M (2009) Trans Indian Inst Met 62:197
30. Lenling WJ, Smith MF, Henfling JA (1990) In: Proceedings of the third thermal spray conference, Long Beach, CA, p 227

31. Ito H, Nakamura R, Shiroyama M, Sasaki T (1990) In: Proceedings of the third thermal spray conference, Long Beach, CA, p 223
32. Stewart A, Shipway P, Maccartney DG (1998) Surf Coat Technol 105:13
33. Khameneh Asl S, Hyderzadeh Sohi M, Hadavi SMM (2004) Mater Sci Forum 465–466:427
34. Khameneh Asl S, Hyderzadeh Sohi M, Hokamoto K, Umera M (2006) Wear 260:1203
35. Guilemany JM, Miguel JM, Vizcaino S, Lorenzana C, Delgado J, Sanchez J (2002) Surf Coat Technol 157:207
36. Stoica V, Ahmed R, Itsukaichi T (2005) Surf Coat Technol 199:7
37. Stoica V, Ahmed R, Itsukaichib T, Tobe S (2004) Wear 257:1103
38. Stoica V, Ahmed R, Golshan M, Tobe S (2004) J Therm Spray Technol 13:93
39. Mateos J, Cuetos JM, Vijande R, Farnandez E (2001) Tribol Int 34:345
40. Bolelli G, Rauch J, Cannillo V, Killinger A, Lusvarghi L, Gadow R (2009) J Therm Spray Technol 18:35
41. Sudaprasert T, Shipway PH, MaCartney DG (2003) Wear 255:7
42. Maiti AK, Mukhopadhyah N, Raman R (2007) Surf Coat Technol 201:7781
43. Berget J, Rohne T, Bardal E (2007) Surf Coat Technol 201:7619
44. Chivavibul P, Watanabe M, Kuroda S, Shinoda K (2007) Surf Coat Technol 202:509
45. Lee CW, Han JH, Yoon J, Shin MC, Kwun SI (2010) Surf Coat Technol 204:2223
46. Karimi A, Verdon C, Barbezat G (1993) Surf Coat Technol 57:81
47. Ozdemir I, Tekmen C, Tsunekawa Y, Grund T (2010) J Therm Spray Technol 19:384
48. Hawthorne HM, Arsenault B, Immarigeon JP, Legoux JG, Parameswaran VR (1999) Wear 225–229:825
49. Levy AV, Wang B (1988) Wear 121:325
50. Barbezat B, Nicoll AR, Sickinger A (1993) Wear 162–164:529
51. Wood RJK, Mellor BG, Binfield ML (1997) Wear 211:70
52. Kim HJ, Kweon YG, Chang RW (1994) J Therm Spray Technol 3:169
53. Roy M, Narkhede BE, Paul SN (1999) In: Proceedings of the 2nd international conference on industrial tribology, Hyderabad, India, 1–4 Dec 1999, p 205
54. Hearley JA, Little JA, Sturgeon AJ (1999) Wear 233–235:328
55. Wang BQ, Verstak A (1999) Wear 233–235:342
56. Smith RW, Mohanty M, Stessel E, Verstak A, Ohmori A (eds) (1995) Thermal spraying-current status and future trends. In: Proceedings of ITSC'95, High Temperature Society of Japan, Osaka, Japan, p 1121
57. Verstak A, Vitiaz P, Lugscheider E (1996) DVS Berichte 175:71, German Welding Society
58. Tayor ML, Murphy JG, King HW (1997) In: Proceedings of ASM symposium on tribological mechanism and wear problems in materials, ASM International, Materials Park, OH, p 143
59. Li CJ, Yang GJ, Ohmori A (2006) Wear 260:1166
60. Tsai PC, Lee JH, Chang C-L (2007) Surf Coat Technol 202:719
61. Kulu P, Hussainova I, Veinthal R (2005) Wear 258:488
62. Uusitalo MA, Vuoristo PMJ, Mäntyla TA (2002) Wear 252:586
63. Wang BQ, Lee SW (1997) Wear 203–204:580
64. Guilemany JM, Cinca N, Fernández J, Sampath S (2008) J Therm Spray Technol 17:762
65. Dent AH, DePalo S, Sampath S (2002) J Therm Spray Technol 11:551
66. Abdel-Samad AA, El-Bahloul AAM, Lugscheider E, Rassoul SA (2000) J Mater Sci 35:3127
67. Gawne DT, Qui Z, Zhang T, Bao Y, Zhang K (2001) J Therm Spray Technol 10:599
68. Liu Y, Fischer T, Dent A (2003) Surf Coat Technol 167:68
69. Habib KA, Saura JJ, Ferrer C, Damra MS, Gimenez E, Cabedo L (2006) Surf Coat Technol 201:1436
70. Nerz JE, Kushner BA, Jr, Rotolico AJ (1991) In: Proceedings of 4th national thermal spraying conference, Pittsburg, PA, 4–10 May 1991
71. Lee HC, Gurland J (1978) Mater Sci Eng 33:125
72. Kumari K, Anand K, Bellaci M, Giannozzi M (2010) Wear 268:1309
73. Stewart DA, Shipway PH, McCartney DG (1998) Surf Coat Technol 105:13

74. Wang Q, Chen ZH, Ding ZX (2009) Tribol Int 42:1046
75. Wirojanupatump S, Shipway PH, McCartney DG (2001) Wear 249:829
76. Zum Ghar KH, Eldis GT (1980) Wear 64:175
77. Thakare MR, Wharton JA, Wood RJK, Menger C (2008) Tribol Int 41:629
78. Stewart DA, Shipway PH, McCartney DG (1999) Wear 225–229:789
79. Qi X, Eigen N, Aust E, Gartner F, Klassen T, Bormann R (2006) Surf Coat Technol 200:5037
80. Kim JH, Baik KH, Seong BG, Hwang SY (2007) Mater Sci Eng A449–551:876
81. Jordan EH, Gell M, Sohn YH, Goberman D, Shaw L, Jiang S, Wang M, Xiao TD, Wang Y, Strutt P (2001) Mater Sci Eng A301:80
82. Gell M, Jordan EH, Sohn YH, Goberman D, Shaw L, Xiao TD (2001) Surf Coat Technol 146–147:48
83. Baik KH, Kim JH, Seong BG (2007) Mater Sci Eng A449–551:846
84. McGrann RTR, Shanley JR (1997) In: Berndt CC (ed) Thermal spray: a united forum for scientific and technological advances. ASM International, Materials Park, OH, p 341
85. Wheeler DW, Wood RJK (2005) Wear 258:526
86. Lavin P (1998) Coating of continuous casting machine components. International Patent, Publication Number WO 98/21379. Monitor Coating and Engineering Limited
87. Sato Y, Midorikawa S, Iwashita Y, Yokogawa A, Takano T (1993) Service life extension technique for cold rolling rolls. Kawasaki Steel Technical Report No. 29, p 74
88. Kasai S, Sato Y, Yanagisawa A, Ichihara A, Onishi H (1987) Development of surface treatment techniques for process rolls in steelworks. Report No. 17, p 81
89. Ren X, Mei X, She J, Ma J (2007) Materials resistance to liquid zinc corrosion of surface of sink roll. In: Proceedings of Sino-Swedish structural materials symposium 2007, p 125
90. Sawa M, Oohori J (1995) In: Ohmri A (ed) Thermal spraying: current status and future tends. Proceedings of 14th international thermal spray conference, Kobe, Japan, 22–26 May 1995. High Temperature Society of Japan, p 37
91. Seong BG, Hwang SY, Kim MC, Kimin KY (2000) In: Berndt CC (ed) Thermal spray: surface engineering via applied research. Proceedings of 1st international thermal spray conference, Montreal QC, 8–11 May 2000. ASM International, Materials Park, OH, p 1159
92. Murakawa M, Watanabe S (1989) In: Proceedings of 2nd international conference on hot isostatic processing, theory and applications, Gaitherburg, MD, 7–9 June 1989
93. Picas JA, Forn A, Matthaus G (2006) Wear 261:477
94. Dong Y, Yan D, He J, Zhang J, Li X (2006) Surf Coat Technol 201:2455
95. Hollis K, Peters M, Bartram B (2003) In: Moreau C, Marple B (eds) Thermal spray 2003: advancing the science and applying the technology. Proceedings of the 2003 international thermal spray conference, Orlando, FL, 5–8 May 2003. ASM International, Materials Park, OH, p 153
96. Fukubayashi HH (2004) In: Proceedings of the international thermal spray conference 2004, Osaka, 10–12 May 2004. ASM International, Materials Park, OH, p 125
97. Sanz A (2004) Surf Coat Technol 177–178:1
98. Wang J, Zhang L, Sun B, Zhou Y (2000) Surf Coat Technol 130:69
99. Wang J, Sun B, Guo Q, Nishio M, Ogawa H (2002) J Therm Spray Technol 11(2):261
100. Fauchais P, Verdelle M (2010) J Therm Spray Technol 19:668
101. Fauchais P (1992) J Therm Spray Technol 1:117
102. Li CJ, Wu T, Li C-X, Sun B (2003) J Therm Spray Technol 12:80
103. Marple BR, Voyer J, Bisson JF, Moreau C (2001) J Mater Process Technol 117:418
104. Landes K (2006) Surf Coat Technol 201:1948
105. Gougeon P, Moreau C (1993) J Therm Spray Technol 2:229
106. Planche MP, Liao H, Coddet C (2004) Surf Coat Technol 182:215
107. Verdelle M, Renault T, Fauchais P (2002) High Temp Mater Process 6:469
108. Doubenskaia M, Bertrand P, Smurov I (2006) Surf Coat Technol 201:1955
109. Xiaa W, Zangb H, Wanga G, Wanga Y (2009) J Mater Process Technol 209:1955

110. Salimijazi HR, Pershin L, Coyle TW, Mostaghimi J, Chandra S, Lau YC, Rosenzweig L, Moran E (2007) J Therm Spray Technol 16:580
111. Kuroda S, Fukushima T, Kitahara S (1988) Thin Solid Films 164:157
112. Kuroda S, Clyne TW (1991) Thin Solid Films 200:49
113. Clyne TW, Gill SC (1996) J Therm Spray Technol 5:401
114. Matejicek J, Sampath S (2003) Acta Mater 51:863
115. Matejicek J, Sampath S, Gilmore D, Neiser R (2003) Acta Mater 51:873
116. Nadeau A, Pouliot L, Nadeau F, Blain J, Berube SA, Moreau C, Lamontagne M (2006) J Therm Spray Technol 15:744

Chapter 2
Nanocomposite Films for Wear Resistance Applications

Manish Roy

2.1 Introduction

A new generation thin films which consists of at least two separate phases with one being nanocrystalline or amorphous phase are termed as nanocomposite coatings. These coatings are usually made from ternary or higher order systems and comprise at least two immiscible phases. The most widely studied nanocomposite coatings are ternary, quaternary or even more complex systems having nanocrystalline grains of hard transition metal nitrides (e.g. TiN, AlN, CrN, Zrn, Bn, etc.), carbides (e.g. VC, WC, ZrC, TiC, etc.), borides (e.g. WB, ZrB_2, TiB_2, CrB_2, VB_2, etc.), oxides (e.g. Al_2O_3, TiO_2, Y_2O_3, ZrO_2, etc.) or silicides (e.g. TiS_2, CrS_2, ZrS_2, etc.) surrounding by amorphous matrices (e.g. Si_3N_2, BN, C, etc.). According to Veprek et al., the nanocrystalline grains should be 5–10 nm in size and separated by 2–5 nm within an amorphous phase [1].

These coatings exhibit enhanced mechanical [2–5], physical [6, 7] and functional [8–11] properties. These enhanced properties are due to large interface and nanoscale effects associated with nanocomposite coatings. In these coatings a large volume fraction of atoms are located in the interface where their behaviour is different from the bulk and crystalline size is smaller than 100 nm. There are many fundamental processes in solids that are associated with length scales of around a few nanometres. In recent times, nanocomposite coatings have attracted attention from researchers due to enormous possibilities of synthesising different varieties of nanocomposite films.

The interest in nanocomposite coating has got impetus due to the fact PVD and CVD systems that are used for deposition of coating, leads to the formation of layers having grain sizes much smaller than that obtained by conventional routes.

M. Roy (✉)
Defence Metallurgical Research Laboratory, Kanchanbagh, Hyderabad 500058, Andhra Pradesh, India
e-mail: rmanish64@rediffmail.com

M. Roy (ed.), *Surface Engineering for Enhanced Performance against Wear*, DOI 10.1007/978-3-7091-0101-8_2, © Springer-Verlag Wien 2013

Further, many of the difficulties for processing the nanocrystalline materials can easily be overcome by these deposition techniques. Vast development of instrumentation techniques allows more precise control of microstructural features and mechanical properties. The possibility to carry out in situ indentation and tribological experiments for both structural and chemical information has propelled design and development of novel coatings in the field of surface engineering.

This chapter reviews the state of art of nanocomposite thin film and their tribological performances and applications. Although it is possible to produce nanocomposite coatings using thermal spray technique, the discussion in this chapter is confined to nanocomposite thin film obtained by CVD or PVD techniques. Thermal spray nanocomposite coatings are discussed in Chap. 1.

2.2 Deposition of Nanocomposite Films

As mentioned earlier, nanocomposite films are usually comprised of minimum two phases, a nanocrystalline phase surrounded by two amorphous phases. Thus synthesis of such coatings depends primarily on co-deposition of both nanocrystalline and amorphous phases such as nc-TiN and nc-TiS_2/a-Si_3N_4 [12], nc-TiB_2/BN [13], nc-TiAlN/a-Si_3N_4 [14], nc-TiC/a-C [15], W_2N/a-Si_3N_4 [16], nc-WC/a-C [17], etc. There are varieties of methods available for deposition of nanocomposite films. The most popular among these methods are plasma-assisted physical vapour deposition (PAPVD) [18], ion-beam and magnetron sputtering [19, 20], combined PVD and PACVD process [1] and also by hybrid process such as magnetron sputtering and laser ablation [21]. However, most widely used method for industrial purpose is magnetron sputtering [22] since this method can be scaled easily for industrial applications and deposition in this method takes place at low temperature. Moreover, any kind of materials can be used as target materials in this method.

Chemical vapour deposition (CVD) technique is another promising method to deposit nanocomposite coatings [23, 24]. Ease of uniform deposition on complicated geometries and fast deposition rate are main advantages of CVD techniques. However, the precursor gases $TiCl_4$, $SiCl_4$ and SiH_4 pose main problem due to their corrosive and fire hazardous nature. The deposition temperature for CVD method is relatively high and thus prevents use of such method in many applications. Further, because of presence of chloride, the protective coating may not withstand high temperature exposure for desired duration.

Plasma-enhanced chemical vapour deposition (PECVD) is also known as glow discharge method. The common PECVD method used at lower pressure such as DC or RF discharges operate at pressure of 10–100 mTorr. In RF–PECVD system, electron and ion densities are at most 10^{11} cm^{-3} at a pressure of 0.1 Torr in which the number density of molecules is of the order of 10^{15} cm^{-3}. Finger et al. [25] employed very high frequency chemical vapour deposition technique (VHF-CVD) to deposit nanocomposite Si films. Such technique results in increase in deposition rate. Use of hot wire (HW) CVD resulted very good quality nanocomposite films [26, 27]. In this deposition method evaporation of the tungsten filament is a problem due to high temperature used and this gives rise to contamination. In this method problems can come due to fragile tungsten filament and there can be inhomogeneity when large area is coated.

Electron cyclotron resonance (ECR) CVD system has attracted interest for low temperature deposition of nanocomposite films [28]. In this system, the energy of the electrons in the plasma can be controlled independent of the generation of the damaging ion species. The formation of higher radical particles which deteriorate the film quality is considerably suppressed and highly crystallised films can be grown at 120 °C by ECR-CVD. Electron cyclotron wave resonance (ECWR) CVD provides better coupling to the plasma. This permits use of this system for large area deposition.

Cathodic arc evaporation (CAE) has been used for production of wear resistance coatings [29], coatings for cutting tools [30] and coatings for automotive applications [31] for more than a decade. This process is extremely versatile due to high deposition rate, high degree of ionisation of evaporation species and high ion energies (20–120 eV). However, only materials having good electroconductivity, reasonable mechanical strength and optimum melting point can be used as target material. Many of these problems can be overcome in sputtering process. The main drawback of sputtering process is low degree of ionisation. This drawback has been overcome by introduction of unbalanced magnetron sputtering (UBM) [22] and high power pulsed magnetron sputtering (HPPMS) [32].

In DC magnetron sputtering, a glow discharge is created by applying negative voltage to the target material under vacuum in low pressure inert gas. The application of magnetic field behind target material allows higher deposition rate during sputtering [33]. The development of unbalanced magnetron sputtering has significantly increased the plasma generation area away from the target [34]. The development of closed field unbalanced magnetron sputtering resulted in much higher ion current densities [34]. One of the most important improvements of magnetron sputtering is reactive deposition of ceramic and compound coatings. However, deposition of insulating materials results in deposition of insulating layer on the anode giving rise to 'disappearing anode' and deposition of insulating film on target giving rise to 'poisoned target'. In order to overcome this problem, radio frequency sputtering (RF) was developed. However, RF sputtering suffers from the setback of low deposition rate, high cost and generation of high temperature. In view of the above, in recent times, pulsed DC magnetron sputtering has been developed. During the normal pulse-on-period, the negative sputtering voltage is applied to the target as in conventional DC sputtering. The target negative potential is periodically interrupted by a positive pulse voltage. In the closed field configuration, two or more magnetrons are used in reactive sputtering. All of them can be run in pulsating mode making different combination of mode. Magnetron potential can be pulsed either in asynchronous bipolar mode or in synchronous bipolar mode giving more variation of deposition conditions. In these deposition techniques a high density plasma is created.

Recently high power pulsed magnetron sputtering in unipolar mode has become widely popular. High density plasma is created in front of the sputtering source by using pulsed high target power density, ionising a large fraction of the sputtered atoms [35]. It is reported that the fraction of ionised species in high power pulsed DC magnetron sputtering can be increased beyond 70 % [36]. This high degree of ionised ions with high energy allows controlling film growth behaviour and thereby produces high quality films. In recent times several nanocomposite coatings are deposited using hybrid coating system where cathodic arc evaporation system is combined with magnetron sputtering technique.

2.3 Important Features of Nanocomposite Films

2.3.1 Microstructural Features of Nanocomposite Films

The main objective of microstructural study of nanocomposite coatings is to determine the structure property relationship so that the gap between atomic structure and the macroscopic properties can be bridged. The most important step in this connection is to ascertain the appropriate length scale where relevant processes take place and the physical basis for such assessment. Although much information can be derived in relation to interfaces because of advent of high resolution transmission electron microscopy (HRTEM), the necessary chemical information at an atomic level is still missing. Field ion microscopy-imaging atom probe is the most important tool for acquiring such information, although other techniques such as an ultrahigh resolution FEG transmission electron microscopy with energy dispersive spectrometry having smaller than 0.5-nm diameter probe can also reveal this information. It can be stated that electron microscopy characterisation accompanied with complimentary techniques serve as a crucial input for deeper understanding for correlating quantitative electron microscopy information to coating properties.

SEM images showing the evolution of the cross-section morphology of W–S–C nanocomposite coatings with increasing C content obtained by reactive sputtering is presented in Fig. 2.1. As can be stated, there is a progressive densification of the coatings with a significant decrease of the number of pores, which results in a decrease of the exposed surface area where O can be adsorbed. With increase in carbon content the film becomes amorphous. The microstructural feature of a nCTiC/a-C:H film deposited using unbalanced magnetron sputter deposition (UMSD) is narrated below. Selected area diffraction (SAD) patterns of four representative samples merged clockwise with decreasing Ti concentration are illustrated in Fig. 2.2 [38]. Diffraction rings point out randomly oriented polycrystalline structure. The upper left diffraction pattern is consistent with hexagonal close-packed lattice of a Ti phase (lattice constants 0.295 and 0.468 nm) with main reflections of the (101), (120), (122) planes. The next pattern in the clockwise direction exhibits that the structure changes with low C content. Simulation of diffraction patterns indicates cubic TiC with lattice constant 0.43 nm and main reflections in the pattern corresponding to (111), (002), (202), (222) planes. The diffraction pattern remains unchanged with increasing C content but becomes more diffuse from about 40 at.% C concentration (sample T303) as TiC nanocrystal grain

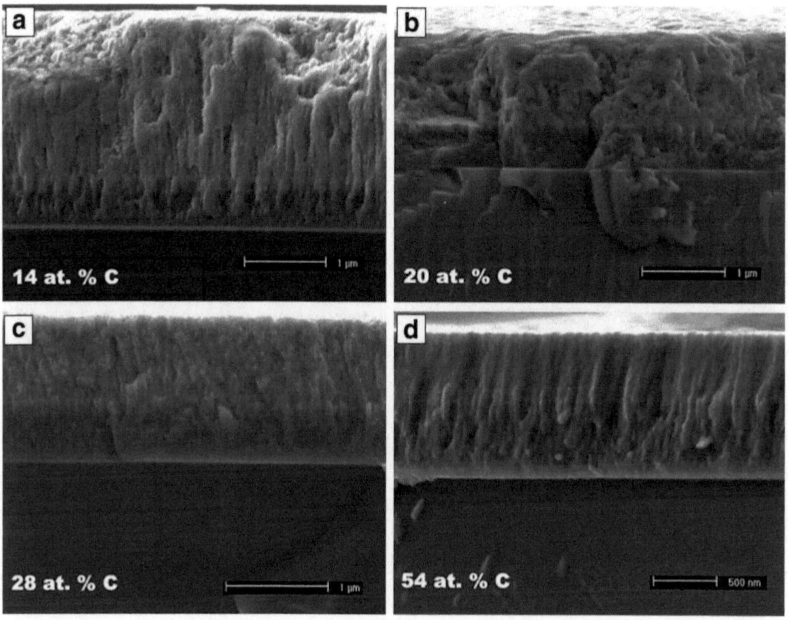

Fig. 2.1 SEM images of a series of W–S–C films containing WC nanograins in the matrix of WS_2 films [37]

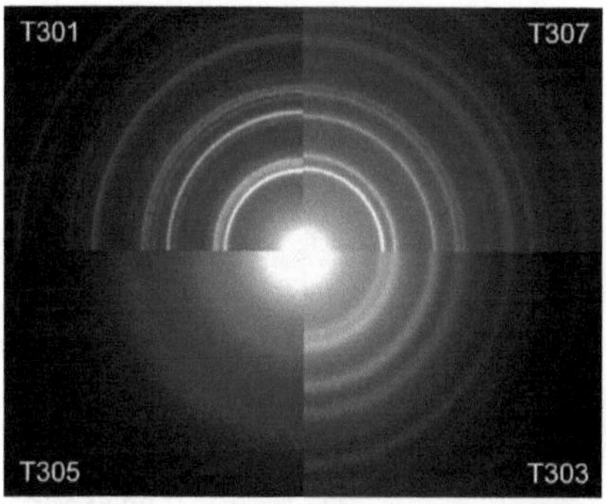

Fig. 2.2 Selected area diffraction (SAD) patterns of four representative samples of a nCTiC/a-C:H film deposited using unbalanced magnetron sputter deposition (UMSD) (clockwise with decreasing Ti concentration) [38]

size decreases and gradually more amorphous a-C:H phase islands are embedded in the TiC phase. At very low Ti concentration (sample T305), no structure was observed in SAD. The highly diffused diffraction pattern corresponds to a-C:H matrix with a decreasing amount of tiny TiC nanocrystals dispersed. Above results correlate well with decreasing grain size in bright field and dark field micrographs as illustrated in Fig. 2.3. TEM images confirm nanometre range grain size.

The XRD patterns of a series of W–S–C films containing WC nanograins in the matrix of WS_2 films are illustrated in Fig. 2.4. The main points worth mentioning

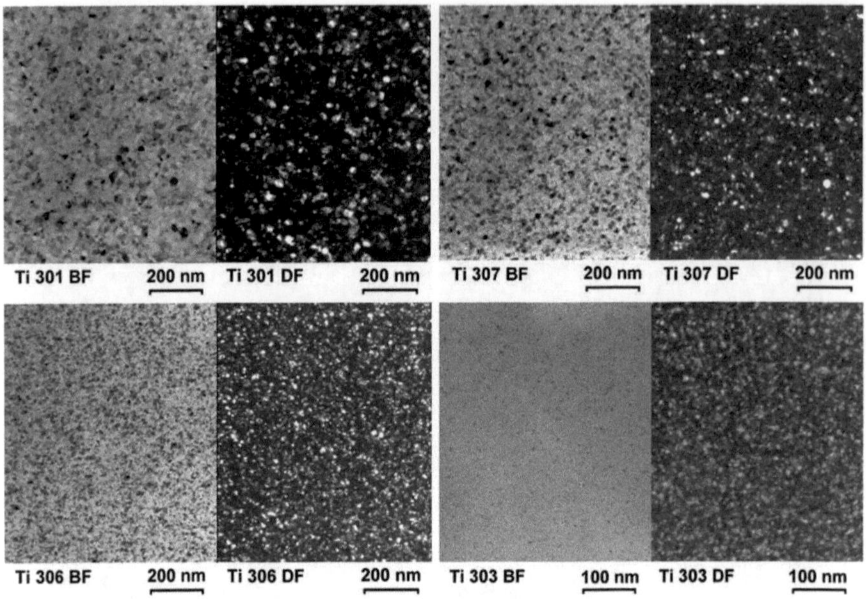

Fig. 2.3 The *bright field* and *dark field* TEM micrographs of a nCTiC/a-C:H film deposited using unbalanced magnetron sputter deposition (UMSD) (with decreasing Ti concentration) [38]

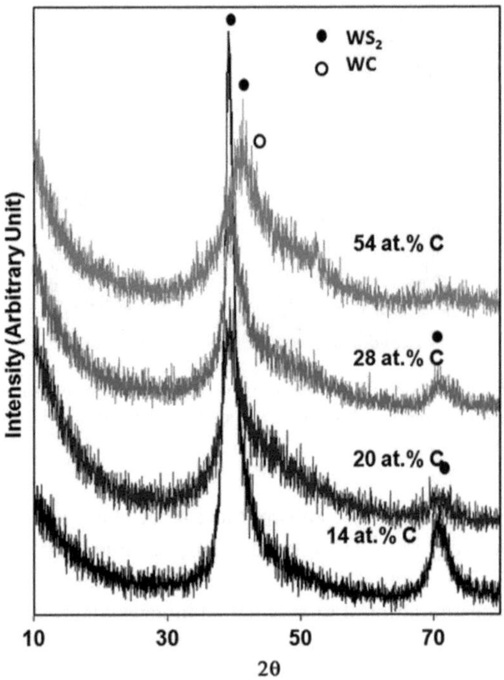

Fig. 2.4 The XRD patterns of a series of W–S–C films containing WC nanograins in the matrix of WS_2 films [37]

are domination of the XRD patterns by main broad asymmetric peaks which are indexed with the general form (10*L*) corresponding to a turbostrating stacking of the family of plans (10*L*) with $L = 0,1,2,3$ [39] of the hexagonal WS_2 phase (International Center for Diffraction data (ICDD), Pennsylvania, Card 84-1398; for hexagonal WS_2 also card 841398 is for the same structure); a small peak close to $2\theta \sim 70°$ can be indexed as the (110) plan of WC and a progressive decrease in the crystallinity degree with increasing C content as shown by either the progressive broadening or the decrease in the intensity of the diffraction peaks. It is difficult to detect traces of any other crystalline phase excepting the small features on the right shoulder of the main broad peak which can be related with the presence of a W–C phase. The microstructure of the films can, thus, be described as a nanocomposite consisting of W–S nanocrystals (and in some cases of W–C nanocrystals) embedded in a C-rich amorphous layer.

Raman spectra obtained from similar nanocomposite films obtained by employing different sputtering procedure are presented in Fig. 2.5. All three films show peaks at wave number 1,360 cm^{-1} corresponding to sp^3 bond and another peak at wave number 1,560 cm^{-1}. The corresponding bond for pure diamond is at 1,330 cm^{-1}. The peak at 1,560 cm^{-1} can be assigned to the G band of microcrystalline graphite. The presence of this band proves existence of sp^2-bonded carbon within the film. Raman measurement with excitation at 532 nm is much more sensitive to sp^2 carbon than sp^3 carbon. As the peak for G band is more diffused for films obtained by co-sputtering of composite and two targets, than the film obtained by reactive sputtering, presence of more defects can be stated for these films. It is to be stated that no peak corresponding to WO_3 could be noted.

The result of the X-ray photoelectron spectroscopy (XPS) analysis obtained from the W–S–C nanocomposite films having nearly 50 at.% C and obtained by reactive sputtering is presented in Fig. 2.6. It shows the survey spectra after some light Ar sputter cleaning. Peaks for C, S, O, W and N can be noticed. The C peak can be seen at around 285 eV for C–C or C–H bonding, S peak can be seen at around 162 eV binding energy, O peak is obtained at a binding energy around 531 eV and peaks for W can be seen at a binding energy of 35 eV as well as 245, 257 and 425 eV. This is to be stated that WO_3 peak could not be noticed in Raman spectra even though spectra obtained from XPS confirm its presence indicating that oxidation of W is merely a surface-related phenomenon.

2.3.2 Mechanical Properties of Nanocomposite Films

It is already mentioned that nanocomposite coatings exhibit outstanding mechanical properties, in particular, superhardness and excellent tribological properties [40–43]. These outstanding properties are related to nanoscale effects of nanocomposite coatings.

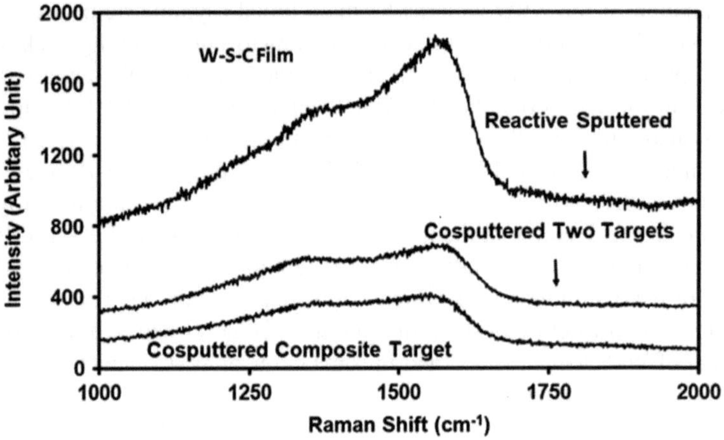

Fig. 2.5 Raman spectra obtained from W–S–C films obtained by employing different sputtering procedure

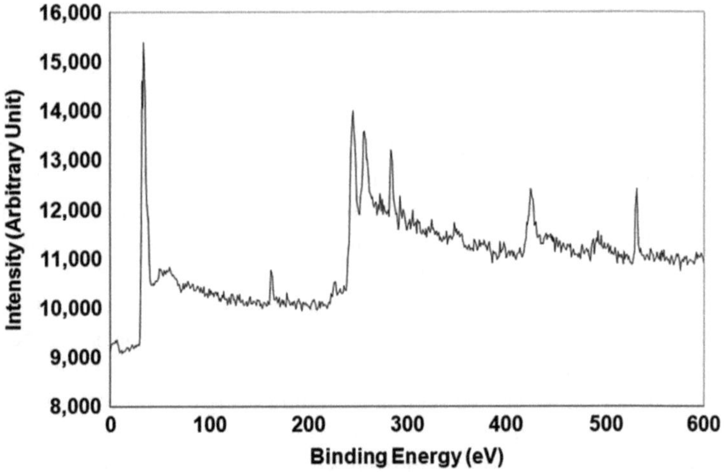

Fig. 2.6 The result of the X-ray photoelectron spectroscopy (XPS) analysis obtained from the W–S–C nanocomposite films having nearly 50 at.% C and obtained by reactive sputtering

2.3.2.1 Deformation Behaviour of Nanocomposite Coatings

Nanocomposite coatings exhibit hardness significantly higher than that given by rule of mixture i.e.

$$H(A_aB_b) = \frac{aH(A) + bH(B)}{a + b} \tag{2.1}$$

where $H(A)$ is the hardness of pure A and $H(B)$ is the hardness of pure B. a and b are the compositions of A and B in the mixture and $H(A_a\ B_b)$ is the hardness of the mixture. In coatings with conventional grain size, grain size is one of the ways to increase the hardness. With decrease in grain size, the multiplication and mobility of dislocations are hindered and hardness of the materials increases according to Hall–Petch equation [44, 45].

$$H_d = H_0 + Kd^{-\frac{1}{2}} \tag{2.2}$$

The very high mechanical characteristics [40–43] of nanocomposite coating should be related to specific deformation mechanisms of nanocomposite coatings. It should be stated that the intrinsic deformation mechanisms operating in nanocomposite coatings are similar to that operating in nanocomposite bulk materials. Differences in external factors such as sample geometry, state of stress and substrate effect may change the macroscopic mechanical properties, but intrinsic deformation behaviour remains unchanged. The deformation behaviour of nanocomposite coating is characterised by abnormal Hall–Petch behaviour which manifests itself as either saturation or decrease of yield stress of the coating with decrease of grain size [46, 47]. Under such condition, conventional dislocation slip dominates in nanocomposite coatings having grain size greater than a critical value. Deformation behaviour controlled by active roles of grain boundaries dominate in coatings having very small grain sizes. Although the mechanisms in this regime is not well understood and is a subject of controversy, based on recent literature [48–54], following deformation mechanisms are expected to operate in nanocomposite coatings having grain size less than critical value.

(a) Lattice dislocation slip
(b) Grain boundary sliding
(c) Grain boundary diffusional creep
(d) Triple junction diffusional creep
(e) Rotational deformation occurring via movement of grain boundary disclinations
(f) Twin deformation conducted by partial dislocations emitted from grain boundaries

Lattice dislocation slip is the dominating deformation mechanism in materials having grain size greater than a critical value about 10–30 nm. In nanocomposite coatings, lattice dislocation slip exhibits some specific features owing to interface and nanoscale effects. When the grain size is very small, there is very dislocation for lattice dislocation slip to occur and hence yield stress deviates from classical Hall–Petch relationship. According to some investigators [55, 56] as the particle sizes become smaller, more and more dislocations are absorbed in the grain boundary, leaving low dislocation density and low flow stress. Although there is good correlation between theoretical and experimental data, the question remains that does lattice dislocation exist and play same role in nanocomposite coating as that of conventional materials? It is pointed out [57] that existence of lattice dislocation is energetically unfavourable in either free nanoparticles or in nanograins.

Grain boundary sliding is caused mainly by grain boundary dislocation having small burgers vector and parallel to grain boundary planes. According to Hahn et al. [58], grain boundary migration is an accommodation mechanism whose combined action with grain boundary sliding gives rise to localization of deformation in nanocomposite coatings. In such case, grain boundary planes become parallel to each other as a result of movement of partial cellular dislocation and defects in arrays of nanograins [59]. In the other case, grain boundary sliding and transformation of defect structures near triple junction of grain boundary result in strengthening. Thus, nanocomposite coatings deformed through grain boundary sliding, either strengthening due to transformation of grain boundary dislocation at triple junctions or softening due to local migration of grain boundary. The competition between strengthening and softening profoundly influences the deformation behaviour of nanocomposite coatings.

Grain boundary diffusional creep or coble creep and triple junction diffusional creep also contribute to plastic flow of nanocomposite coating having very small grains. As the volume fraction of grain boundaries and triple junction phases increases, the importance of these mechanisms in plastic deformation increases. Coatings where grain sizes are very small and thus grain boundary sliding is suppressed, coble creep becomes important deformation mechanism. Some nanocomposite coatings do not contain mobile dislocation grain boundary responsible for grain boundary sliding. In such coatings, coble creep and triple junction diffusional creep are considered important operating mechanism.

Rotational deformation in nanocomposite coating is plastic deformation accompanied by crystal lattice rotation. The main carriers of rotational plastic deformation are believed to be dipoles of grain boundary disinclinations [60]. Grain boundary disinclination is a line defect that separates two grain boundary fragments having different misorientations. Movement of disinclination dipoles causes plastic deformation accompanied by crystal lattice rotation. This mode of deformation is very intensive at times in nanocomposite coatings [61]. The elastic energy of disinclination dipoles diverges rapidly on increasing the distances between disinclinations [62]. Thus dipoles that are close to each other are energetically favourable for grain boundary disinclinations. In nanocomposite coatings, interspacing between neighbouring grains are extremely small. Further, the number of triple junctions where the crossover from conventional grain boundary to rotational deformation occurs is extremely high in nanocomposite coatings.

2.3.2.2 Hardness, Fracture Toughness and Residual Stresses of Nanocomposite Coatings

Hardness of thin films are determined using a nanoindenter equipped with a Berkovich three-sided pyramidal diamond indenter with a nominal angle equal to

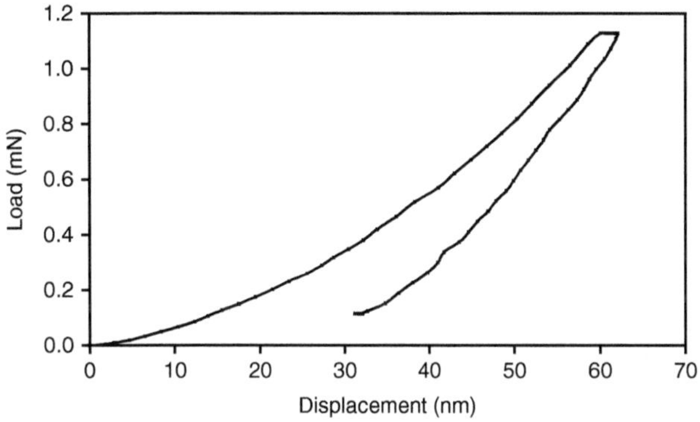

Fig. 2.7 Load vs. displacement diagram obtained for Ti–C:H nanocomposite film

63.5°. In this method, displacement (i.e. penetration depth) is continuously recorded during the total loading and unloading cycle and hence a typical load vs. displacement diagram obtained for Ti–C:H nanocomposite film as presented in Fig. 2.7. The contact depth h_c is estimated from load displacement curve as given below:

$$h_c = h_{max} - \varepsilon \frac{P_{max}}{S} \tag{2.3}$$

where h_{max} is the maximum indentation depth, P_{max} is the peak indentation load, S is the contact stiffness and ε is a constant which depends on indenter geometry. For Berkovich, $\varepsilon = 0.75$. The contact area A is estimated by evaluating an indenter shape function at contact depth h_c as given below:

$$A = f(h_c) \tag{2.4}$$

The shape function $f(h_c)$ is related to the cross-sectional area of the indenter to the distance from its tip. For the Berkovich indenter, the shape function is given by

$$f(h_c) = 24.56{h_c}^2 \tag{2.5}$$

Once the contact area is determined from load displacement curve, the hardness is given by

$$H = \frac{P_{max}}{A} \tag{2.6}$$

The elastic modulus was determined using a procedure enumerated elsewhere [63]. Elastic modulus was calculated employing equation as given below:

$$\frac{1}{E_r} = \frac{(1-\nu^2)}{E} + \frac{(1-\nu_i^2)}{E_i} \tag{2.7}$$

and E_r is given by

$$E_r = \frac{0.89S}{\sqrt{A}} \tag{2.8}$$

where S is the slope of the initial part of the unloading curve (in N/m), A is the contact area between the indenter and the substrate (in m^2). E and E_i are the elastic moduli and ν and ν_i are the Poisson ratios of the film and the indenter, respectively. The hardness and elastic modulus of a series of nanocomposite film containing WC grains in WS_2 matrix is presented in Fig. 2.8 [37]. Both the hardness and elastic modulus are improved with increasing at.% of carbon up to a threshold value and then drop down with a further increase of C. The maximum value of the hardness is attained for 54 at.% of carbon, whereas the maximum elastic modulus is noted at 28 at.% of C. As discussed in previous publication [64], such a variation can be attributed to the fact that the film becomes denser with increasing carbon content. Moreover, the possible formation of W-carbide nanograins in coatings with C contents in the range from 40 to 55 at.%, as demonstrated by XPS measurements [64], can justify the maximum of hardness for that value. Once the amount of carbon reaches a saturation point, a further increase in carbon results only in a decrease on the relative amount of W-carbide nanocrystals. This phenomenon is reflected by a decrease of both the hardness and the elastic modulus.

Fracture toughness is another important property for nanocomposite films. According to the theoretical analysis of Li and Bhushan [65], the fracture toughness of nanocomposite film is given by

$$K_{1C} = \left[\left(\frac{E}{(1-\nu^2 2\pi C_R)}\right)\left(\frac{U}{t}\right)\right]^{1/2} \tag{2.9}$$

where E is elastic modulus, v is Poisson's ratio, $2\pi C_R$ is the crack length in the film plane, U is the strain energy difference before and after cracking and t is the thickness of the film layer. It is well known that the work performed by the indenter during elastic plastic deformation of the film/substrate system is given by the area under the load displacement curve. The strain energy release can be calculated from the loading curve. U is estimated from the peak load of loading and unloading curve. Based on such calculations, the fracture toughness of some films is reported to be between 1.74 and 3.90 MPa $m^{1/2}$ [66].

The residual stress (σ) of the nanocomposite films are the results of three different types of stresses such as thermal stress (σ_T), growth-induced stress (σ_g) and structure mismatch-induced stress (σ_m) as given below:

$$\sigma = \sigma_T + \sigma_g + \sigma_m \tag{2.10}$$

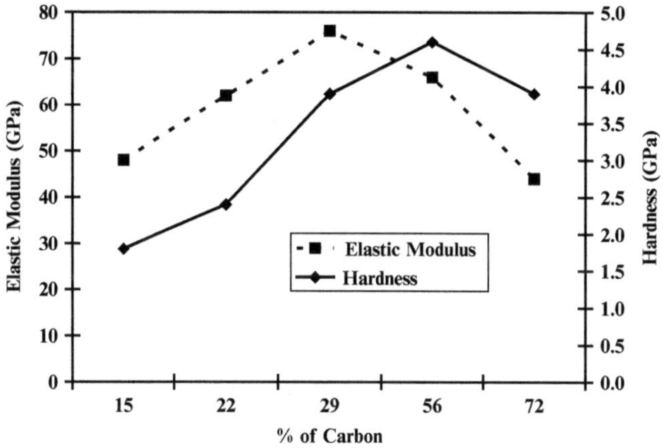

Fig. 2.8 The hardness and elastic modulus of a series of nanocomposite W–S–C film containing WC grains in WS_2 matrix [37]

As most of the nanocomposite films are having amorphous matrix, no stress is generated for matrix mismatch resulting in $\sigma_m = 0$.

Thermal stress is generated due to the mismatch between the CTE of the film and the substrate. Thus the stress in the film due to thermal mismatch can be given as [67, 68]

$$\sigma_T = \frac{E_f}{1 - \nu_f} \int_{T_{dep}}^{T_{rm}} (\alpha_s(T) - \alpha_f(T))dT \qquad (2.11)$$

where T_{rm} and T_{dep} are room temperature and deposition temperature, E_f and ν_f are the elastic modulus and Poisson ratio of the film. $\alpha_f(T)$ and $\alpha_s(T)$ are CTE of the film and the substrate respectively.

The growth stress depends on the substrate temperature, gas pressure and kinetic energy of the ions. In general, increase in power density increases the kinetic energy which in turn increases the strain. Increase in power density results in increase in residual stress from 0 to 150 MPa. Sometimes the stress is compressive and sometimes this stress is tensile.

2.4 Tribological Properties of Nanocomposite Films

2.4.1 Friction Behaviour

While engineering the surface of a material for improved performances, attention should be paid in the bulk temperature of the wearing surface. This bulk temperature (T_b) during sliding wear is given as [69, 70]

$$T_{\mathrm{b}} = T_0 + \frac{\alpha\mu F v l_b}{A_{\mathrm{n}} K_{\mathrm{m}}} \tag{2.12}$$

where T_0 is room temperature, α is a constant known as heat distribution coefficient, μ is the friction coefficient, F is applied load, v is sliding velocity, l_b is mean heat diffusion distance, A_{n} is nominal area of contact and K_{m} is thermal conductivity of the metal.

The contact temperature at the asperities, known as flash heating temperature is given by [69, 70]

$$T_{\mathrm{f}} = T_{\mathrm{b}} + \frac{\mu v r}{2K_{\mathrm{e}}} \left(\frac{H_{\mathrm{o}} F}{N A_{\mathrm{n}}} \right)^{\frac{1}{2}} \tag{2.13}$$

where H_{o} is the hardness of the oxide scale, N is the total no of the asperities, r is the radius of the contacting area and K_{e} is the equivalent thermal conductivity. Other symbols are as above.

From both the equation it is clear that for a given tribological condition, the bulk temperature or the flash temperature is controlled by friction coefficient i.e. these temperatures increase with increase of friction coefficient. If the flash temperature increases beyond some values, the mechanism of wear will change from plasticity-dominated wear to mild oxidational wear. Further if the bulk temperature is high, the hardness of the degrading layer tends to come down leading to increased wear rate in plasticity dominated wear. One way of reducing the temperature is to lower the friction coefficient. In view of these, most nanocomposite films contain an amorphous matrix phase. These amorphous phases are known for their low friction coefficient [71, 72]. This is more so with a-C phase which transform to graphitic structure and helps in reducing friction coefficient by acting as lubricant.

In support of above statement, the observation of Sánchez-López et al. [73] can be presented. Their observation indicates decrease of friction coefficient with increasing volume fraction of amorphous a-C phases. The tribological behaviour of nanocomposite films composed by metallic carbides mixed with amorphous carbon (a-C) was assessed by comparing TiC/a-C, WC/a-C and TiBC/a-C series prepared by magnetron sputtering technique with variable carbon contents. The friction coefficients appear to correlate with the overall carbon content. By estimation of the percentage of carbon atoms situated in the amorphous lubricant phase, a comparative analysis was done in selected samples providing similar friction coefficient (~0.2). The Raman analysis of the ball counterfaces demonstrates the formation of a graphitic-like material for the more lubricant films (i.e. those containing higher fraction of a-C) during friction. The relative amount of a-C [xa-C] needed to reduce friction below 0.2 is diminished in the nanocomposite films formed by metastable or non-stoichiometric carbides (TiBC/a-C and WC/a-C) from 55–60 (case of TiC/a-C) to 30–40 %. This decrease was attributed to a partial C release into the contact due to the decomposition or transformation of these compounds under friction. Figure 2.9 plots the evolution of friction coefficients

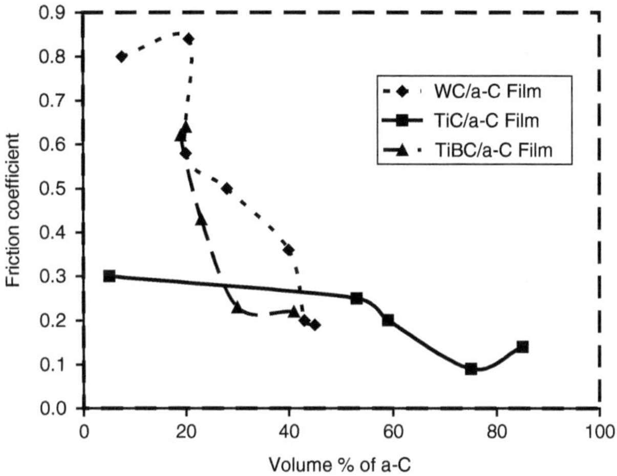

Fig. 2.9 The evolution of friction coefficients with lubricant fraction, xa-C [73]

vs. lubricant fraction, xa-C. The friction coefficient tends to decrease rapidly for the WC/a-C and TiBC/a-C systems from 0.6–0.8 to values below 0.2 for a-C contents above 40 %. For obtaining a similar low friction coefficient with the TiC/a-C nanocomposite films, it is necessary the a-C fraction rises up to 60 %. Further reduction of the friction, even below 0.1, can be achieved by increasing the fraction lubrication to a higher extent but always smaller percentages are needed for the WC/a-C and TiBC/a-C films. Feng et al. [74] deposited Ti–C:H coatings by inductively coupled plasma (ICP)-assisted hybrid chemical vapour deposition-physical vapour deposition (CVDPVD). These Ti–C:H coatings consist of nanocrystalline TiC clusters embedded in an a-C:H matrix and are thin film TiC_y amorphous hydrocarbon (a-C:H) nanocomposite materials. Results point to the fact that at Ti compositions 20 at.%, the friction coefficients stay relatively constant at 0.1. This low friction coefficient is characteristic of hard carbon-based coatings. At Ti compositions greater than 30 at.%, the friction coefficients are also relatively constant. However, they now show a greater than twofold increase in the friction coefficient to 0.25. A friction coefficient in this range is typical for unlubricated sliding between ceramic materials, such as metal carbides and metal nitrides [75].

A comparative study of nc-TiC/a-C(Al) coatings ($C_{56}C_{31}Al_{13}$) and nc-TiC/a-C coating ($C_{64}Ti_{36}$) was carried out against 100Cr6 steel balls. It was noted that the friction coefficient of nc-TiC/a-C(Al) coatings is lower than the friction coefficient of nc-TiC/a-C coating in spite of having lower a-C content as nc-TiC/a-C coating was harder and rougher than nc-TiC/a-C(Al) coatings. The friction coefficients of these nanocomposite coatings are higher than that of pure a-C coatings due to presence of less amount of graphitic phase and high hardness. However, friction coefficients of these nanocomposite coatings are lower than that of currently used industrial coatings such as TiN, CrN, TiC, etc. [76–81]. Similar observation related to lower friction coefficient of nanocomposite coatings due to presence of

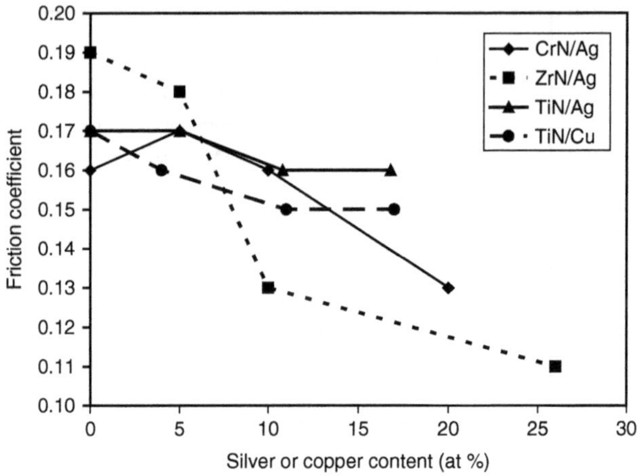

Fig. 2.10 Variation of friction coefficient as function of Ag or Cu content of a series of CrN/Ag. ZrN/Ag, TiN/Ag, TiN/Cu nanocomposite coatings deposited using dual pulsed magnetron sputtering system [83]

amorphous a-C phase and decrease of friction coefficient as a result of increase of amorphous a-C phase is also noted by Lindquist et al. [82] and Stuber et al. [83]

Another reason for reduction of friction coefficient is increase of elastic modulus and hardness of these coating. The increase in elastic modulus and hardness results in decrease of the plowing component of the friction force as per given equation [84]:

$$\mu_P = C\frac{{K_{1C}}^2}{E\sqrt{HN}} \tag{2.14}$$

where C is a constant, K_{1C} is fracture toughness for mode one fracture, E is elastic modulus, H is hardness and N is normal load. Kelly et al. [85] deposited CrN/Ag, ZrN/Ag, TiN/Ag and TiN/Cu nanocomposite coatings using dual-pulsed magnetron sputtering system. These coatings were subjected to unlubricated thrust washer wear testing [86] against 100Cr6 steel washer at ambient condition at an rpm of 30 at applied load of 100 N. This kind of test provides conformable geometry that is encountered in mechanical seal, pumps and sideways. Their results are shown in Fig. 2.10. The friction coefficient varies between 0.16 and 0.19, this reduced friction coefficient is attributed to enhanced film growth condition obtained due to pulsed DC mode [87]. Further the friction coefficient was found to decrease with increasing silver or copper content. The largest reduction is obtained for the film ZrN/Ag which decreased from 0.19 to 0.11 at 25.9 % Ag primarily due to increase of hardness and elastic modulus as described in Eq. (2.14).

Cheng et al. [88] deposited TiSiN nanocomposite coatings with a thickness of 2.5 μm using a large area filtered arc deposition (LAFAD) technique with TiSi targets having different Si content. The tribological behaviours of the TiSiN

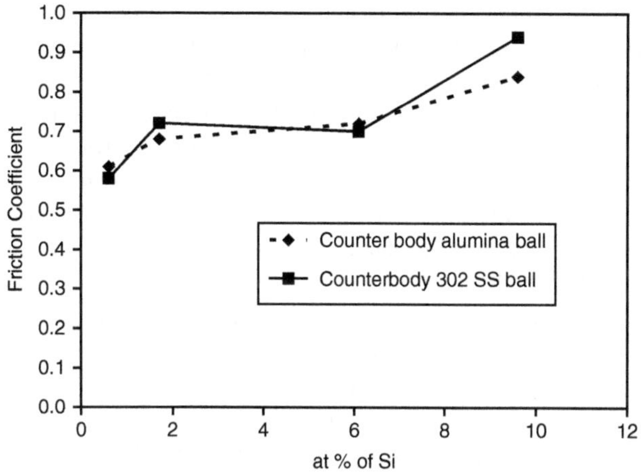

Fig. 2.11 Variation of friction coefficient with Si content of TiSiN coating against Al_2O_3 and 302 SS ball as counterbody [88]

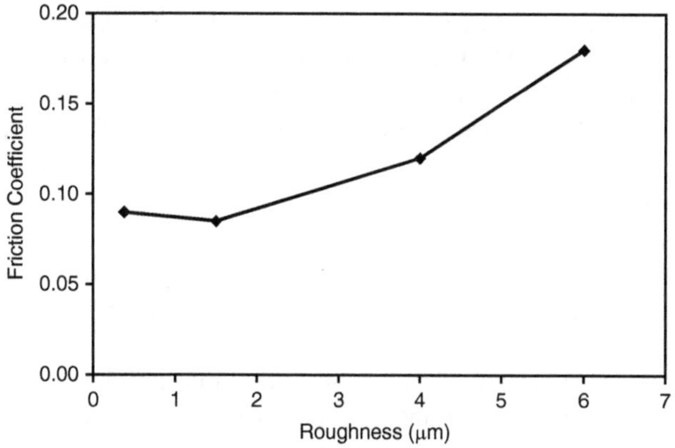

Fig. 2.12 The influence of roughness on the friction coefficient of nanocomposite TiC/a-C coating [89]

coatings are strongly dependent on the Si content in the coatings and the testing ball material. TiSiN coatings exhibit similar friction coefficient when tested against Al_2O_3 and 302 stainless steel balls as shown in Fig. 2.11. However, increase of Si content causes an increase in the friction coefficient of the TiSiN coatings. With the increase in the Si content in the coatings, the wear rate of the TiSiN coatings decreases when tested against Al_2O_3 balls, but increases significantly when tested against 302 stainless steel balls. The capability of forming a transfer layer on the ball surface contributes to the change in the friction coefficient and wear rate with Si content in the coating and ball materials.

Roughness is an important parameter that governs the friction coefficient. The work due to Shaha et al. [89] is presented in Fig. 2.12. They deposited TiC/a-C film by simultaneous DC sputtering of Ti and pulsed DC sputtering of graphite. Their work exhibits increase of friction coefficient with increasing roughness. The increase of friction coefficient with increasing roughness is in line with the mechanism proposed by earlier investigator [90, 91]. Higher roughness increases the asperities slope angle which in turn increases friction coefficient [90]. Thus the ratcheting mechanism where relative motion between two surfaces is achieved by asperities riding over each other is responsible for increase of friction coefficient. The second possible mechanism related to energy loss mechanism where asperities push each other also holds good here

Yttria-stabilized zirconia (YSZ) has a unique combination of mechanical properties. Young's modulus and thermal expansion coefficient of this material are close to metals. Hardness is 11–15 GPa. Fracture toughness is one of the highest among oxide ceramics. They have excellent thermal stability [92]. The tribological properties of YSZ have been intensely studied [93–95]. At high contact loads, the friction coefficient quickly increases (over a period of a few hundred cycles) to 0.6–1.2 due to the accumulation and entrapment of hard wear debris, causing abrasive wear [96]. Wear and friction mechanisms of thin YSZ coatings have not been investigated very intensely [97–99]. A common problem reported for these coatings is brittle fracture in friction contacts due to substrate compliance and coating residual stress [100]. One of the most successful uses of YSZ protective coatings has been as 20–50-nm overcoats on magnetic recording media [101]. Voevodin et al. [102] improved the friction performances of YSZ film by adding gold, making the film a nanocomposite, where hard and brittle nanograins of YSZ were surrounded by amorphous gold matrix. Their observation is illustrated in Fig. 2.13. As can be seen from Fig. 2.13 that addition of 10 % Au results in nanocomposite structure which in turn helps in reducing the friction coefficient to as low as 0.2 even at 773 K suppressing abrasion wear as commonly observed in YSZ coatings.

In addition, a large variety of nanocomposite films exhibited outstanding tribological properties. An excellent antiwear performance is reported by Wang et al. [103] for CrAlN nanocomposite films with various Al contents prepared by a reactive magnetron sputtering technique using Cr and Al targets in the reactive gas mixture. Scharf et al. [104] noted that Co-sputtering at room temperature with any amount of Ti induced a dramatic change in the microstructure. Ti prevented the formation of crystalline WS_2, making it amorphous with well-dispersed nanocrystalline (1–3 nm) precipitates. A longer thin film lifetime was exhibited when the thin films were doped with low amounts of Ti (5–14 at.%) in comparison to pure WS_2, but there was no change in friction coefficient (0.1). According to Low et al. [105], nanocomposite nickel coatings reinforced with regularly shaped nanotubular titanates (multilayered wall structure with 5 nm internal diameter and 30–500 nm tube length), electrodeposited from a modified Watts nickel electrolyte have shown (1) 22 % reduction in surface friction against a spherical diamond tip, (2) 29 % enhancement in wear resistance in a three-body slurry abrasive wear test (steel counter body and 5 μm SiC particles), when compared with a nickel coating containing irregularly shaped nanosized titanium dioxide particles. Ma et al. [106]

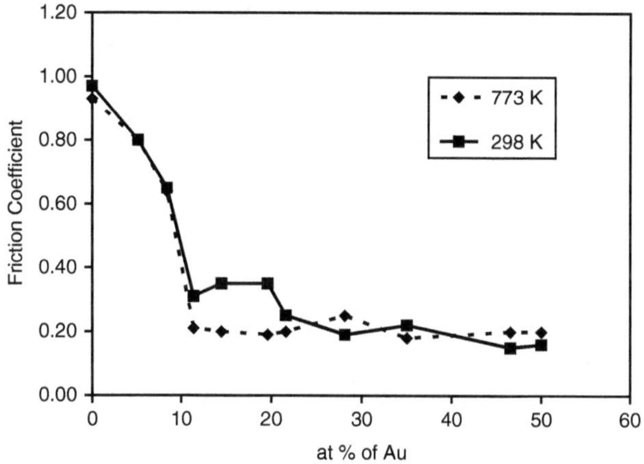

Fig. 2.13 The effect of Au content on the friction coefficient of YSZ nanocomposite film containing gold [100]

studied a series of quaternary Ti–Si–C–N nanocomposite coatings deposited on HSS substrate at 823 K using a pulsed direct circuit plasma-enhanced chemical vapour deposition (PECVD) equipment with a gas mixture of $TiCl_4/SiCl_4/H_2/N_2/CH_4/Ar$. The results show that the nanocomposite Ti–Si–C–N coatings with low Si and high C contents have a lower friction coefficient of 0.17–0.35 at room temperature. The Ti–Si–C–N nanocomposite coating containing 12 at.% Si and 30 at.% C shows excellent tribological properties with a low friction coefficient of 0.30 and a low wear rate at 773 K. Neidhardt et al. [107] successfully synthesised wear-resistant Ti–B–N nanocomposite coatings by reactive arc evaporation from partially metallic Ti/TiB_2 cathodes. Higher N_2 fractions in this film result in a thermodynamically stable nanocrystalline TiN phase embedded in an amorphous BN matrix. The amount of boundary phase can be adjusted with the Ti/B ratio of the cathodes. At lower N_2 partial pressures, a TiN–TiB_x structure is observed, with strong indications for a substitutional solid solution of B in TiN, giving rise to a maximum in hardness (>40 GPa) and wear resistance regardless the Ti/B ratio in the cathodes. All coatings show coefficients of friction of 0.7 ± 0.1 against alumina balls. The dry wear behaviours of reactive unbalanced dc-magnetron sputtered Ti–B–N thin films deposited on silicon (100) at room temperature were investigated by Lu et al. [108]. Both friction coefficient and specific wear rate decreased with incorporation of B into TiN or increasing B content. With increasing applied load, the friction coefficient increased. However the specific wear rate exhibited an inverse trend, accompanying with an increase in amount of deformed wear debris. Increasing sliding velocity yielded similar results to those of increasing loading, which might be due to strain rate effect. Oxygen of H_2O in air participated in and reacted with films and formed mainly Ti oxide consisting of TiO, TiO_2 and Ti_2O_3, accompanying with minor amount of FeO in debris. The formation of oxide in debris was believed to have an effect on increasing wear resistance.

Nanocomposite TiBCN coatings with different nitrogen contents were synthesised by sputtering a TiBC (80 mol.% TiB_2 to 20 mol.% TiC) compound target in various Ar/N_2 mixtures by Lin et al. [109]. When the nitrogen content was less than 5 at.%, the coatings were characterised as a mixture of nano-columnar grains of the Ti(B,C) crystalline phase embedded in a small fraction of amorphous-free carbon and CN. These coatings exhibited superhardness (>45 GPa), but poor adhesion, toughness and wear resistance due to the prominent TiB_2-type structure and the deficiency of a sufficient amorphous phase. When the nitrogen content was increased to 8–15 at.%, the substitution of N for B atoms in the Ti(B,C) lattice led to a gradual decrease in the Ti(B,C) phase and an increase in the volume fraction of the Ti(N,C) and amorphous BN phases. The microstructure of these coatings was characterised as nanocrystalline solid solution Ti(B,C) and Ti(N,C) compounds embedded in an amorphous BN, free carbon and CN matrix. These coatings showed a slight decrease in the coating hardness (37–45 GPa), but significant improvements in the adhesion and wear resistance. However, excessive incorporation of N into the TiBCN coatings (>20 at.%) resulted in a great reduction of the hard Ti(B,C) phase accompanied with the formation of a large volume fraction of amorphous BN, which resulted in a decrease in both hardness and wear resistance.

Friction coefficient can be reduced if it is possible to increase the number of slip systems available for deformation. This can be achieved by adding Si to some systems. Addition of Si results change of orientation of surfaces, increasing the number of available slip system and also formation of amorphous phase a-SiN giving rise to lower friction coefficient. Both phenomena result in reduction of friction coefficient. Pilloud et al. [110] reduced the friction coefficient of ZrN film by adding Si on it. Zr–Si–N films have been deposited on steel and silicon substrates using a reactive magnetron process. ZrN coatings exhibit a [111] preferred orientation, whereas Zr–Si–N ones with low silicon content (<5.7 at.%) exhibit a [100] preferred orientation. Furthermore, the nanocomposite films grow with a columnar structure and their roughness decreases when the silicon content increases. The resultant nanocomposite film exhibits reduced friction coefficient. Friction tests conducted against steel and alumina balls show that the Zr–Si–N films do not suffer any wear when sliding against steel. Only material transfer from the ball onto the film has been detected. When the alumina ball was used, the coatings suffer an significant wear except the amorphous film containing the highest amount of silicon (7.6 at.%).

Similarly ZrN and ZrN–Ag films were deposited using reactive unbalanced magnetron sputtering by Aouadi et al. [111]. The films formed a microstructure that consisted of dense and homogeneous nanocrystals of Me evenly distributed throughout the ZrN matrix. The films became more textured with the (111) preferred orientation for larger substrate bias. Nanoscratch tests suggested that the films with the largest silver content had the lowest wear rate probably because of the high pressures associated with these measurements. Wear testing using the more conventional ball-on-disc technique indicated improvements in the tribological properties of the films that may be correlated to the improvements in the resistance to plastic deformation established from nanoindentation measurements. Overall, nanoindentation, nanowear, and ball-on-disc techniques proved that ZrN–Ag films had superior mechanical properties than ZrN. Further, Cr–Si–N coatings with a nanocomposite microstructure were deposited using closed unbalanced magnetron sputtering ion plating system by Bao et al. [112]. The hardness and modulus of Si-doped Cr–N coatings have been improved with the rise of Si concentrations. The improvement of hardness of Cr–Si–N coatings is attributed to both solid solution strengthening and higher density of the doped coatings. Si-doped Cr–N coatings have a lower friction coefficient which is due to the non-crystal Si compounds with a low friction coefficient.

2.4.2 *Wear Behaviour*

In recent time, it is demonstrated that not only hardness or elastic modulus but also the ratio *H*/*E* is a very important material property [113]. The ratio *H*/*E* multiplied by geometric factor, which is ratio of the diameter of plastic zone to total deformed zone, gives the plasticity index. The plasticity index describes the deformation properties of contacting surfaces. This quantity also appears in various expressions for fracture toughness. It was recognised by several authors that the ranking of materials according to their *H*/*E* ratio can provide extremely close agreement to their ranking in terms of wear [114]. Various authors have accepted that the *H*/*E* ratio as a ranking parameter since Oberle's early study [115, 116].

The support of above hypothesis can be found in the work of Lin et al. [109] as given in Fig. 2.14. TiBCN nanocomposite coatings were deposited in a closed field unbalanced magnetron sputtering system using pulsed magnetron sputtering of a TiBC compound target with various Ar/N_2 mixtures. Improvements in the coating adhesion, *H*/*E* ratio and wear resistance were achieved together with a decrease in the coating hardness to 35–45 GPa as the N content in the coatings was increased from 8 to 15 at.%. The microstructure of the coatings changed from a nano-columnar to a nanocomposite structure in which 5–8 nm nanocrystalline Ti(B,C) and Ti(N,C) compounds were embedded in an amorphous matrix consisting of BN, free carbon and CN phases. With a further increase in the N content in the coatings to levels greater than 20 at.%, the inter-particle spacing of the nanocrystalline compounds increased significantly due to the formation of a large amount of the amorphous BN phase, which also led to low hardness and poor wear resistance of the TiBCN coatings.

The Ti–Al–B–N phase system offers a broad range of possibilities for designing coating structures with optimal properties for different applications. Further, the addition of aluminium to TiN/TiB_2-based nanocomposite structure provides scope for lowering the elastic modulus of the coating (E_{Al}—70 GPa). The presence of boron allows the possibility to produce a soft, low-modulus (and potentially dry-lubricant) h-BN phase, superhard (or ultrahard) and stiff c-BN or an amorphous phase, depending on the deposition conditions [117–119]. Furthermore, the 'doping' of titanium-based vapour-source materials with several atomic percent

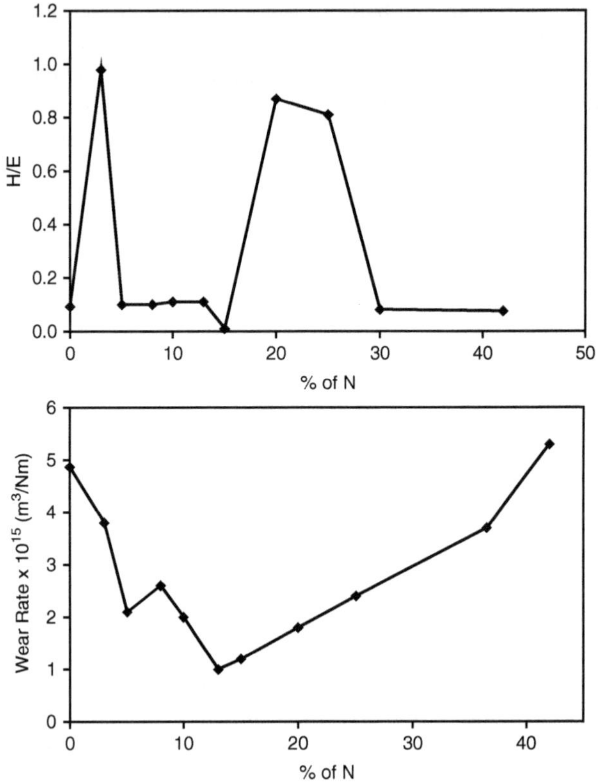

Fig. 2.14 Variation of *H*/*E* ratio and wear resistance of TiBCN nanocomposite coating as function of nitrogen content [109]

of boron can be used to produce a nanocrystalline material with very similar hardness and elastic properties to those of tool steel, which makes an excellent metallic interlayer for ceramic coatings [120]. The incorporation of 2–3 at.% B into a TiN film creates a coating which is significantly harder (i.e. >30 GPa) than 'conventional' TiN and exhibits dramatically reduced sliding wear [119]. The use of interstitial doping to create metallic coatings which has lower elastic moduli than most ceramic coatings and are hard and wear resistant, has been observed in systems other than Ti–Al–B–N (e.g. W–C [121] and Cr–N [122]), with distinct benefits in each case.

There exist two main practices for the production of nanocomposite coatings. In one of the practices, the nanocomposite structure is obtained by the combination of two hard and immiscible phases in one coating (e.g. nc-TiN/a-Si_3N_4; a-$TiSi_2$-nc-$TiSi_2$) while in the other practice, the mixture of one hard and one soft immiscible phases (e.g. Zr–Cu–N, Cr–N–Cu, Ti–Cu–N and Mo–Cu–N) are used. Nanocomposite coatings based on hard/hard phases has been extensively studied [123–126]. Less attention has been paid to those composite coatings that are based

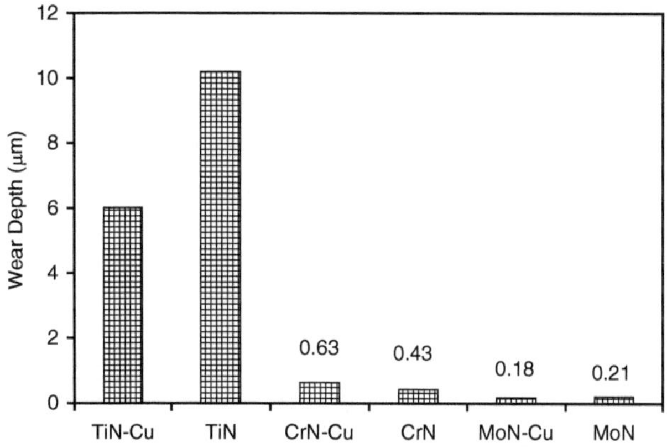

Fig. 2.15 Wear depth of various nanocomposite coatings showing the beneficial effect of Cu addition in improving wear [127]

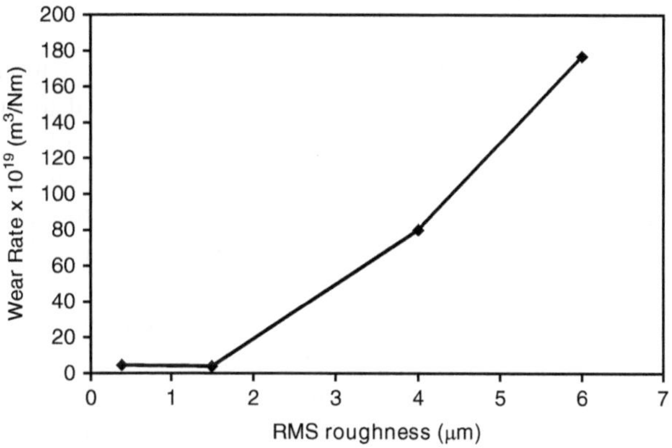

Fig. 2.16 The influence of roughness on wear rate of TiC/a-C coating [89]

on hard/soft phases. Ozturk et al. [127] investigate tribological behaviours of Cu-doped TiN, CrN, and MoN coatings using hard soft phase combination. The results of their investigation confirmed that the addition of Cu into TiN, CrN and MoN coatings has indeed modified the grain size and morphology, but beneficial effect is experienced only on the friction and wear behaviour of MoN. Their results are shown in Fig. 2.15. Similar to the case of friction coefficient, the wear rate also increases with increase of surface roughness as can be shown in Fig. 2.16 by Shaha et al. [89].

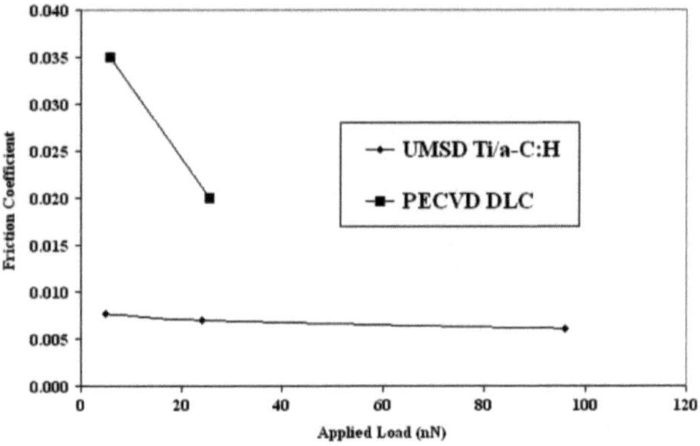

Fig. 2.17 The effect of applied load on the friction coefficient of Ti/a-C:H nanocomposite and DLC film [38]

2.4.3 Nanotribology

The possibilities of using these coatings for tribological applications at low load such as for microelectromechanical system, micromechanical assembly, etc. are explored [71, 128]. Work due to Kvasnica et al. [38] on tribology of n-TiC/a-C nanocomposite film at nN load is presented in Fig. 2.17. It is noted that the friction coefficient is independent of scan speed and applied load in this load range for n-TiC/a-C nanocomposite film. In a separate study by Koch et al. [37], using constant load scratch indenter, it is noted that the friction coefficient of W–S–C nanocomposite film decreases with increase of carbon content. Their observation is presented in Fig. 2.18. WS_2 has an excellent friction behaviour in dry atmospheres contrary to humidity containing ones. Some kinds of C-phases behave in reverse way. Thus, when testing in ambient condition, the coatings rich in C-phases exhibit lower friction coefficient than the others with lower C content. Further such behaviour specially the friction behaviour at very low load may also be attributed to presence of oxygen in the film. Oxygen either in adsorbed state or by forming a thin complex layer increases the friction coefficient. The problem of the humidity in W–S–C system is the fact that O can establish strong bonds between the basal planes impeding the easy sliding and increasing the friction coefficient. Of course if one increases the porosity, the surface where O/H_2O can be adsorbed is increased facilitating the establishment of these bonds. Both porosity and the tendency to react with O_2 increases with decreasing carbon content in these films. This results in lower friction coefficient at higher carbon content. Further, it is also found that the friction coefficients exhibit a maximum value at the lowest applied load. Such behaviour is clearly related to the fact that at lower applied load the friction is powerfully governed by the surface forces. Furthermore, the friction coefficient also increases with the sliding distance. As the scratch tip moves along the surface, the material scratched out piles up in the front of the tip, giving rise to

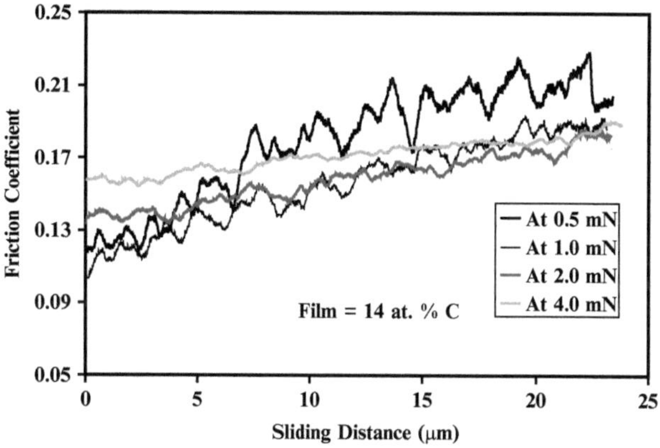

Fig. 2.18 Variation of friction coefficient with sliding distance for W–S–C film containing 14 at. % C [37]

an enhanced resistance to the motion of the tip and hence a higher friction force is realised. Further, it is reported that during sliding in WS_2 system, propagation of pre-existing cracks breaks covalent bonds creating more unsaturated bonds [129]. When sliding takes place in ambient condition, such unsaturated bonds react with the moisture in air forming WO_3. The formation of WO_3 tends to increase the friction force.

Finally, it is to be stated that the scratch test is mostly concerned with scratching a specimen with a V-shaped indenter in general a diamond indenter. Such scratching simulates an ideal single contact abrasive process. However, such a single asperity contact has several shortcomings. Single asperity contact does not simulate possible mechanical interlocking between asperities. Such experiment also fails to simulate the distribution of loads over several asperities as encountered in practical applications. On the other hand self-lubricating films are used mostly in sliding wear applications. The state of stress and the thermodynamic nature of the deformation of the material under an abrasive process or a sliding process differ considerably. For example, the strain rate during abrasive wear is around 1–10 s^{-1} whereas the strain rate during sliding wear is around 10^{-2} to 1 s^{-1}. During abrasive wear, material deformation is constrained and partially adiabatic [130]. In contrast, during sliding wear, material deformation is characterised by constrained isothermal deformation. Thus friction study using a microtribometre with ball-on-disc configuration simulates ideal sliding wear conditions. It is possible to cover the entire contact area uniformly with carbonaceous graphitic material during scratching. This may not take place during reciprocating sliding action.

In order to overcome the above-mentioned shortcomings, low load tribological study is carried out for n-TiC/a-C nanocomposite film. The effect of the stroke length on the friction coefficient of such film at 500 mN load is shown in Fig. 2.19 [131]. Clearly the friction coefficient reaches a minimum at 500 μm stroke length. The highest friction coefficient is noted for 1,000 μm stroke length. Friction

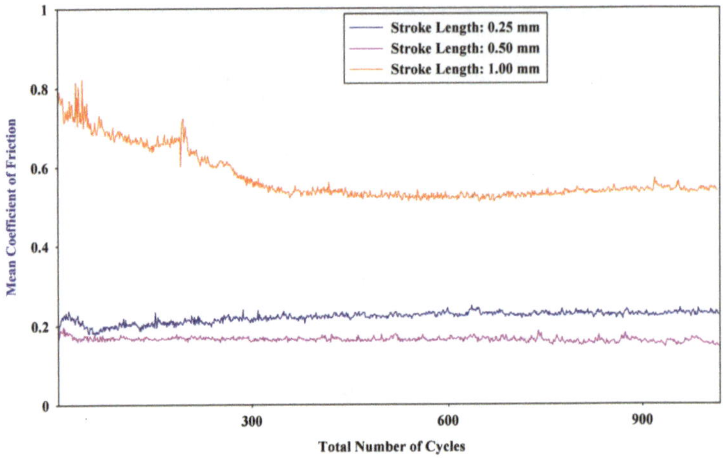

Fig. 2.19 The effect of the stroke length on the friction coefficient of such film at 500 mN load of TiC/a-C film [137]

coefficients for 500 and 250 μm stroke length remain constant through out the test. The friction coefficient at 1,000 μm stroke length decreases from an initial high value and attains a constant value.

The friction force is found to be minimum at an intermediate stroke length. The friction coefficient is very high at higher stroke length of 1,000 μm. As the frequency of reciprocation (0.8 Hz) remains constant, the average velocity of the interacting surfaces increases with increase in stroke length. This results in increase of temperature at the contact point giving rise to the formation of a higher amount of oxide debris. This oxide debris subsequently acts as an abrasive giving rise to higher friction coefficients at higher stroke length. It is also noted that the friction coefficient at higher stroke length decreases from an initial high value to a low constant value. This may be attributed to the fact that the contact temperature during reciprocating motion with high stroke length is high. Hence carbon tends to get graphitised as the process of wear continues and the friction coefficient decreases. After some time the friction coefficient goes down and the contact temperature does not attain a high enough value to allow graphitisation to take place. Consequently the friction coefficient does not decrease further but rather assumes a steady state value. The above observation can also be related to the fact that as the sliding continues, the real contact area increases, resulting in a decrease of the mean pressure and consequently ploughing action decreases.

Topographic images obtained using SPM from the worn surfaces of W–S–C nanocomposite films obtained by employing different sputtering procedure and rubbed against 52100 steel ball at very low load are illustrated in Fig. 2.20. In all films, extensive ploughing can be noted. This ploughing forces the material to the ridges along the sides of the grooves. These displaced material subsequently chipped out leading to material loss. These grooves are due to ploughing action

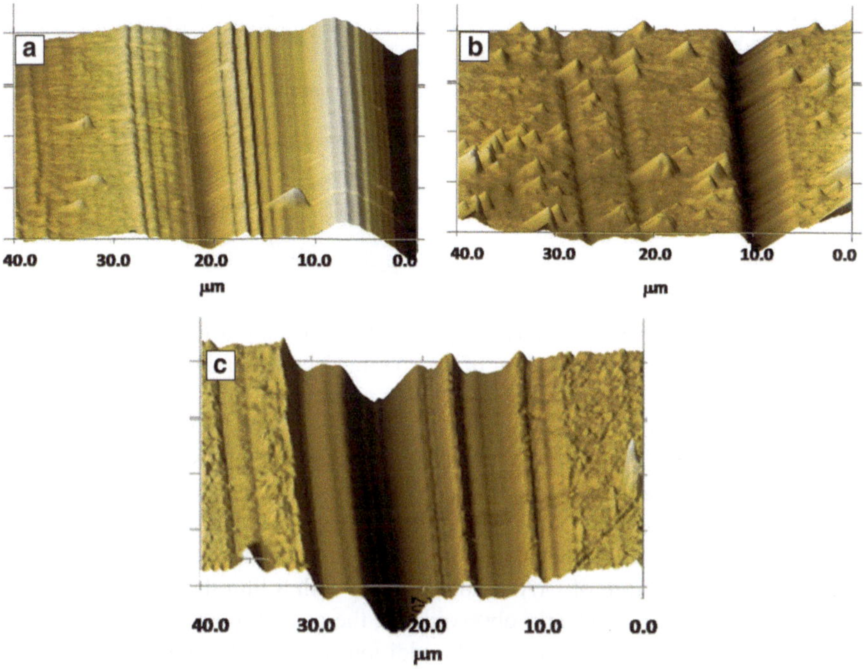

Fig. 2.20 Topographic images obtained using SPM from the worn surfaces of W–S–C films obtained by employing different sputtering procedure and rubbed against 52100 steel ball at very low load, (**a**) reactively sputtered, (**b**) co-sputter with two different targets and (**c**) co-sputter with composite target

of the asperities. Within the wear track, the depth and width of the grooves varied from place to place and from films to films.

2.5 Applications of Nanocomposite Films

Piston ring of internal combustion engine is considered to be a potential area for application of nanocomposite nc-TiC/a-C(Al) film [132]. Pure a-C film is not considered for such application as high residual stress prevents thick film to be deposited. Thickness below 1.5 µm or even a few microns does not meet the requirement of acceptable mileage. Further pure a-C coating exhibits poor thermal stability, oxidation resistance and brittle behaviour. In contrast, the important features of nc-TiC/a-C(Al) film indicated that this will be excellent coating for such application. The low residual stresses of the film allow thicker coating to be deposited. The piston ring of a two-stroke gasoline engine having capacity 41 cc, cylinder bore of 39.8 mm and stroke of 32.55 mm is coated with nc-TiC/a-C(Al) film. The piston ring is 1.65-mm thick, having inside diameter of 36.32 mm and

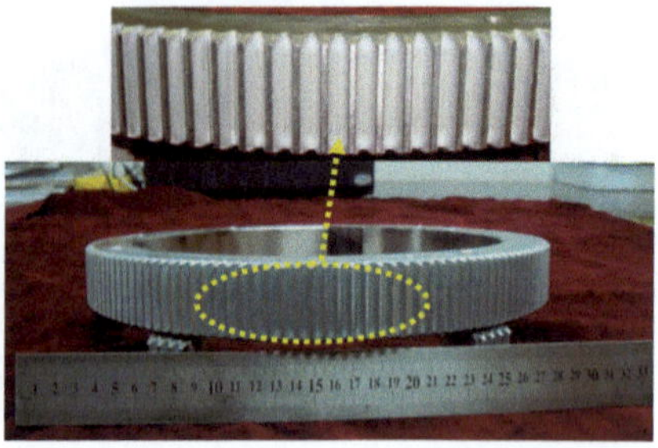

Fig. 2.21 Image of CrAlN coated gears with a film thickness of approximately 15 μm [132]

outside diameter of 39.37 mm. The engine test is carried out for 610 h. It is noted that with nc-TiC/a-C(Al) film, wear of the coating is considerably lower and around 3 % fuel is saved in first 30 h and 2 % fuel is saved in next 580 h.

As mentioned earlier, nanocomposite films are considered as potential protective surfaces for moving parts such as gears, cutting tools and shafts used in engines or other industries to replace the traditional CrN films. Such kinds of nanocomposite films have now been successfully deposited on large-scale gears (nearly 300 mm in diameter, stainless steel AISI 304) used in polishing machines [133]. Figure 2.21 shows the whole picture of CrAlN-coated gears with a film thickness of approximately 15 μm. It can be clearly observed that the coated tooth surface of gears is very uniform, smooth and compact. In addition, the CrAlN films exhibited very excellent bonding with the gears. The servicing life and failure process of such nanocomposite CrAlN-coated gears are found to improve significantly. Images of several components such as turbine blades, cutting tools, etc. coated with nanocomposite film (Dextron Research Ltd., Russia) using UniCoaT 600 and UniCoat 700 systems are given in Fig. 2.22.

The nanocomposite nc-$(Al_{1-x}Ti_x)N$/a-Si_3N_4 film is developed for the purpose of metal cutting operation [134, 135]. The cutting performance of this coating in dry and minimum lubrication condition is 2–4 times better than the state-of-the-art $(Al_{1-x}Ti_x)N$ coating. During dry drilling of tough steel, the life time of drill bits made of cemented carbide coated with this film is four times higher than that coated with state-of-the-art $(Al_{1-x}Ti_x)N$ coating [136, 137].

Fox et al. [138] reported MoS_2/Ti composite coating by co-deposition of small amount Ti in MoS_2 matrix. The properties of this coating are available elsewhere [139]. Drilling tests were carried out. High-speed steel M6 tap drills have been coated with TiN, TiCN (low carbon content), TiAlN (high aluminium content), TiN + MoS_2/Titanium composite and TiCN (low carbon content) + MoS_2/titanium composite. These tap drills were tested under dry conditions and lubricated (mixed oil and water soluble by 20 %) condition to drill 5.5 mm through holes into AISI 400 stainless steel (11.7 mm thick, HRB 60–70). These tests were performed at 530 rpm and at 1,061 rpm. The number of holes produced was recorded. Use of MoS_2/titanium composite in dry conditions on the top of hard coatings such as TiN and TiCN has increased the number of holes produced by twice in the case of TiN and by 4.1–4.8 times for TiCN [140]. Carbide end mills have been coated with

Fig. 2.22 Images of several components coated with nanocomposite film (Dextron Research Ltd., Russia)

TiCN and TiCN + MoS_2/titanium composite and were tested. These end mills were tested under dry conditions and lubricated (Trim Sol: mixed oil and water-soluble by 5 %) at a cutting speed of 150 m/min and a feed rate of 0.04 mm/turn onto AISI 304 stainless steel. The cutting depth was 4 mm, while the radial depth cut was 3 mm and the axial depth cut was 5 mm. Average milling force, tool wear and surface finishing were recorded. It has also been observed that MoS_2/titanium composite coatings running in dry conditions deposited on top of a TiCN hard coating offer an increase in milling distance and a reduction in average milling force during the test and improvement of surface finish [140]. Punches were coated with MoS_2/titanium composite (1.2 μm) on top of TiCN (3–3.5 μm of TiCN produced by arc evaporation) coating and used to perforate 12 mm HSLA steel with water soluble lubricant (20 %). In the piercing application, an uncoated tool was able to pierce 15,000 parts before tool failure. The use of a TiCN coating increased the number of parts to 50,000 and depositing MoS_2/titanium composite on top of TiCN further increased the number of parts to 200,000 [140].

2.6 Summary and Perspectives

In this chapter, the state of art of the research of nanocomposite coatings and their tribological performances is reviewed. Over the last decade, significant efforts are being put for research and development of nanocomposite coatings and this has yielded wealth of information on structure property relation and tribological performances of these coatings. There is a growing interest in nanocomposite wear-resistant coating, in particular, transition metal nitrides with increased thermal stability because of their wide spread applications. These coatings are finding other applications such as nuclear fuel materials, diffusion barrier, laser diodes electrical contacts, etc.

The generic concept of design of super hard nanocomposite coatings is based on spinodal phase segregation resulting in formation of stable nanostructure by self organisation. The deformation behaviour of nanocomposite coatings is having typical features controlled by interfaces and nanoscale effects in these coatings. For tribological performances the ratio of hardness to elastic modulus of the coating has been proved to be an important property. These nanocomposite coatings lend themselves to incorporate 'adaptive' and 'chameleon'-type attributes.

One important challenge for nanocrystalline coating is to develop a process which can produce nanocomposite coating with controlled grain size, shape, crystallographic orientation and lattice structure. Such nanocomposite coatings are expected to have enhanced optical, electrical, magnetic, electronic and photocatalytic properties.

Doping of oxides by metal is expected to produce new generation nanocomposite coating with self-healing functionality. Considerable effort has been provided to produce TiO_2-based film with photocatalytic, self-healing, antifogging and antibacterial properties [141, 142]. The present research is concerned with doping this film with an element in order to shift activation of these properties from ultraviolet to visible light region by forming a nanocomposite coating.

Industrial applications of nanocomposite coatings expose them to elevated temperature and oxygen. However, detail study on oxidation mechanism is rare although such data will help designing and tailoring the coatings for applied thermal load. Further, research in the area of diffusion related phenomena controlling recovery, recrystallization, segregation, grain growth interdiffusion, etc. are to be investigated in much more detail for development of next generation nanocomposite coatings.

Another important area is to have enhanced understanding of the nonlinear elastic behaviour and mechanism of plastic deformation of nanocomposite coatings. Many nanocomposite coatings are macroscopically brittle but microscopically ductile and this may be due to plastic instability in the material. It is important to understand and control plastic instability of these materials.

References

1. Veprek S, Nesladek P, Niederhofer A, Glatz F, Jilek M, Sima M (1998) Surf Coat Technol 108–109:138
2. Veprek S, Reiprich S (1995) Thin Solid Films 268:64–71
3. Budrovic Z, Van Swygenhoven H, Derlet PM, Van Petegem S, Schmitt B (2004) Science 304:273–276
4. Van Swygenhoven H, Specer M, Caro A (1999) Acta Mater 47:3117–3126
5. Provenzano V, Holtz RL (1995) Mater Sci Eng A204:125–134
6. Gleiter H (1989) Prog Mater Sci 33:223–315

7. Musil J, Vicek J (2001) Surf Coat Technol 142–144:557–566
8. Yip S (1998) Nature 391:532–533
9. Mazaleyrat F, Varga LK (2000) J Magn Magn Mater 215–216:253–259
10. Gusev AI (1998) Physics-Uspekhi 41(1):49–76
11. Maya L, Allen WR, Grover AL, Mabon JC (1995) J Vac Sci Technol B13(2):361–365
12. Veprek S, Niederhofer A, Moto K, Bolom T, Mannling HD, Nesladek P, Dollinger G, Bergmeier A (2000) Surf Coat Technol 133–134:52
13. Park IW, Kim KH, Kumrath AO, Zhong D, Moore JJ, Voevodin AA, Levashov EA (2005) J Vac Sci Technol B23(2):588
14. Park IW, Kim KH, Suh JH, Park CG, Lee MH (2003) J Korean Phys Soc 42(6):783
15. Roy M, Kvasnica S, Eisenmenger-Stittner C, Vorlaufer G, Pauschitz A (2005) Surf Eng 21 (3):257
16. Cavaleiro A, Louro C (2002) Vacuum 64(3):211
17. Voevodin AA, O'neil JP, Prasad SV, Zabinski JS (1999) J Vac Sci Technol A17(3):986
18. Veprek S (1999) J Vac Sci Technol A17:1521
19. Musil J (2000) Surf Coat Technol 125:322
20. Kelly PJ, Arnell RD (2000) Vacuum 56:159
21. Voevodin AA, Zabinski JS (1998) Diam Relat Mater 7:463
22. Seo SC, Ingram DC, Richardson HH (1995) J Vac Sci Technol B13(2):2856
23. Safi I (2000) Surf Coat Technol 127:203
24. Choy KL (2003) Prog Mater Sci 48:57
25. Finger F, Hapke P, Luysberg M, Carius R, Wagner H, Scheib M (1994) J Appl Phys Lett 65:2588
26. Alpium P, Chu V, Conde JP (1999) J Appl Phys 86:3812
27. Heintze M, Zedlitz R, Wanka HN, Schubert MB (1996) J Appl Phys 79:2699
28. Wang KC, Hwang HL, Leong PT, Yew TR (1995) J Appl Phys 77:6542
29. Knoteck O, Prengel HG (1991) Surf Modif Technol 4:507
30. Tanaka Y, Ichimiya N, Onishi Y, Yamada Y (2001) Surf Coat Technol 146–147:215
31. Merlo AM (2003) Surf Coat Technol 174:21
32. Ehiasarian AP, Hopsipian PE, Hartman L, Helmersson U (2004) Thin Solid Films 457:270
33. Kelly PJ, Arnell RD (1998) Surf Coat Technol 108–109:317
34. Arnell RD, Kelly PJ (1999) Surf Coat Technol 112:170
35. Kouznetsov V, Macak K, Schneider JM, Helmersson U, Petrov I (1999) Surf Coat Technol 122:190
36. Ehiasarian AP, New R, Munz WD, Hartman L, Helmersson U, Kouznetsov V (2002) Vacuum 65:147
37. Koch T, Pauschitz A, Roy M, Evaristo M, Caveleiro A (2009) Thin Solid Films 518:185
38. Kvasnica S, Schalko J, Benardi J, Eisenmenger-Sittner C, Pauschitz A, Roy M (2006) Diam Relat Mater 15:1743
39. Regula M, Ballif C, Moser JH, Lévy F (1996) Thin Solid Films 280:67
40. Veprek S, Argon AS (2002) J Vac Sci Technol B20:650
41. Patscheider J (2003) MRS Bull 28:173
42. Veprek S (2003) Rev Adv Mater Sci 5:6
43. Veprek S, Argon AS (2001) Surf Coat Technol 146–147:175
44. Hall EO (1951) Proc Phys Soc Lond Sec B64:747
45. Petch NJ (1953) J Iron Steel Inst 174:25
46. Mohamed FA, Li Y (2001) Mater Sci Eng A298:1
47. Padmanavan KA (2001) Mater Sci Eng A304:200
48. Ovid'ko A (2002) Science 295:2386
49. Ovid'ko A (2005) Int Mater Rev 50:65
50. Kim HS, Estrin Y, Bush MB (2000) Acta Mater 48:493
51. Liao XZ, Zhao F, Lavernia E, He DW, Zhu YT (2004) Appl Phys Lett 84:632
52. Chen MW, Ma E, Hemkar KJ, Sheng HW, Wang YM, Cheng XM (2003) Science 300:1275

53. Koch CC (2003) Scr Mater 49:657
54. Ovid'ko A (2005) Rev Adv Mater Sci 10:89
55. Malygin GA (1995) Phys Solid State 37:3
56. Evans AG, Hirth JP (1992) Scr Metall 26:1675
57. Romanov AE (1995) Nanostruct Mater 6:125
58. Hahn H, Mondal P, Padmanabhan KA (1997) Nanostruct Mater 9:603
59. Ovid'ko IA (2003) Philos Mag Lett 83:611
60. Seefeldt M (2001) Rev Adv Mater Sci 2:44
61. Yu Gutkin M, Ovid'ko IA (2003) Rev Adv Mater Sci 4:79
62. Ramanov AE, Valdimirov VI (1992) In: Nabarro FRN (ed) Dislocations in solids, vol 9. North Holland Pub, Amsterdam, p 191
63. Oliver WC, Pharr GM (1992) J Mater Res 7:1564
64. Nossa A, Cavalerio A (2004) J Mater Res 19:2356
65. Li X, Diao D, Bhushan B (1997) Acta Metall 45:4453
66. Ding J, Meng Y, Wen S (2000) Thin Solid Films 371:178
67. Fu YQ, Du HJ, Sun CQ (2003) Thin Solid Films 424:107
68. Slack GA, Bartram SF (1975) J Appl Phys 46(1):89
69. Lim SC, Ashby MF (1987) Acta Metall 35:11
70. Roy M (2009) Trans Indian Inst Met 62:197
71. Roy M, Koch T, Pauschitz A (2010) Adv Surf Sci 256:6850
72. Sánchez-López JC, Martínez-Martínez D, López-Cartes C, Fernández A (2008) Surf Coat Technol 202:4011
73. Sánchez-López JC, Martínez-Martínez D, Abad MD, Fernández A (2009) Surf Coat Technol 204:947
74. Feng B, Cao DM, Meng WJ, Rehn LE, Baldo PM, Doll GL (2001) Thin Solid Films 398–399:210
75. Hutchings IM (1992) Tribology, friction and wear of engineering materials. CRC, Boca Raton, FL
76. Singer ILS, Fayeulle S, Ehni PD (1991) Wear 149:375
77. Myake S, Kaneko R (1992) Thin Solid Films 212:256
78. Jamal T, Nimmaggada R, Bunshah RF (1980) Thin Solid Films 73:245
79. Zhang S, Bui XL, Li X (2006) Diam Relat Mater 15:972
80. Panjan P, Navisek B, Cekada M, Zalar A (1999) Vacuum 53:127
81. Zhang S, Fu YQ, Du HJ (2002) Surf Coat Technol 162:42
82. Lindquist M, Wilhelmsson O, Jansson U, Wiklund U (2009) Wear 266:379
83. Stuber M, Leiste H, Ulrich S, Holleck H, Schild D (2002) Surf Coat Technol 150:218
84. Briscoe J, Stolarski TA (1993) In: Glaeser WA (ed) Friction in characterisation of tribomaterials. Butterworth-Heinemann, Boston, MA, p 44
85. Kelly PJ, Li H, Benson PS, Whitehead KA, Verran J, Arnell RD, Lordanova I (2010) Surf Coat Technol 205:1606
86. Kelly PJ, Li H, Whitehead KA, Verran J, Arnell RD, Lordanova I (2009) Surf Coat Technol 204:1137
87. Kelly PJ, von Braucke T, Liu Z, Arnell RD, Doyle ED (2007) Surf Coat Technol 202:774
88. Cheng YH, Browne T, Heckerman B, Meletis EI (2010) Surf Coat Technol 204:2123
89. Shaha KP, Pei YT, Martinez-Martinez D, De Hosson JTM (2010) Surf Coat Technol 205:2624
90. Hayward IP, Singer IL, Seitzman LE (1991) Wear 157:215
91. Bull SJ, Chalkar PR, Johnston C, Moore V (1994) Surf Coat Technol 68:603
92. Wachtman JB (1996) Mechanical properties of ceramics. Wiley, New York, p 161, 392
93. Lancaster JK, Mashal Y, Salher R (1992) J Phys D 25:A205
94. Fischer TE, Anderson MP, Jahamir S, Salher R (1988) Wear 124:133
95. Stachowiak GW, Stachowiak GB (1989) Wear 132:151
96. Lee SW, Hsu SM, Shen MC (1993) J Am Ceram Soc 76:1937

97. Kao AS, Hwang C (1990) J Vac Sci Technol A 8:3289
98. Dugger MT, Chung YW, Bhushan B, Rothschild W (1993) Tribol Trans 36:84
99. Yamashita T, Chen GL, Shir J, Chen T (1988) IEEE Trans Magn 24:2629
100. Moulzolf SC, Lad RJ, Blau PJ (1999) Thin Solid Films 347:220
101. Theunissen GSAM (1998) J Mater Sci Lett 17:1235
102. Voevodin AA, Hu JJ, Fitz TA, Zabinski JS (2001) Surf Coat Technol 146–147:351
103. Wang L, Zhang G, Wood RJK, Wang SC, Xue Q (2010) Surf Coat Technol 204:1517
104. Scharf TW, Rajendran A, Banerjee R, Sequeda F (1999) Thin Solid Films 347:220
105. Low CTJ, Bello JO, Wharton JA, Wood RJK, Stokes KR, Walsh FC (2010) Surf Coat Technol 205:1856
106. Ma D, Ma S, Dong H, Xu K, Bell T (2006) Thin Solid Films 496:438
107. Neidhardt J, Czigány Z, Sartory B, Tessadri R, Mitterer C (2010) Int J Refract Met Hard Mater 28:23
108. Lu YH, Shen YG, Zhou ZF, Li KY (2007) Wear 262:1372
109. Lin J, Moore JJ, Mishra B, Pinkas M, Sproul WD (2010) Acta Mater 58:1554
110. Pilloud D, Pierson JF, Takadoum J (2006) Thin Solid Films 496:445
111. Aouadi SM, Bohnhoff A, Sodergren M, Mihut D, Rohde SL, Xu J, Mishra SR (2006) Surf Coat Technol 201:418
112. Bao M, Yu L, Xu X, He J, Sun H, Teer DG (2009) Thin Solid Films 517:4938
113. Musil J, Jirout M (2007) Surf Coat Technol 201:5143
114. Oberle TL (1951) J Met 3:438
115. Lancaster JK (1963) Br J Appl Phys 14:497
116. Halling J (1982) Tribologia 1(2):15
117. Rebholz C, Ziegele H, Leyland A, Matthews A (1998) J Vac Sci Technol A 16(5):2851
118. Rebholz C, Schneider JM, Voevodin AA, Steinebrunner J, Charitidis C, Logothetidis S, Leyland A, Matthews A (1999) Surf Coat Technol 113:126
119. Rebholz C, Leyland A, Matthews A (1999) Thin Solid Films 343–344:242
120. Rebholz C, Leyland A, Larour P, Charitidis C, Logothetidis S, Matthews A (1999) Surf Coat Technol 116–119:648
121. Rebholz C, Schneider JM, Leyland A, Matthews A (1999) Surf Coat Technol 112:85
122. Rebholz C, Ziegele H, Leyland A, Matthews A (1999) Surf Coat Technol 115:222
123. Musil J, Zeman P (1999) Vacuum 52:269
124. Abadias G, Tse YY, Michel A, Jaouen C, Jaouen M (2003) Thin Solid Films 43:166
125. Musil J, Zeman P, Hruby H, Mayrhofer PH (1999) Surf Coat Technol 120–121:179
126. He JL, Setsuhara Y, Shimizu I, Miyake S (2001) Surf Coat Technol 137:8
127. Ozturk A, Ezirmik KV, Kazmanlı K, Urgen M, Eryılmaz OL, Erdemir A (2008) Tribol Int 41:49
128. Tomala A, Gebeshuber IC, Pauschitz A, Roy M (2010) Int J Mater Res 101(7):845–851
129. Prasad S, Zabinski J (1997) Nature 387:761
130. Roy M (2006) In: Proceedings of the international conference on advances in materials and materials processing, Kharagpur, India, 3–5 Feb, p 202
131. Pauschitz A, Jisa R, Kvasnica S, Bernardi J, Koch T, Roy M (2008) Diam Relat Mater 17:2010
132. Zhang S, Bui XL, Sun D (2007) In: Zhang S, Ali N (eds) Nanocomposite thin films and coatings. Imperial College, London
133. Wang L, Zhang G, Wood RJK, Wang SC, Xue Q (2010) Surf Coat Technol 204:3517
134. Halubar P, Jilek M, Sima M (2000) Surf Coat Technol 133–134:145
135. Veprek S, Mannling HD, Jilek M, Halubar P (2004) Mater Sci Eng A366:202
136. Jilek M, Cselle T, Halubar P, Morstein M, Veprek-Heijman MGJ, Veprek S (2004) Plasma Chem Plasma Process 24(4):493
137. Cselle T (2003) Werkzeug Technick No 77, March 2–7
138. Fox VC, Renevier NM, Teer DG, Hampshire J, Rigato V (1998) In: Proceedings of the PSE conference, Garmisch Partenkirchen, 14–18 Sept

139. Teer DG, Hamphire J, Fox V, Bellido-Gonzalez V (1997) Surf Coat Technol 94(95):572
140. Renevier NM, Lobiondo N, Fox VC, Teer DG, Hampshire J (2000) Surf Coat Technol 123:84
141. Herman JM, Disdier J, Pichat P (1984) Chem Phys Lett 108:618
142. Asahi R, Morikawa T, Ohawaki T, Aoki K, Taga Y (2001) Science 293:269

Chapter 3
Diamond Films and Their Tribological Performances

Manish Roy and Roland Haubner

3.1 Introduction

Diamond is the hardest known material. Diamond was initially considered as organic material. Discovery of X-ray diffraction technique and formulation of Brags law helped scientist and technologist to realise that carbon can built different structures like cubic, hexagonal and amorphous. However, diamond is exclusively face-centred diamond with cubic structure. This structure consists of eight corner atoms, six face-centred atoms and four other atoms from adjacent interpenetrating lattices spaced on quarter of a cube diagonal of diamond cubic. Every atom is tetrahedrally coordinated to four other atoms via σ bonds emanating from sp^3 hybrid atomic orbital. The four (111) directions in cubic diamond are in bond directions. The phase diagram provided in Fig. 3.1 exhibits stability ranges of diamond, graphite and other forms of diamonds [1].

Hard coatings with combined functional and structural properties have long been used extensively for improving the tribological performances of mechanical components such as cutting or forming tools [2]. Diamond film is well known for several extraordinary features such as high mechanical strength, excellent thermal conductivity [3], outstanding wear and friction properties [4, 5], high chemical inertness [6], etc. This makes potential of diamond film for tribological application very high, although tribological study of diamond film is very complicated. Usually diamond coatings wear mating surfaces and they do not wear themselves. However, these films exhibit high surface roughness because of their columnar growth [7]

M. Roy (✉)
Defence Metallurgical Research Laboratory, Kanchanbagh, Hyderabad 500058, Andhra Pradesh, India
e-mail: rmanish64@rediffmail.com

R. Haubner
Institute for Chemical Technologies and Analytics, Vienna University of Technology, Vienna, Austria

M. Roy (ed.), *Surface Engineering for Enhanced Performance against Wear*,
DOI 10.1007/978-3-7091-0101-8_3, © Springer-Verlag Wien 2013

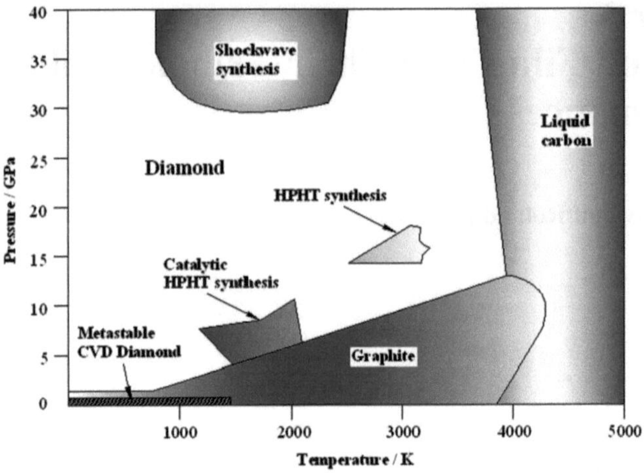

Fig. 3.1 The phase diagram showing the stability ranges of diamond graphite and other forms of diamonds [1]

making it unsuitable for many applications such as in microelectronics and as optical components [8, 9]. One exciting way of getting rid of roughness problem of these films is to use smooth nanocrystalline diamond (NCD) film. NCD is fast becoming serious candidate material for high performance mechanical and tribological surfaces such as micro-electromechanical systems (MEMS), biosensors [10, 11] and other fields.

Until now, diamond film has been deposited on several substrates, but chemical interactions had to be considered [12]. Without application of additional interlayer [13], only substrates can be used showing less reactivity. Ceramic materials such as WC, SiC [14] are more or less stable and metals like, W, Mo and Si [15, 16] quickly form a carbide layer stabilising the surface. For conventional tools, hardmetal bodies (WC–Co) are most important and so intensive research was necessary to solve the adhesion problems of diamond coatings on hardmetal substrates [17, 18]. As adhesion of diamond film with the substrate plays crucial role for tribological performances of the film, numerous studies have been undertaken to improve adhesion characteristics by employing interlayer [19–21], by meticulous optimisation of surface pre-treatment [22–24] or by careful choice of deposition technology [25]. The adhesion characteristics of diamond film with the substrate is determined by number density of interfacial bonds between atoms from interface and carbon atoms from nucleated diamond together with corresponding chemical bond strength. These properties are governed by substrate surface phases and corresponding film growth conditions.

The main objective of this chapter is to give a brief outline of the tribological characteristics of diamond film. This chapter highlights the present state of art of diamond film as tribological coating. It is also the purpose of this chapter to bring out the important application of diamond film under demanding tribological conditions. The reason for using this film in spite of high cost of the film is also put forward. This chapter is divided into five sections.

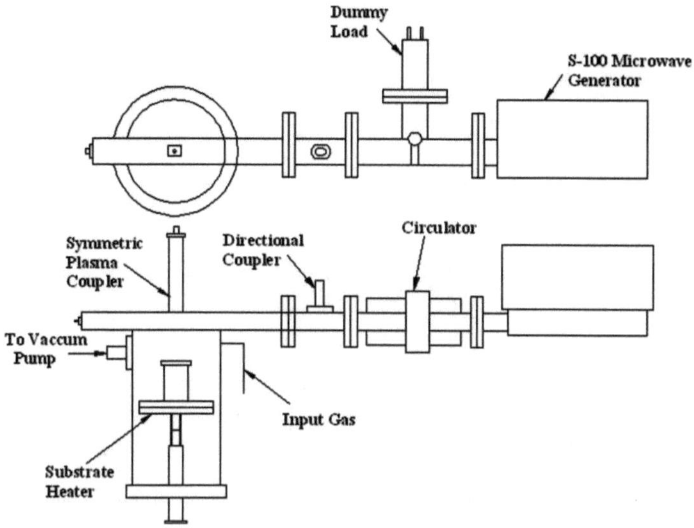

Fig. 3.2 A schematic diagram of ASTeX type of reactor

3.2 Deposition of Diamond Film

There are several techniques that can be employed to deposit high quality diamond film. These are plasma enhanced CVD, hot filament CVD, microwave CVD, DC arc jet, combustion flame, laser-assisted CVD [7, 26–30]. All of these methods produce atomic hydrogen, which is necessary for the diamond growth.

Hot filament CVD was the first method to produce CVD diamond. This method was first employed by national institute for research in inorganic material (NIRIM), Japan. As this method heats the entire substrate to deposition temperature, no additional heating of the substrate is necessary. Microwave plasma-enhanced CVD was designed shortly after the hot filament method also at NIRIM. One major drawback of such a reactor is its inability to control temperature of the substrate and the gases independently. Development of ASTeX type of reactor has eliminated this problem for MPECVD. A schematic diagram of ASTeX type of reactor is presented in Fig. 3.2. Hot filament CVD (HFCVD) provides advantages like low cost, simplicity, ease of scaling and ability to deposit diamond film over a large area. In laser-assisted CVD, light is used as heating source. Such deposition technique allows low temperature deposition, enables selected area or localised deposition and avoids film damage during deposition due to low energy deposition [31].

In all these methods some common sequences occur. A gaseous phase, which contains carbon-containing species such as hydrocarbons, alcohols, etc. must be activated either by plasma or by elevated temperature. Further, small additions of H_2O or CO_2 are possible. Under good deposition conditions, if there is enough atomic hydrogen and low carbon supersaturation, well-faceted high quality

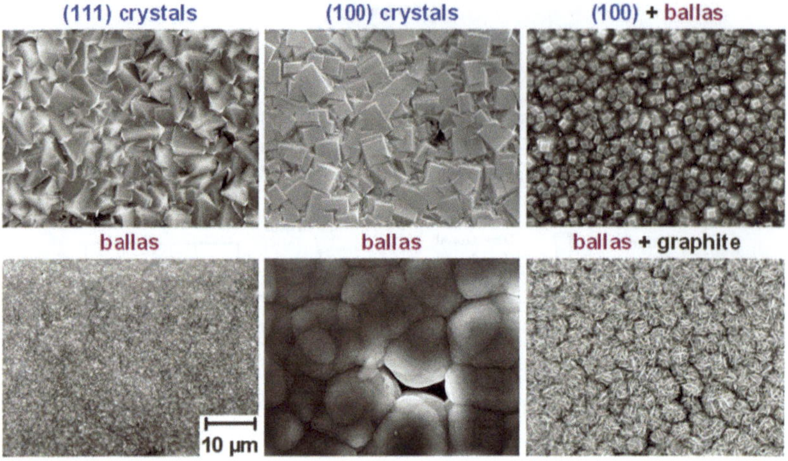

Fig. 3.3 Different morphologies of diamond films deposited under different conditions

diamond coatings can be deposited [1, 32]. When deposition conditions become conducive, the amounts of defect in the diamond increase and surface morphology changes [33]. At least a nanocrystalline diamond called ballas is deposited [34, 35].

Nanocrystalline diamond films (NCD) can be deposited by microwave or hot filament CVD techniques. To decrease the grain size of diamond, the gas phase has to be changed and additional biased enhanced nucleation (BEN) can be applied [36–39]. Various argon, methane and hydrogen mixtures are commonly used for deposition of NCD films. Increasing the methane content in methane/hydrogen plasma, increased nucleation rate can be achieved. NCD films have been synthesised from microwave assisted hydrogen poor plasmas ($CH_4/H_2/Ar$) or ($C_{60}/H_2/Ar$) with Ar concentration up to 97 % and sometimes even without addition of molecular hydrogen [40–42].

3.3 Characterisation of Diamond Films

Diamond films are extensively characterised using various techniques. Scanning electron microscopy (SEM) is a useful technique to characterise diamond film. Various morphology of diamond films can be differentiated based on the micrographs obtained from SEM. Figure 3.3 exhibits different types of morphologies of diamond films deposited under different conditions [43]. At best deposition conditions, the (111) facets are formed changing to (100) facets and unfaceted ballas morphologies. If the deposition conditions are really bad, graphite mixtures with ballas can be observed [33]. In order to examine the nature of growth and the nature of the interface between the substrate and the film, cross-section SEM is carried out. Figure 3.4 represents the cross section of a nanocrystalline

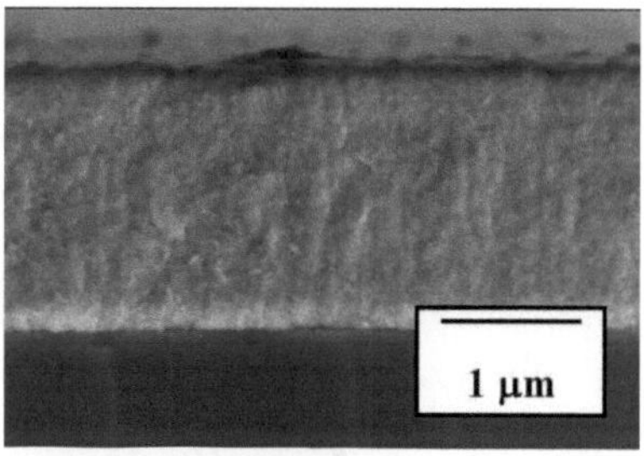

Fig. 3.4 SEM image showing the cross section of diamond film

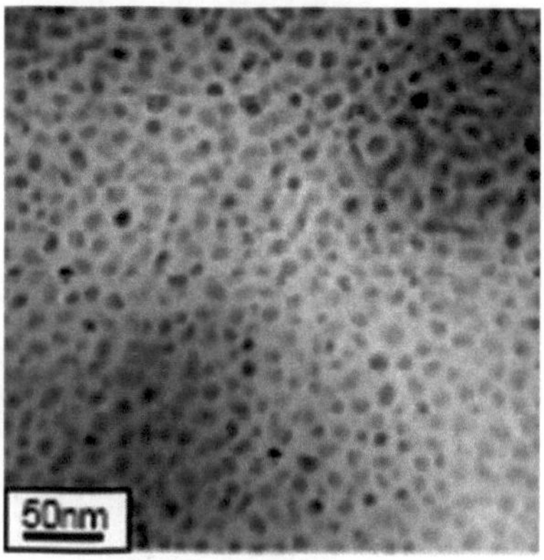

Fig. 3.5 TEM images showing nanocrystalline structure of diamond film

diamond film showing columnar structure and excellent interface between the substrate Si and the film. This type of structure can be explained by van der drift growth mechanism [44]. Transmission electron microscopy (TEM) is another useful technique for characterising diamond film mainly because of its high spatial resolution. One powerful tool to separate out nanocrystalline diamond film from conventional microcrystalline diamond film is to use TEM micrograph. TEM images showing nanocrystalline structure of diamond film can be seen in Fig. 3.5 [45]. Due to small aperture, conventional TEM is not able to get much accurate results for grain boundaries of diamond film with nm size grains. In order to overcome this problem, high resolution electron microscopy (HRTEM) has been introduced. Cross-section TEM analysis was also performed by Jiao et al. [46]. It is to be stated that amorphous carbon layers can be deposited by physical vapour deposition (PVD) methods. These coatings are called diamond-like carbon (DLC) and show very different properties depending on its deposition conditions. If the crystal size in nanocrystalline diamond decreases and the amount of amorphous carbon increases it ends up being DLC.

Atomic force microscopy (AFM) is an important tool to study topographies of diamond films. An AFM image showing the differences between microcrystalline and nanocrystalline diamond films is presented in Fig. 3.6 [47]. Figure 3.6 also shows differences between ballas and faceted morphologies of microcrystalline

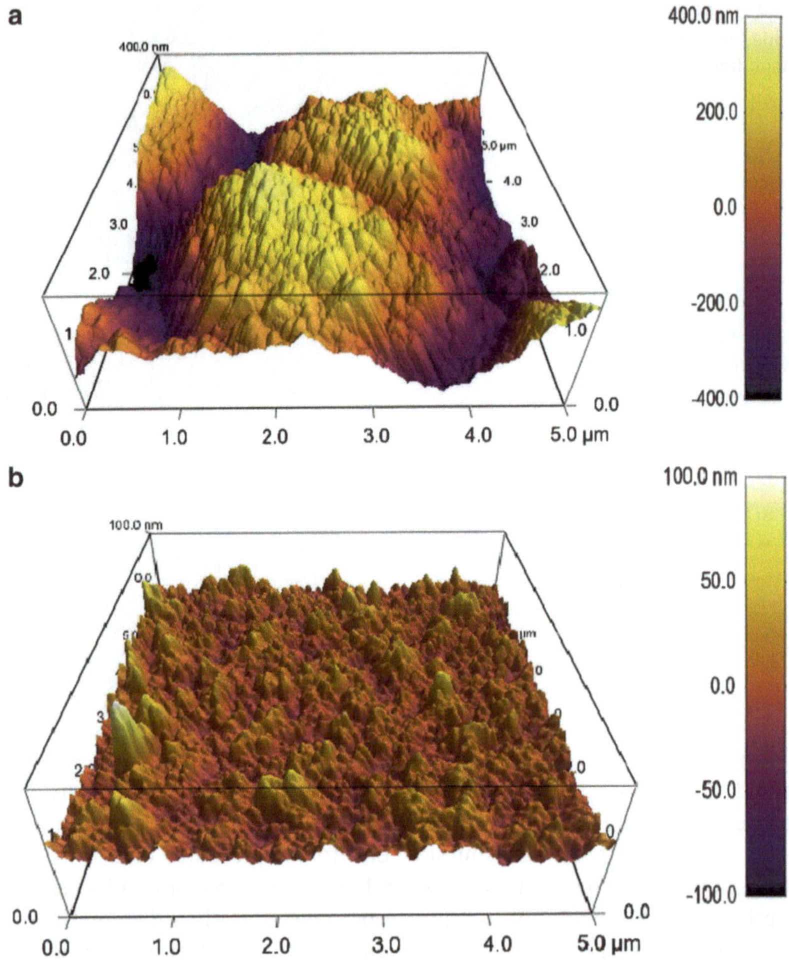

Fig. 3.6 AFM images showing the differences between (**a**) microcrystalline and (**b**) nanocrystalline diamond films

diamond film; these topographic images are useful to measure various roughness parameters of films. The large differences of topographic parameters of these films are summarised in Table 3.1.

Raman spectroscopy, which is extensively used in various chemical pharmaceutical and semiconductor industries, is the most powerful tool for characterisation of diamond film. In this system, the sample is illuminated with a laser and scattered light is collected. The wavelength and intensities of the scattered light are the measures of functional group in a molecule. Raman scattering occurs with a change in vibrational, rotational and electronic energy of a molecule. Raman scattering is about 50 times more sensitive to p-bonded amorphous carbon and graphite than to the photon band of diamond. Hence Raman spectroscopy can be used to establish

Table 3.1 Topographical parameters of the diamond films having fine ballas, coarse ballas and faceted morphology

Diamond films	Parameters Average height (nm)	R.M.S. (nm)	Minimum height (nm)	Maximum height (nm)
Fine ballas	−66.0 ± 1.6 9	66.0 ± 1.3	−67.6	−64.3
Fine ballas	151.0 ± 9.	151.4 ± 7.8	152.2	209.6
Faceted	29.9 ± 2.4	30.0 ± 1.7	27.9	33.0

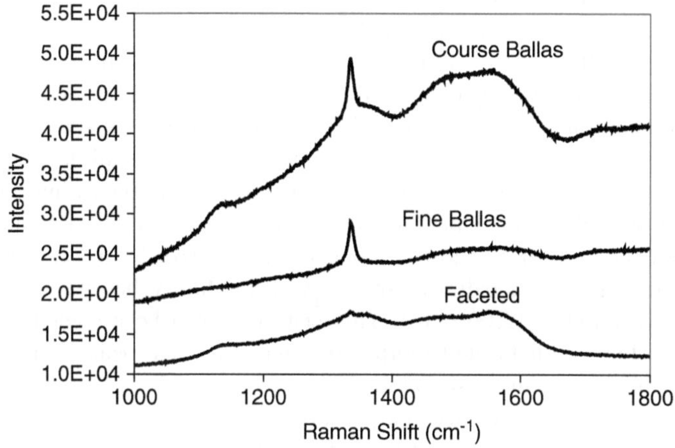

Fig. 3.7 Raman spectra obtained from films having three different types of morphologies [43]

crystalline quality of diamond film by estimating the amount of sp^2-bonded carbon in the film. In order to quantify the quality of diamond film, diamond quality factor as given below is introduced [48–50]:

$$q_D = \frac{75 \times \int I_D}{75 \times I_D + \sum_{ND} \int I_{ND}} \times 100\,\% \tag{3.1}$$

In this equation, I_D is the diamond peak intensities and I_{ND} are the peak intensities of all non diamond carbon phases such as graphite, amorphous carbon, etc. From Eq. (3.1) the diamond quality of smooth nanocrystalline diamond film is estimated to be 75 %. This is due primarily to presence of large amount of grain boundary and preferences of most non diamond phases to settle preferentially at the grain boundary.

Raman spectra obtained from films having three different types of morphologies namely faceted, fine ballas and coarse ballas are presented in Fig. 3.7 [43]. All three films show peaks at wave number (1,340 cm^{-1}) corresponding to sp^3 bond. The corresponding bond for pure diamond is at 1,330 cm^{-1}. This peak is very prominent for the films having course ballas and faceted morphology and very weak for the film with fine ballas morphology. The Raman spectra of the film with fine ballas morphology are characterised by a weak peak at 1,160 cm^{-1}, a weak and broad peak in the band of 1,300–1,400 cm^{-1} and another broad peak in the band from 1,400 to 1,650 cm^{-1}. These peaks at 1,160 cm^{-1} and in the band from 1,400 to 1,650 cm^{-1} belong to diamond nanocrystals, G band of microcrystalline graphite and diamond nanocrystal respectively. The broad peak in the band of 1,300–1,400 cm^{-1} can be deconvoluted in to two peaks at 1,330 and 1,370 cm^{-1}. While the peak at 1,330 cm^{-1} corresponds to diamond, the peak at 1,370 cm^{-1} will certainly pertain to the D band of microcrystalline graphite. The spectra of the film with course ballas morphology consist of a broad band between 1,400 and 1,650 cm^{-1}. This can be deconvoluted in to two peaks at 1,470 and 1,560 cm^{-1}. The peak at 1,470 cm^{-1} is commonly observed in nanocrystalline diamond films [51–53]. Ferrari and Robertson [54] however, assigned this peak to transpolyacetylene situated at the grain boundaries of diamond nanocrystals. The peak at 1,560 cm^{-1} can be assigned to the G band of microcrystalline graphite. The presence of this band proves existence of sp^2 bonded carbon within the film. At this point, it should be stated that due to resonance effect [55, 56], Raman measurement with excitation at 532 nm is much more sensitive to sp^2 carbon than sp^3 carbon. Film with course ballas morphology also exhibits the presence of a weak peak at wave number 1,160 cm^{-1}. This peak can also be assigned to diamond nanocrystal. Film with faceted morphology has only one peak. Thus, film with faceted morphology is the best diamond film.

X-ray core level reflectance and absorption spectra are site and symmetry selective and magnitudes and energy positions of spectral features contain bonding information of the film. The near edge region of core level photoabsorbtion has been used to relatively quantify sp^2 and sp^3 bonding of thin films. It is sensitive to local bond order of the material because of dipole-like electronic transition from core states with well-defined orbital angular momenta into empty electronic (e.g. antibonding) states [57]. This technique can be used to determine the symmetry of the final state and the difference between sp^2 and sp^3 can be distinguished. In addition NEXAFS with a photon beam of the size of millimetre has the ability to investigate the film over a large area.

Electron energy loss spectroscopy (EELS) is a powerful technique to study electronic structure of the materials [58, 59]. The energy loss near edge structure (ELNES) varies with crystal structure. The π^* states can be observed between 282 and 288 eV and σ^* states can be observed between 290 and 320 eV for sp^2-bonded carbon. For sp^3-bonded carbon, σ^* states can be observed between 289 and 320 eV. Thus employing EELS and ELNES, it is possible to separate out diamond and non-diamond phases. EELS spectra obtained from nanocrystalline diamond film are presented in Fig. 3.8. Characteristic σ^* onset at 289 eV is evident. The sp^2-related peak at 285.4 eV cannot be seen. The film exhibits a minimum at the spectral shape near 302 eV indicating a low density of accessible unoccupied states.

Diamond has got exceptional mechanical properties. However, because of high roughness it is difficult to measure mechanical properties of microcrystalline diamond films. The mechanical properties of nanocrystalline diamond film are

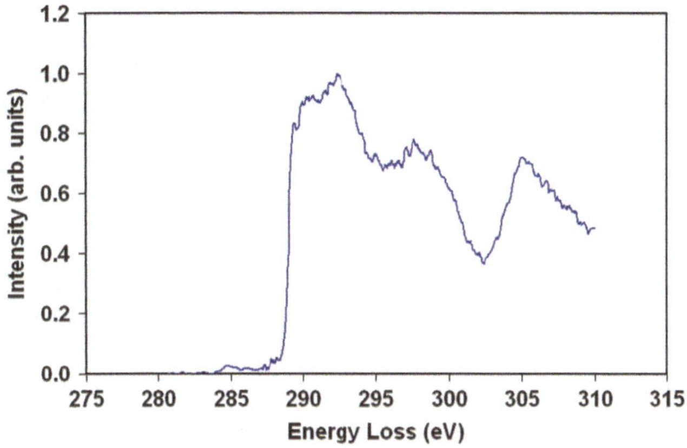

Fig. 3.8 EELS spectra obtained from nanocrystalline diamond film [45]

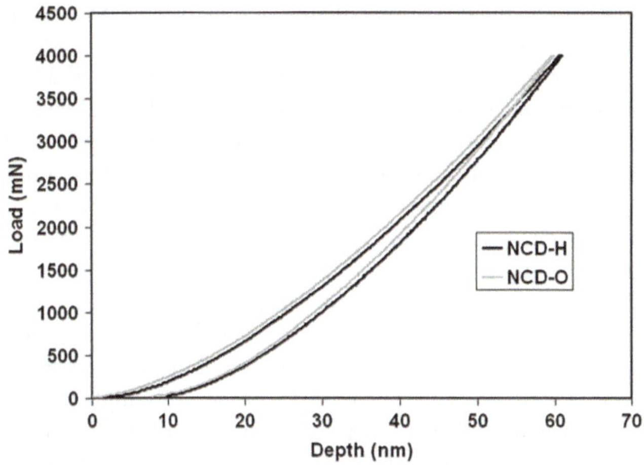

Fig. 3.9 The load vs. displacement curves of the NCD films having H-terminated and O-terminated surfaces [60]

measured by nanoindentation technique. Hardness, elastic modulus and at times, indentation toughness can be reported from indentation test. The load vs. displacement curves of the NCD films having H-terminated and O-terminated surfaces are given in Fig. 3.9 [60]. The highly elastic nature of the curves is evident. The maximum depth of indentation is 60.7 nm where the residual depth of indentation is 9.5 nm for the H-terminated film. The corresponding values for the O-terminated NCD film are 59.5 nm and 7.8 nm respectively. The hardness and elastic modulus of the film increase marginally after an oxidation treatment. In recent times various micro tensile test techniques are employed to measure fracture toughness of internal stresses of thin films [61, 62]. Espinosa et al. [62] carried out membrane deflection experiment to (MDE) to investigate the strength of free standing nanocrystalline diamond film.

Fig. 3.10 Diamond coated hardmetal drill bit for processing wood (Boehlerit GmbH, Austria)

3.4 Tribological Applications of Diamond Films

Diamond film has numerous applications. However, in spite of their excellent tribological properties, it has found limited application in tribology. Tribological application of diamond films depends on whether these films are microcrystalline or nanocrystalline. Some important applications of microcrystalline diamond films are given below.

Although there are quite new and very attractive fields of usage for CVD diamond films, they are still used as low friction and wear-resistant coating. This outstanding property allows fabrication of single-sealing systems with only the working fluid as lubricant and cooling media. This avoids the need of more complex and particularly more expensive double-working sealing with separate cooling and lubrication systems [63]. Even if the working fluid is not available temporarily, diamond-coated ceramic seals are able to sustain the mechanical and thermal strain in the dry running sealing system. Without diamond coating, the unlubricated seal faces generate an enormous amount of friction heat leading to sealing ring deformation and direct contact of the very fast rotating seal faces. Consequently complete seal system breaks down due to brittle fractures of the ceramic counter parts.

Cutting and forming tools coated with thin diamond film are well suited for extreme conditions (high speed cutting, dry turning) machining and forming operation of abrasive work piece materials, e.g. metal matrix composites (MMC) containing high amount of ceramic particles, graphite, copper, brass, refractory metals [64–66]. Diamond with its intrinsic properties—high hardness, thermal conductivity, wear resistance and chemical stability—results in tools with a long life time and good performance if the quality (adhesion, crystallinity, thickness, roughness) of the coating is high. Diamond-coated hardmetal drill bit for processing wood (Boehlerit GmbH, Austria) is illustrated in Fig. 3.10.

Many moving mechanical assemblies are made of such materials having high hardness, fracture toughness and flexural strength in order to withstand the bending, shear, torsional and contact stresses. In many occasion they are made to operate in demanding tribological condition making it imperative to put hard wear resistance coating such as diamond film [67]. However, coating such system requires deposition of interlayer and upscaled laboratory reactor to be able to coat a multitude of complex shaped components.

Bearing are such type of mechanical components which are subjected to Hertzian contact stresses, surface traction force subsurface shear stresses [67]. These severe conditions demand hard thin coating on bearing surfaces. The involute teeth of gears present a still more demanding test than bearing. At the dedendum and the addendum circle, the sliding component is the highest. Diamond coating

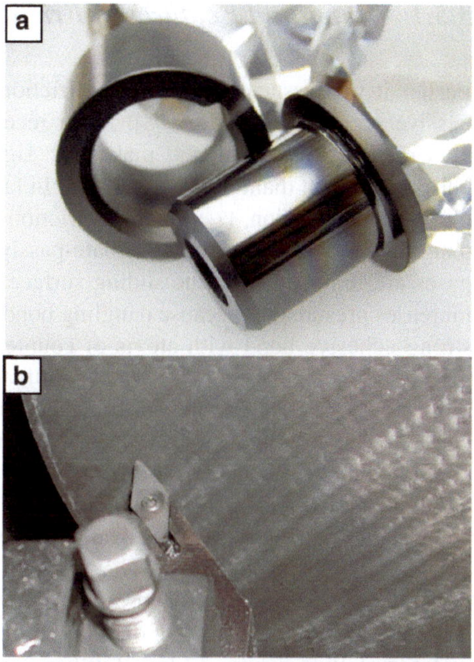

Fig. 3.11 (**a**) Ultrananocrystalline diamond coated bearing (Advanced Diamond Technologies, USA) and (**b**) diamond coated hardmetal cutting tool during machining of CFC (carbon fibre composite; Boehlerit GmbH, Austria)

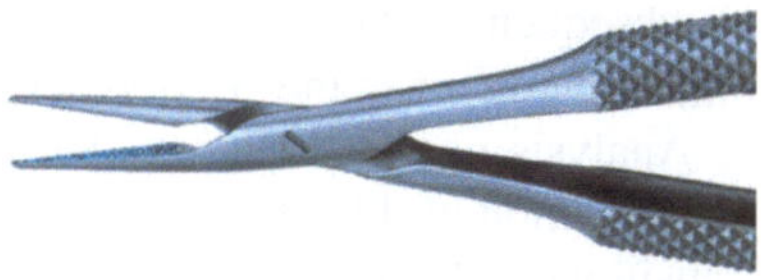

Fig. 3.12 Diamond-coated micro scissors of Ti (KLS Martin Group, Germany)

is expected to offer improved performances for such components. Ultrananocrystalline diamond-coated bearings (Advanced Diamond Technologies, USA) and hardmetal cutting tool for machining of carbon fibre composite (Boehlerit GmbH, Austria) are shown in Fig. 3.11a and Fig. 3.11b respectively.

Diamond-coated parts are used in medical component due to excellent wear and corrosion resistance of diamond films. The cages of heart pumps are coated with diamond film. Diamond coating on micro scissors of titanium shown in Fig. 3.12 (KLS Martin Group, Germany) resulted in increased service life and enhanced functionality. Many knives and other biomedical components are coated with diamond films for improved tribological performances.

3.5 Tribological Behaviour of Diamond Film

Studies in recent times confirmed that there are several factors that can adversely affect the wear and friction performances of diamond films.

3.5.1 Important Features of Friction Behaviour of Diamond Film

Earlier it was assumed that the low friction of diamond film is due to inert and passive nature of the film [68]. More recent studies however revealed that the presence of contaminants makes a significant difference in the friction performances of diamond film [69–71]. In fact in the absence of any shear-induced phase transformation, coefficient of friction is essentially governed by number of dangling, reconstructed or absorbate-passivated surface bonds. When absorbed gases are removed from the sliding surfaces of diamond film, friction coefficient increases presumably because dangling bonds are reactivated and available to form strong adhesive bond with atoms of counterface material. In contrast, when diamond surface is exposed to the contaminants, dangling bonds are repassivated reducing the friction coefficient. It is noted that among various absorbates, friction coefficient is minimum with H [72]. New and effective method to passivate the dangling bond has been developed. Addition of F has reduced the wear and friction coefficient of diamond and diamond film [73–75]. With increased temperature, some dangling bonds become available for reconstruction leading to increased friction coefficient [76, 77]. The friction behaviour of diamond film is also governed by the roughness and morphology of the film [78–80]. In addition, the wear and friction behaviour of diamond film is influenced by growth orientation of microcrystalline diamond films. Further, type of counterface materials can affect significantly the wear and friction response of diamond films [81, 82] as discussed in subsequent section. Finally it is also to be stated that phase transformation plays a major role in wear and friction performances of diamond coatings [79, 83].

Analysis of the evolution of friction coefficient with sliding distance leads to identification of different tribological stages and the associated sliding mechanism transitions [84]. The friction coefficient consists of an adhesion and a ploughing component. The adhesion component can be expressed by the following relation [85]:

$$\mu_{\mathrm{ad}} = \frac{\tau_{\mathrm{m}}}{H} \tag{3.2}$$

where H is the hardness of the softer material and τ_{m} the shear strength of the contacting asperities, which depends on the work of adhesion of the mating materials [86]. The ploughing component is associated with the surface plastic deformations exerted by the hard asperities through a softer counterface. If the abrading asperities are modelled by a cone with an included angle θ, μ_{pl} may be expressed as [85]

$$\mu_{\mathrm{pl}} = \tan\theta \tag{3.3}$$

If we consider that the roughness of the diamond asperities yields an average attack angle of about 30°, Eq. (3.3) predicts the contribution of the ploughing term to friction is nearly 0.6. This value of friction coefficient has been observed many times in diamond films particularly in the initial run-in period. Once the film gets smoothened due to continuous sliding, roughness component decreases and adhesion component starts playing influential role. In this regime, friction coefficient increases from a low value due to increase of ploughing components. Many times the friction coefficient of diamond coating decreases with sliding distance. Liu et al. [87] and Erdemir et al. [88] attributed this decrease to a decrease in τ_m of Eq. (3.2) connected with the formation of a graphite-rich transfer layer. Sometimes a high increase of friction coefficient is noted due to spallation of coating. In general, spallation can be due to tensile or compressive stresses. In the case of compressive stresses, spallation can be due to buckling or wedging [89]. Spallation by buckling develops through the nucleation of a decohesion area at the interface between the coating and the substrate and its propagation along the interface. In order to have the onset of buckling, the following condition (simply obtained from the Euler relation) must be satisfied [89]:

$$\sigma_b = \frac{E_c}{3}\left(\frac{t\pi}{a}\right)^2 \tag{3.4}$$

where σ_b is the compressive stress in the coating, t the thickness of the coating and a is the length of the area of decohesion. There are occasions when spallation occur by wedging. Coating damage occurred by wedging is the typical fracture process which takes place in brittle materials subjected to a compressive stress. Wedging is initiated when the acting (compressive) shear stress σ_{sh} (which is maximum in a plane at 45° with the surface) reaches a critical value, i.e. when [89]

$$\sigma_{\mathrm{sh}} = \frac{K_{\mathrm{IIC}}}{Y\sqrt{\pi a}} \tag{3.5}$$

where K_{IIC} is the critical stress intensity factor of the coating for mode II of fracture, Y a constant of the order of 1 and a is the critical dimension of a defect in the coating.

3.5.2 Sliding Wear

Sliding wear of diamond film has been studied most extensively among various wear mechanisms. Important characteristics of sliding wear of diamond films are described below.

3.5.2.1 Influence of Roughness

It is mentioned earlier, polycrystalline diamond films are very rough. In addition to environmental characteristics and surface chemistry, the wear and friction

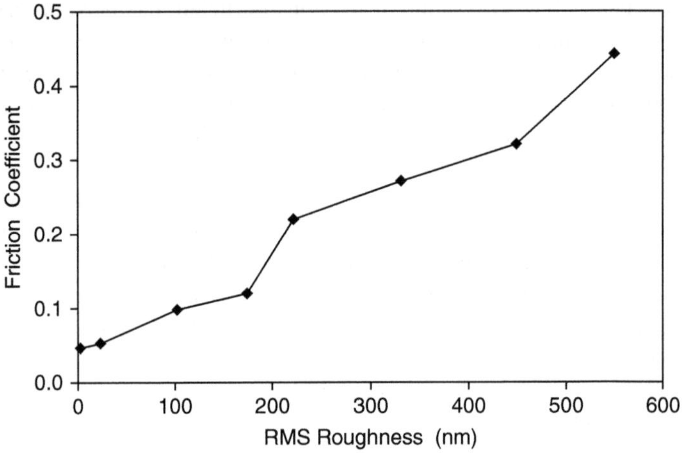

Fig. 3.13 The influence of roughness on the friction coefficient

properties of diamond film are largely governed by the surface roughness of the film. The friction coefficient of diamond film is found to increase with surface roughness. Several investigators carry out a systematic study on the roughness of diamond film on friction coefficient. Their observation is compiled in Fig. 3.13. Higher roughness increases the asperities slope angle, which in turn increases friction coefficient [90]. There are two well-established mechanisms responsible for higher friction coefficient with film having higher roughness. The first one is ratcheting mechanism where relative motion between two surfaces is achieved by asperities riding over each other. The second one is energy loss mechanism where asperities push each other. In the initial stage of sliding, there can be other operative mechanisms.

In general, the roughness of the coating decreases with time and so is the friction coefficient. If the coating is smoothened by polishing either mechanically or chemically [91], a lower friction coefficient is obtained at the onset of sliding [92]. If the counterface material is other than diamond, the effect of roughness on friction coefficient is even more pronounced. Higher coating roughness leads to much higher wear of counterbody [93, 94]. The friction and wear properties of polycrystalline diamond coatings as well as the wear of the counterface are dependent significantly on the surface roughness, the morphology and crystalline structure of diamond coatings as well as the counterface materials. For (111)-textured diamond coatings with rough surface and sharp asperities sliding with Al_2O_3 balls, the coefficient of friction is much higher than that of (100)-textured coatings, and the wear of the counterface material is quite high [95]. After polishing the diamond coating, the surface roughness, coefficient of friction and wear of counterface decreases significantly.

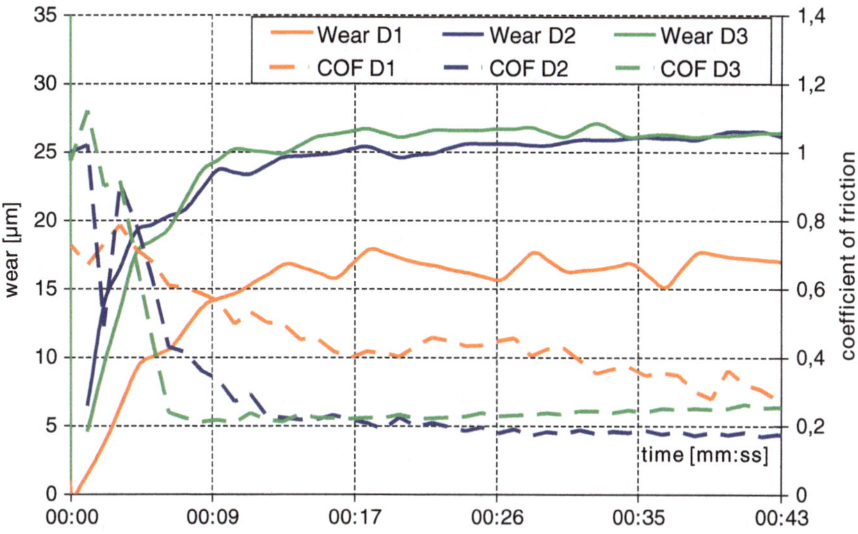

Fig. 3.14 The influence of morphology on the friction coefficient and wear of diamond film

3.5.2.2 Influence of Morphology

Morphology is another important parameter that influences wear and friction behaviour of diamond film. Although it is difficult to isolate morphology from roughness, a general trend can still be observed. Synthesis of diamond film aims at faceted diamond layers even though hemispherical polycrystalline diamonds (ballases) are also found under non-optimal growth conditions [96, 97]. The CVD ballas films are also known as cauliflower-like, ball-shaped diamond layers. Faceted diamond layers can be obtained for a wide range of depositing parameter. These methods are based on activation of CH_4/H_2 gas mixtures in order to generate large amounts of atomic hydrogen and carbon radicals [98]. If there is super saturation of carbon in the gas phase, growth conditions of faceted diamond becomes poor and a high density of twins and stacking faults is introduced [34]. This finally, leads to radial growth of polycrystalline unfaceted diamonds of "ballas-type" [99].

SEM images showing the morphologies of different types of diamond films are presented in Fig. 3.3. The morphologies and the structures of the films differ considerably. Wear and friction of diamond films having three different morphologies such as fine ballas (D1), coarse ballas (D2) and faceted (D3) have been characterised. Film with fine ballas morphology is relatively featureless. Film D2 exhibits radially grown coarse ballas morphology and is obtained when diamond is grown under conditions similar to regimes of faceted diamond [100]. The film D3 grown at 2 % CH_4 consists of crystallites with faceted morphologies having cubo-octahedral and icosahedral symmetries. This ballas morphology can be seen when the film is grown close to graphitic region. The variations of friction coefficient as function of time for all three films are presented in Fig. 3.14. The friction coefficient decreases from an initial high value to a low steady state value in all the films. This behaviour is consistent with what is noted for nanocrystalline diamond film [101] and it is in contrast to what is observed with some microcrystalline diamond film [102]. This behaviour is often associated with the severe plastic deformation of the asperities and creation of third body [103]. Among these films, film D2 having coarse ballas morphology exhibits lowest friction coefficient. In contrast, film D3 with faceted morphology has highest friction coefficient. Friction coefficients of all the films are quite low in the range of 0.2–0.3. Such low value is result of transfer of carbonaceous materials from the film to the counterbody as reported by others [104]. Thickness loss of the tribosystem containing the film and the ball was monitored on line with an eddy current sensor system. The results of the online wear investigations are also illustrated in Fig. 3.14. Clearly the rate of change of total wear height with respect to time reaches a lower value after an initial high value for all three films. The higher friction coefficient of film D3 compared to other films can be attributed to the fact that film D3 with faceted morphology breaks during sliding, giving rise to abrasive particles. These abrasive particles in turn increase friction coefficient and wear rate by "third body" abrasion. In contrast, a graphitic carbonaceous material of the films D1 and D2 with ballas morphology is responsible for low friction. Low friction coefficient and low wear rate of the film recommends use of diamond film with ballas morphology for tribological application at high load even though diamond film with faceted morphology is best quality of diamond film and has the best topographical parameter [105].

3.5.2.3 Influence of Environment

There is significant reduction of the friction of diamond film when made to slide against diamond in presence of lubricant. It is however still not clear whether it is due to formation of slippery debris or due to change of surface chemistry. According to Gardos and co-workers [106, 107], the friction coefficient of diamond coated or uncoated SiC at low gas pressure is very low. The friction coefficient increases with increase of temperature and then decreases again with decrease of test temperature. They attributed such observation on presence of dangling bond. Few researchers also observe a decrease of friction coefficient with increase of temperature due to formation of graphitic layer. Other investigators found [108–110] diamond to get oxidise and graphitise gradually at elevated temperature making them unsuitable for elevated temperature applications.

3.5.2.4 Influence of Counterface Material

Influence of counterface materials on the friction coefficient of diamond film can be noted in the work of Habig [81, 111]. Bar diagram showing the friction coefficient of rough diamond film against various sliding pins is depicted in Fig. 3.15. As

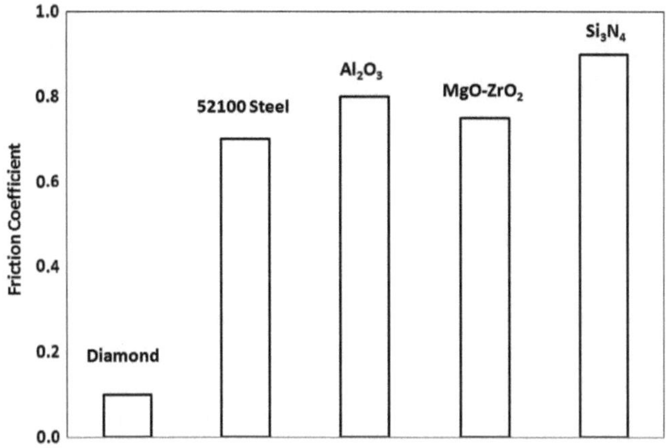

Fig. 3.15 Bar diagram showing the friction coefficient of rough diamond film against various sliding pins

expected, diamond film sliding against diamond film has less friction. Against other counterface material, friction coefficient is significantly high. In commensurate with friction coefficient, the wear of diamond against diamond film is also low. However, the wear of Si_3N_4 pin is also very low even though the friction coefficient is very high. The friction coefficient against all counterface material increases with roughness. Mohrbacher et al. [112] however, noted that friction coefficient of diamond film can be classified under two different heads. The friction coefficient against brittle materials such as diamond or Al_2O_3 is lower than 0.1 whereas the friction coefficient of diamond film against metallic materials such as steel or WC–Co is always above 0.5. The above response was attributed to nature of transfer film and its shear strength tm as described in Eq. (3.2).

3.5.2.5 Influence of Transfer Film and Debris

It is well established that the adhesion component of friction is related to the chemical nature of the contacting surfaces. If the nature and composition of the debris generated is different from the original diamond film, friction behaviour is altered. When diamond is made to slide against diamond the debris formed is not diamond rather it is either graphite or amorphous carbon phases. It also contains more oxygen than the original diamond [69]. Studies with Raman spectroscopy, electron energy loss spectroscopy, electron diffraction and transmission electron microscopy reveals that most of the debris particles are highly disordered graphitic microstructure. During friction test most diamond films exhibit a decrease in friction coefficient after a running in period. This is due primarily to formation of lower shear strength transfer film or tribolayer with low adhesion strength [113]. As the counterbody is normally softer than the diamond film, generally material is transferred to the diamond film [114, 115]. If Al_2O_3 is made to slide against diamond film in moist environment, the resultant debris is altered forming a self-lubricating transfer film on the diamond surface and thus, reduces the friction coefficient [78]. Similar reduction of friction coefficient is obtained by applying a self-lubricating layer of MoS_2 on diamond film [116].

3.5.3 Abrasive Wear

Diamond films exhibit exceptional performances under abrasive wear condition due to their very high hardness [117, 118]. In many occasion, the film fails due to poor mechanical properties of the substrate [119]. Sometimes the friction coefficient during abrasion can be as high as 0.3 leading to failure of the film substrate interface [120]. Work due to Alahelisten [121] indicates that during abrasive wear with SiC or Al_2O_3 abrasives, the life of diamond coatings increases with increase of coatings thickness. Diamond coatings display a superior wear resistance to hard metals (WC–Co) and Ti (C,N) and TiC coatings in abrasion tests with diamond grains as abrasives [122, 123]. The wear of the diamond coatings starts with blunting of protruding asperities and continues until the surface is flat. Tolkowsky has shown that when diamond is abraded with a disc with diamond powder, the amount of worn material is proportional to the total number of revolutions and independent of the rotational speed [124]. This shows that the abrasion process is not thermally activated. The abrasion process is dependent of the abrasion direction [117].

Roughness of the diamond film also plays an influential role in abrasive wear behaviour of counterface material. The roughness can be controlled during post-coating processes such as by polishing or during coating processes by changing the morphology, etc. The roughness of microwave-assisted chemical vapour-deposited coating was varied for different methane concentration. The wear test indicated that the wear rate of the alumina counterface material decreased with increase of methane concentration.

Buijnsters et al. [13] investigated the effect of substrate temperature and methane concentration on the adhesion of polycrystalline diamond films grown by hot-filament CVD on molybdenum substrates by means of indentation, scratch and sand abrasion tests From the indentation and scratch adhesion tests, it is concluded that higher deposition temperatures lead to stronger adhesion, whereas increasing methane concentrations result in a decrease of the adhesion. However, for sand abrasion, the lifetime of films grown at a methane concentration of 4.0 % is about three to eight times higher than that of films grown for lower methane concentrations as shown in Fig. 3.16.

The effect of high biaxial stress on the abrasive wear rate of diamond coating worn by diamond abrasive has been evaluated using AFM by Gahlin et al. [125]. According to them, wear rates of highly stressed coating is 5–20 % of stress free (free standing) coatings. The difference in wear mechanism lies in the direction of crack propagation. Crack seems to propagate in plans more parallel to the surface in highly stressed coatings. The resistance of diamond faces to abrasion is much

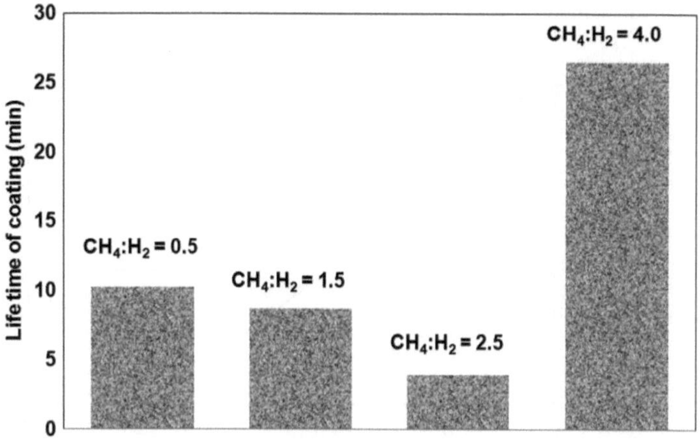

Fig. 3.16 Influence of methane concentration on the abrasive wear rate of diamond film [13]

greater for {111} diamond faces than for {100} faces [126]. Thus, diamond film with <111> texture where {111} diamond faces are parallel to the sliding planes should have better wear resistance than an <100> textured diamond film [127]. The wear coefficient is an important parameter giving severity of the prevailing wear mechanisms. The wear coefficient during abrasion of diamond coating against diamond particle is very low of the order of 10^{-5}. The corresponding value for other PVD coatings such as TiN and CrN is 10^{-2}. This value is typical for severe abrasion. In severe abrasion regime, material removal is by micro chip formation and microcracking. Since the wear coefficient for diamond film is of the order of 10^{-5}, micro chip formation and microcracking is expected to be rear. In such case, possible wear mechanism should be surface fatigue caused by repeated contact against abrasives. This mechanism will involve nucleation and propagation of crack due to cyclic high stress contact with abrasives.

3.5.4 Erosive Wear

Much less result is available about erosion of diamond films. Erosion behaviour is dominated by fracture and delamination. Wheeler and Wood [128] noted that erosion rate of diamond coating increases with increase of coating thickness. Work due to Kral et al. [129] is presented in Fig. 3.17. It is noted that diamond-coated Si exhibited highest erosion resistance and diamond-coated Ti–6Al–4V had better erosion resistance compared to uncoated substrate. Coating was mostly failed by delamination due to high residual stress and stress concentration at the coating substrate interface.

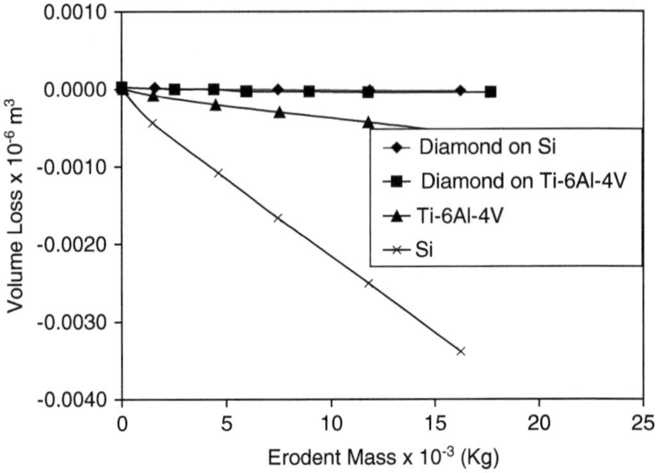

Fig. 3.17 The erosion rate of diamond-coated and -uncoated Si and Ti–6Al–4V substrate [129]

Head and Harr [130] developed a model for the wear rate of brittle materials. According to his model, parameter D = erosion resistance per unit volume is given by

$$D = H_t - \alpha \left\{ \frac{H_t - T}{\pi/2} \right\} \tag{3.6}$$

where T is modulus of toughness and is given by

$$T = \frac{\varepsilon_u}{2} (\sigma_y + \sigma_u) \tag{3.7}$$

Regression analysis gives the erosion rate (W) of brittle materials as

$$W = \frac{V^{3.06} \alpha^{2.69} {H_P}^{2.08} D^{0.03}}{{H_t}^{3.64}} \tag{3.8}$$

where ε_u is ultimate tensile strain, σ_y 2 % yield strength, σ_u is ultimate tensile strength α is the impact angle, H_t is hardness of the target and H_P is hardness of the erodent. The theoretical erosion rate as predicted by Head and Harr [130] expression was compared with the experimental result. The experimental results were lower than theoretical value by a factor of 0.2×10^{-3} to 7×10^{-3}.

Feng et al. [131] investigated 6–15 μm coatings on Si and Si_3N_4 at impact velocities of 34 and 59 ms^{-1} at an impact angle of 90°. It was noted that the CVD diamond behaved in a manner similar to natural diamond. The main difference is the manner in which the material is removed. In the case of diamond films, delamination of the coating from the substrate occurs, whereas in natural diamond,

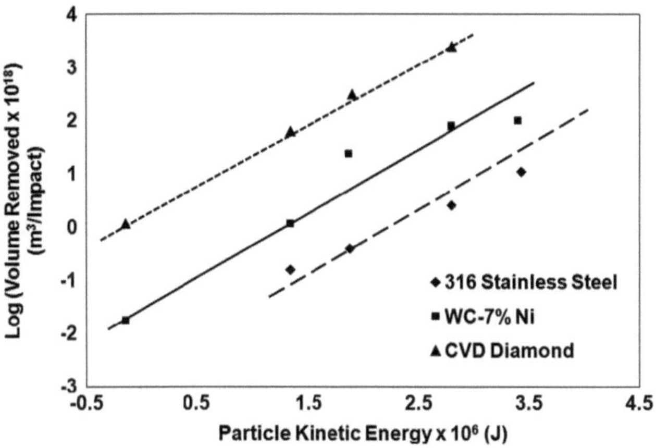

Fig. 3.18 Influence of particle kinetic energy on the erosion rate of lapped diamond coating on W, sintered WC–7 % Ni and 316 Stainless Steel [134]

material is removed when ring or cone cracks intersect to separate small volumes of diamond from the bulk [132]. Alahelisten et al. [133] used particle velocities of between 46 and 72 ms^{-1} and found that the erosion rate increased with coating thickness; however, these results were achieved on very thin coatings of 1.5–6 μm. Work due to Wheeler and Wood [134] indicates that CVD diamond films with a thickness of 15 mm have exhibited a substantial increase in erosion resistance compared with uncoated SiC. The CVD diamond film deposited on SiC has a 200 times erosion resistance than that of polycrystalline diamond. It also showed that at 268 ms^{-1}, the erosion rate of 39–47 μm CVD diamond on tungsten is six times lower than cemented WC–7Ni. Their result is illustrated in Fig. 3.18 which shows CVD diamond film offers considerable improved erosion resistance compared to other materials.

Wheeler and Wood [128] also noted that the erosion rates of thick CVD diamond coatings having a lapped surface are approximately an order of magnitude lower than cemented tungsten carbide at a particle velocity of 268 ms^{-1}. They have also shown that a coating with a lapped surface has an erosion performance approximately three to four times better than an as-grown surface of comparable coating thickness. The study suggests that for lapped coatings between 60 and 160 μm the erosion rate is essentially independent of thickness. Another work by Wheeler and Wood [135] revealed presence of circumferential cracks and pin-holes are seen at all the angles at which the tests were conducted. Moreover, the diameter and aspect ratio of the circumferential cracks do not appear to be dependent on impact angle. The circumferential crack diameters appear to be dependent on coating thickness with larger diameter cracks observed on the 60 μm coatings than on the 37–43 μm coatings. This provides further evidence that Hertzian theory is not applicable in attempting to explain the mechanism by which the circumferential cracks are formed. The coating thickness dependent erosion rate is presented in Fig. 3.19.

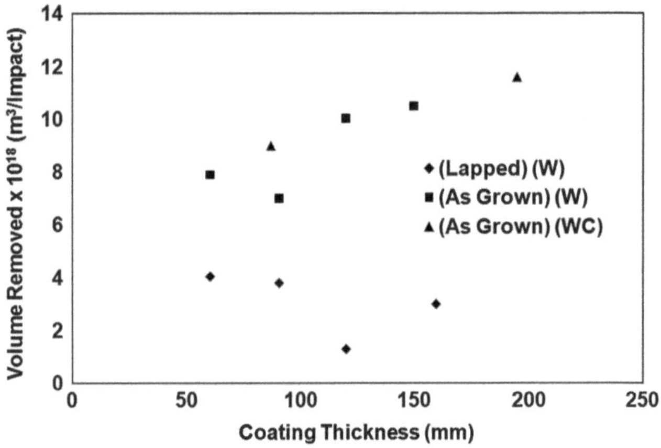

Fig. 3.19 Influence of coating thickness on the erosion rate of lapped and as-grown diamond coatings on W and WC substrates [135]

3.5.5 Oscillating Wear

Most application of diamond films simulates fretting wear. In spite of this, fretting wear of diamond film is studied to a limited extent. Fretting experiment of diamond coating against spherical steel ball indicates that the under fretting condition dominant wear mechanism is fatigue. According to Alzoubi et al. [136], the wear rate of diamond coating increases with increase of coating thickness. As softer steel balls cannot cause abrasion of harder diamond film, wear mechanism should be fatigue. Thus, it can be concluded that if fatigue is the wear mechanism, thicker film deteriorates faster than thinner coatings. When both the counterbodies are diamond film, the wear mechanism under fretting condition is abrasive [137, 138]. In general, sliding against brittle non-metallic materials friction coefficient is low. However, against metallic materials friction coefficient under oscillating sliding is high. According to Liu et al. [139], at the beginning of oscillating contact the friction coefficient is high and this is related to the morphology, structure of the coating and size and orientation of the diamond crystal. The observation of Miyoshi [140] is summarised in Table 3.2. As reported by Miyoshi, as-deposited and carbon-implanted diamond coating can be used as self-lubricating film. The carbon ion-implanted, fine-grain diamond films can be effectively used as wear-resistant, self-lubricating coatings not only in humid air and in dry nitrogen but also in ultrahigh vacuum. This is to be stated that as-received diamond film has relatively high friction coefficient in vacuum making it unsuitable to use in vacuum. Skopp and Klaffke [141] noted that during running-in, system wear is mainly caused by abrasive wear and material transfer to the diamond coating occurs. After this initial wear phase, adhesive wear is probably the dominating mechanism against 100Cr6, and a tribochemical wear mechanism with formation of effective "triboreaction layers" is assumed against Al_2O_3. Depending on test conditions, test duration and

Table 3.2 Summary of tribological properties of diamond film [140]

Environment	Properties	Fine grain	Coarse grain
As-deposited diamond film			
In air and dry nitrogen	Friction coefficient in running in	0.15	>0.4
	Friction coefficient at 30,000 passes	0.03–0.04	0.03–0.04
	Wear rate (m^3/Nm)	10^{-17}	10^{-17}–10^{-16}
In ultrahigh vacuum	Friction coefficient in running in	1.2	1.2
	Friction coefficient at 30,000 passes	1.7	1.7
	Wear rate (m^3/Nm)	10^{-13}	1–4×10^{-13}
Carbon ion-implanted diamond film			
In air	Friction coefficient in running in	0.1	0.3
	Friction coefficient at 30,000 passes	0.08	0.1
	Wear rate (m^3/Nm)	10^{-16}	10^{-16}
In dry nitrogen	Friction coefficient in running in	0.06	0.12
	Friction coefficient at 30,000 passes	0.05	0.08
	Wear rate (m^3/Nm)	10^{-16}	10^{-16}
In ultra high vacuum	Friction coefficient in running in	0.1–0.15	0.2
	Friction coefficient at 30,000 passes	<0.1	0.35
	Wear rate (m^3/Nm)	10^{-15}	10^{-15}

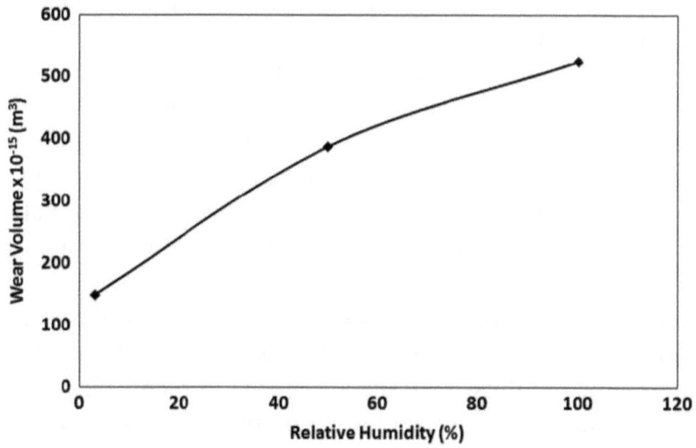

Fig. 3.20 The influence of humidity on the wear volume of diamond film

counterbody material friction coefficients between 0.03 and 0.7 and system wear coefficients below 1×10^{-7} mm^3/Nm as well as those in the order of 1×10^{-5} mm^3/Nm were determined for the same diamond coating, same test rig and same test configuration. Relative humidity also play crucial role on the wear rate. Work by Skopp and Klaffke [141] presented in Fig. 3.20 indicates wear volume of diamond film increases with relative humidity. Such behaviour was attributed to formation of hydrated layer which can be sheared easily.

3.6 Tribology of Nanocrystalline Diamond Films

The interest on nanocrystalline diamond (NCD) coating was increasing rapidly in the last few years. Specific properties, like high hardness, stiffness, thermal conductivity, low friction coefficient, chemical resistivity and imperviousness make it an optimal material for many engineering applications, including cutting tools and mechanical assembly devices, etc. [10, 142]. The potential of NCD-coatings in reducing friction and wear on different ceramic material is also proven and mentioned in few literatures [143–147]. NCD-coatings deposited on spherical hard metal substrates have been tested in dry and lubricated (water, oil) sliding against hard metal, ball-bearing steel, stainless steel, titanium and aluminium [144]. In general, this investigation shows that smooth NCD display exceptionally low friction and wear in dry and water-lubricated sliding against metallic materials. It is noted [145] that hydrogen plasma pre-treatment of silicon nitride substrates has a beneficial effect on the delamination threshold loads under tribological stress of nanocrystalline diamond coatings. Adequate adhesion between the coating and the substrate is crucial for tribological applications. Si_3N_4 and SiC ceramics are materials with high potential for depositions of stress-free highly adherent NCD-coating due to their low thermal expansion coefficient [146]. For this reason the tribological investigation of NCD-coated materials are mostly done with SiC and Si_3N_4 as substrate material for specimens.

In general, use of NCD film resulted in significant reduction of friction and wear properties [148–150]. Such reduction in friction and wear is noted in ambient lubricated, unlubricated condition and also in physiological solution. In most of the work, a short run-in period with high friction followed by low steady state friction is observed. Such behaviour is in general attributed to formation of carbonaceous layer. Further the wear is primarily found by micro abrasion in most investigation.

The friction coefficient and wear of counterbody during wear study of NCD coated specimens having surface roughness Ra = 0.02–0.04 μm in presence of lubricant are presented in Figs. 3.21 and 3.22 [151]. It seems that the nature of the used lubricants has no relevant effect on the friction coefficient and counteracting body wear. Lubricant however reduces the friction coefficient marginally in both varieties of films. The friction values in most cases are higher than 0.1 (0.11–0.17), and this is acceptable for the selected severe test condition. In this case there is no significant difference between H- and O-terminated specimens. However the best results were achieved with O-terminated coating specimens. The specific pressure is high at the beginning of the test (calculated Hertzian pressure about 1,800 N/mm^2). At the end of the test the specific contact pressure can be estimated to be 50–60 N/mm^2 and 80–82 N/mm^2 for H-terminated surfaces and O-terminated surfaces respectively. There is no measurable wear on the coated surfaces. It seems that, in general the O-terminated NCD-coated specimens achieve the better tribological behaviour.

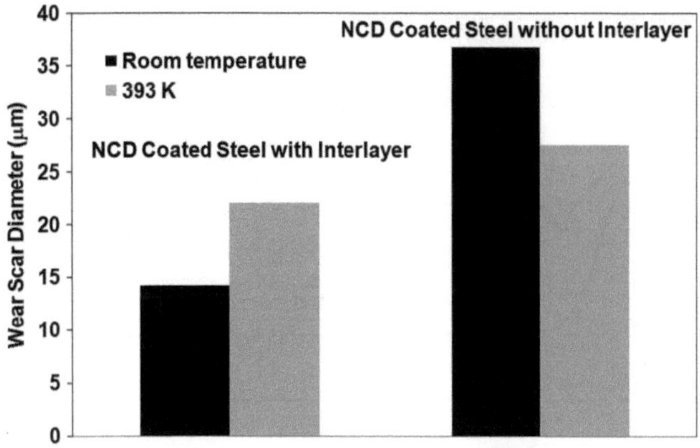

Fig. 3.21 The wear of counterbody during wear study of NCD-coated specimens [151]

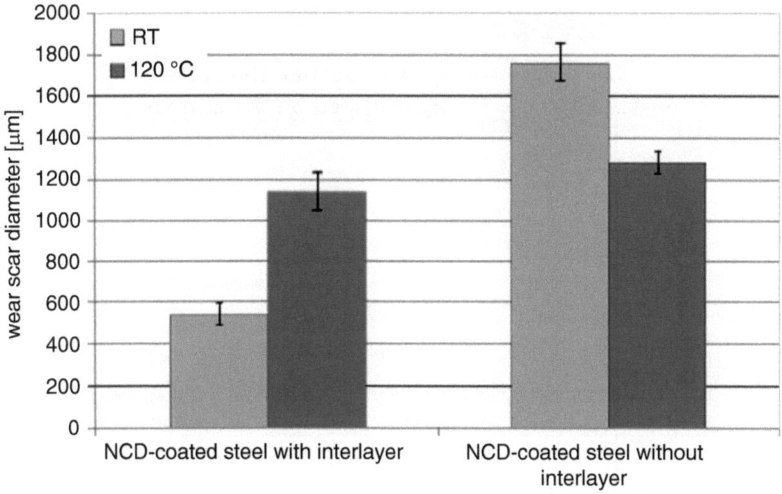

Fig. 3.22 The wear of counterbody during wear study of NCD-coated specimens [151]

3.7 Tribology of Diamond Film in Micro and Nanoscale

As microcrystalline diamond (MCD) film is very rough, micro or nanotribological behaviour of MCD films are not studied. Rather, such study of diamond film is confined with NCD films. Development of MEMS and NEMS has made micro and nanotribology relevant. Development of new equipments and superfast computer has made such study possible. Figure 3.23 shows the variation of friction coefficient of H-terminated and O-terminated NCD films slide against steel ball at a load of 200 mN [45]. The friction coefficient decreases from an initial high value to a steady state value. The friction coefficient of O-terminated film is always higher than the friction coefficient of H-terminated films.

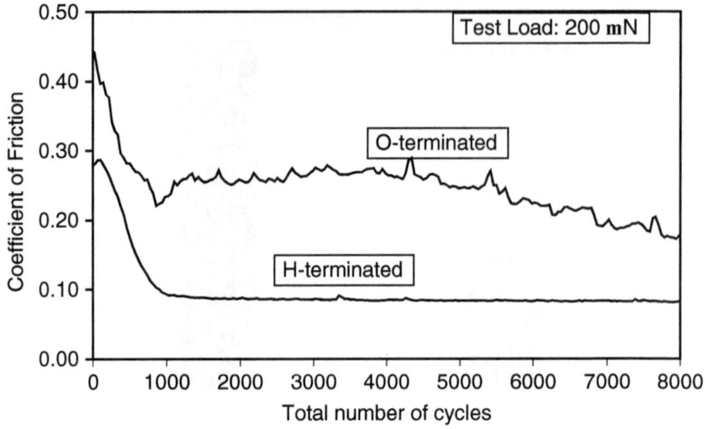

Fig. 3.23 The variation of friction coefficient of H-terminated and O-terminated NCD films slide against steel ball at a load of 200 mN [45]

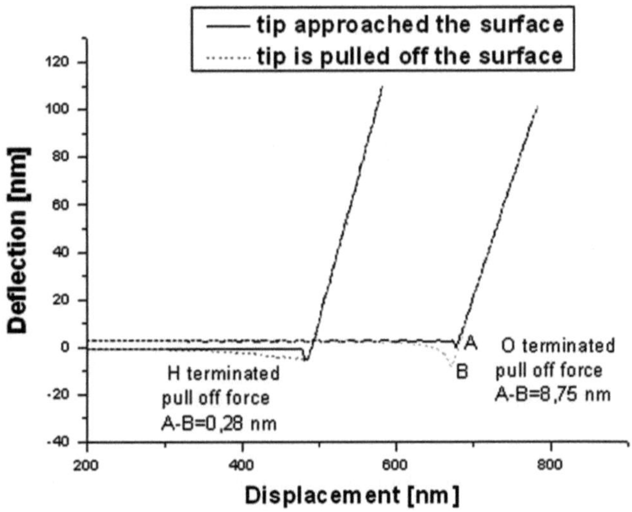

Fig. 3.24 The deflection of the cantilever tip as a function of the distance from the surfaces of H-terminated and O-terminated diamond films

Nanotribological response of nanocrystalline diamond film can be studied using AFM. To understand the nature of interaction between the cantilever tip of an AFM and the coating, the deflection displacement curves are recorded. The deflection of the cantilever tip as a function of the distance from the film surfaces is presented in Fig. 3.24 for H-terminated and O-terminated NCD films [60]. In all cases, the darker line indicates the tip approach and the lighter line represents the tip being pulled away. The vertical separation between the point where the tip is touching the

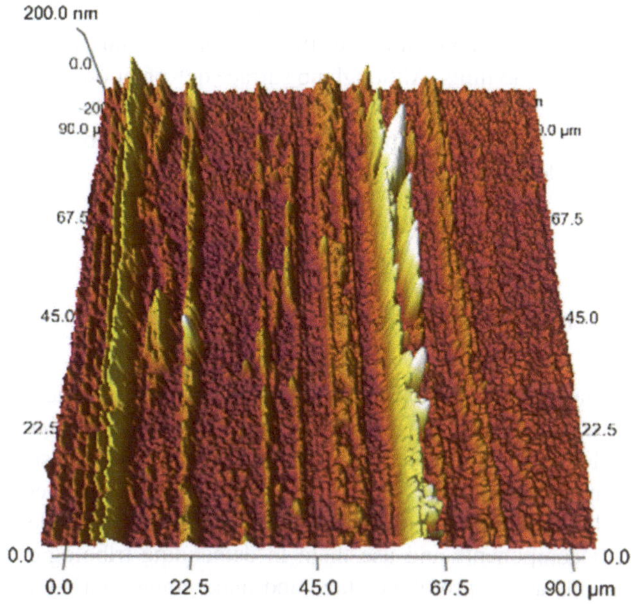

Fig. 3.25 Topographic AFM image obtained from the worn surfaces of the H-terminated nanocrystalline films

film and the point where the tip is pulled off the film is a measure of the pull-off (adhesive) force. The product of this horizontal distance of separation and the spring constant of the cantilever (0.1 nN/nm) gives the pull-off forces [152]. The pull-off forces of the H-terminated and the O-terminated films are 12.8 nN and 406.9 nN respectively. Thus, the pull-off force is significantly higher on the O-terminated NCD surfaces than on the H-terminated surfaces.

Topographic images obtained using AFM from the worn surfaces of the H-terminated nanocrystalline films are illustrated in Fig. 3.25 [45]. Extensive ploughing can be noted. This ploughing forces the material to the ridges along the sides of the grooves. These displaced materials subsequently chipped out leading to material loss. These grooves are due to ploughing action of the asperities and debri. The metallic oxide debri that formed due to rubbing action is primarily responsible for such prominent grooves formation. These grooves are more closely spaced for the O-terminated films whereas these grooves are uniformly distributed for the H-terminated films. The heights of these grooves are higher for the O-terminated film than that of the H-terminated films.

Liang et al. [153] deposited undoped and boron-doped nanocrystalline (NCD) diamond films on mirror polished Ti–6Al–4V substrates in a microwave plasma-assisted chemical vapour deposition system. It was found for diamond/diamond sliding, coefficient of friction decreases with increasing normal loads. It was also found that the wear rate of boron-doped NCD films is about ten times higher than that of undoped films. A wear rate of ~5.2 $\times$ 10^{-9} mm^3/Nm was found for undoped NCD films. This value is comparable to the best known value of that of polished polycrystalline diamond films. Although no surface deformation, film delamination or microcracking was observed for undoped films, boron-doped NCD film undergoes a critical failure at a normal stress of 2.2 GPa, above which surface deformation is evident.

3.8 Direction for Future Research

Systematic studies on the influence of microstructural features such as grain size, ballas or faceted morphology, on the tribological behaviour of diamond film are hardly available and it deserves dedicated attention. Over the last few decades, great strides have been made in development of deposition technology and utilisation of diamond films. With present state of art, it is possible to coat large number of tribocomponents and use them in demanding tribological conditions. Many applications are in prototype stage and undergoing continuous refinement. However, many are being evaluated for reliability and endurance. Relatively high cost, poor adhesion and less reliability are the main reasons for their delayed applications. As these hard coatings have much to offer for tribocomponents, their large-scale utilisation is yet to be realised.

Because of highly inert character of diamond, the intrinsic adhesion characteristics of the film are very poor with most materials. One major reason for failure of diamond film is their poor adhesion with the substrate. Significant effort is directed towards deposition of diamond film on various substrates. The adhesion characteristics of different film will vary depending on the substrate materials and interlayer. The adhesion characteristics can also be altered by employing different deposition technologies and processes. However, a comprehensive study to evaluate and improve the adhesion behaviour of diamond coatings on various substrates needs to be undertaken.

High roughness is another stumbling block for large scale application of diamond film particularly in microelectronics. Attempts are directed to deposit CVD diamond film with improved roughness. Although the success in direction is far from satisfactory, improved process parameter, process control and innovative process may see interesting development in this direction.

Free standing diamond films are increasingly being used in microelectro mechanical systems. Development of free standing films with higher degree of reliability and improved mechanical and physical properties also draws significant attention of researchers in recent times. This development of diamond film should accompany improved post processing techniques such as micromachining, micro finishing, etc.

References

1. Bundy FP (1980) J Geophys Res 85(B12):6930
2. Lux B, Haubner R (1998) CVD diamond for cutting tools. In: Dischler B, Wild C (eds) Low pressure synthetic diamond. Springer, Berlin, p 223
3. Sigals I, Caveney RJ (eds) Diamond materials and their applications, Verlag GmbH, p 479
4. Donnet C (1996) Surf Coat Technol 80:139
5. Zehnder T, Patscheider J (2000) Surf Coat Technol 133–134:138
6. Wu RLC, Miyoshi K, Vuppuladhadium R, Jackson HE (1992) Surf Coat Technol 54(55):576
7. Haubner R, Lux B (1993) Diam Relat Mater 2:1277
8. Kohzaki M, Higuchi K, Noda S, Uccida K (1992) J Mater Res 7:1769
9. Miyoshi K, Wu RLC, Garscadden A (1992) Surf Coat Technol 54(55):428
10. Olszyna A, Smolik J (2004) Thin Solid Films 459:224
11. Gerbi JE, Birrel J, Sardela M, Carlisle JA (2005) Thin Solid Films 473:41
12. Köpf A, Lux B, Haubner R (2001) New Diam Front Carbon Technol 11:11
13. Buijnsters JG, Shankar P, van Enckevort WJP, Schermer JJ, ter Meulen JJ (2005) Thin Solid Films 474:186
14. Deuerler F, Lemmer O, Frank M, Pohl M, Heising C (2002) Int J Refract Met Hard Mater 20:115
15. Wang WL, Liao KJ, Fang L, Esteve J, Polo MC (2001) Diam Relat Mater 10:383
16. Zeiler E, Klaffke D, Hiltner K, Grogler T, Rosiwal SM, Singer RF (1999) Surf Coat Technol 116:599
17. Haubner R, Schubert WD, Lux B (1998) Int J Refract Met Hard Mater 16:177
18. Haubner R, Kalss W (2010) Int J Refract Met Hard Mater 28:S. 475
19. Glozman O, Halperin G, Etsion I, Berner A, Schectman D, Lee GH, Hoffman A (1999) Diam Relat Mater 8:859
20. Glozman O, Berner A, Schectman D, Hoffman A (1998) Diam Relat Mater 7:595
21. Haubner R, Lux B (2006) Int J Refract Met Hard Mater 24:380
22. Avigal Y, Hoffman A (1999) Diam Relat Mater 8:127
23. Iijima S, Aikawa Y, Baba K (1990) Appl Phys Lett 57:2646
24. Wolter SD, Okuzume F, Prater JT, Sitar Z (2003) Thin Solid Film 440:145
25. Haubner R, Köpf A, Lux B (2002) Diam Relat Mater 11:555
26. Tachibana T, Ando Y, Watanabe A, Nishibayashi Y, Kobashi K, Hirao T, Oura K (2001) Diam Relat Mater 10:1569
27. Ahmed W, Sein H, Ali N, Garcio J, Woodswards R (2003) Diam Relat Mater 12:1300
28. Matsumoto S, Sato Y, Kamo M, Setaka N (1982) Jpn J Appl Phys 21:L183
29. Butler JE, Windischman H (1998) MRS Bull 23:22
30. Lux B, Haubner R, Holtzer H, Devaries RC (1997) Int J Refract Met Hard Mater 15:263
31. Choy KL (2003) Prog Mater Sci 48:57
32. Lindlbauer A, Haubner R, Lux B (1992) Int J Refract Met Hard Mater 11:247
33. Joksch M, Wurzinger P, Pongratz P, Haubner R, Lux B (1994) Diam Relat Mater 3:681
34. Lux B, Haubner R, Holzer H, De Vries RC (1997) Int J Refract Met Hard Mater 15:263
35. Haubner R, Lux B (2002) Int J Refract Met Hard Mater 20:93
36. Lee ST, Pang HY, Zhou XT, Wang N, Lee CS, Bello I, Lifshitz Y (2000) Science 287:104
37. Zhou XT, Lee Q, Meng FY, Bello I, Lee CS, Lee ST, Lifshitz Y (2002) Appl Phys Lett 80:3307
38. Chen HG, Chang L (2005) Diam Relat Mater 14:183
39. Liu H, Dandy DS (1995) Diam Relat Mater 4:1173
40. Popov C, Kulisch W, Gibson PN, Ceccone G, Jelinek M (2004) Diam Relat Mater 13:1371
41. Wu K, Wang EG, Cao ZX, Wang ZL, Jiang X (2000) J Appl Phys 88:2967
42. Zhou D, McGauley TG, Qin LC, Krauss AR, Gruen DM (1998) J Appl Phys 83:540
43. Bogus A, Gebeshuber IC, Pauschitz A, Roy M, Haubner R (2008) Diam Relat Mater 17:1998
44. Drift A (1967) Philips Res Rep 22:267

45. Roy M, Ghodbane S, Koch T, Pauschitz A, Steinmüller-Nethl D, Tomala A, Tomastik C, Franek F (2011) Diam Relat Mater 20:573
46. Jiao S, Sumant A, Kirk MA, Gruen DM, Krauss AR, Auciello O (2001) J Appl Phys 90:118
47. Roy M, Vorlaufer G, Pauschitz A, Haubner R (2010) J Tribol Surf Eng 1:1
48. Schade A, Rosiwal SM, Singer RF (2007) Surf Coat Technol 201:6197
49. Griesser M, Grasserbauer M, Kellner R, Bohr S, Haubner R, Lux B (1995) Fresenius J Anal Chem 352:763
50. Bühlmann S, Blank E, Haubner R, Lux B (1999) Diam Relat Mater 8:194
51. Gruen DM, Pan X, Krauss AR, Liu S, Luo J, Foster CM (1994) J Vac Sci Technol A 12:1491
52. Nemanich RJ, Glass JT, Lucovsky G, Shroder RE (1988) J Vac Sci Technol A 6:1783
53. Prawer S, Nugent KW, Jamieson DN, Orwa JO, Bursill LA, Peng JL (2000) Chem Phys Lett 332:93
54. Ferrari AC, Robertson J (2001) Phys Rev B 63:121405
55. Night DS, White WB (1989) J Mater Res 4:385
56. Okada K, Kanda K, Komatsu S, Matsumoto S (2001) J Appl Phys 88:1674
57. Gruen DM, Krauss AR, Zuiker CD, Scsencsits R, Terminello LJ, Carlise JA, Jimenez I, Sutherland DGJ, Shuh DK, Tong W, Himpsel FJ (1996) Appl Phys Lett 68:1640
58. Kimoto K, Sekiguchi T, Aoyama T (1997) J Electron Microsc 46:369
59. Bridson R (2001) Electron energy loss spectroscopy. Springer, New York
60. Roy M, Koch T, Steinmuller D, Pauschitz A (2009) In: Proceedings of the 3rd Vienna conference in nanotechnology, Vienna, 18–20 Mar 2009
61. Sharpe WN, Jackson KM, Hemkar KJ, Xie Z (2001) J Microelectromech Syst 10:3
62. Espinosa HD, Porok BC, Peng B, Kim KH, Moldovan N, Auciello O, Carlise JA, Gruen DM, Mancini DC (2003) Exp Mech 43:256
63. Hallman P, Bjorkman O, Alahelisten A, Hogmark S (1998) Surf Coat Technol 105:169
64. Collins JL (2001) In: Kneringer G, Rodhammer P, Wildner H (eds) Proceedings of the 5th international Plansee seminar, vol 2, HM 94, Reutte. Plansee AG, pp 711–725
65. Layendecker T, Lemmer O, Jurgens A, Esser S (1999) Surf Coat Technol 48:253
66. Schafer L, Gabler J (2000) In: Batz W (ed) Proceedings of the 12th international colloquium tribology 2000 plus, vol III, 11–13 Jan 2000, Esslingen
67. Braza JF, Sudarshan TS (1992) Mater Sci Technol 8:574
68. Bowden FP, Hanwell AE (1966) Proc R Soc Lond A295:233
69. Gardos MN (1999) Surf Coat Technol 113:183
70. Miyoshi K, Wu RLC, Garscadden A, Barnes PN, Jackson HE (1993) J Appl Phys 74:4446
71. Chandraseka S, Bhushan B (1992) Wear 153:79
72. Gardos MN, Gebelich SA (1999) Tribol Lett 6:103
73. Miyake S (1994) Appl Phys Lett 65:1109
74. Smentkowski VS, Yates JT Jr (1996) MRS Symp 416:293
75. Molian PA, Janvrin B, Molian AM (1993) Wear 165:133
76. Dugger D, Peebles E, Pope LE (1992) In: Chung YW, Homolo AM, Street GB (eds) Surface science investigation in tribology, ACS symposium series 485, American Chemical Society, Washington, DC, p 72
77. Gardos MN (1994) In: Spears KE, Dismukes JP (eds) Emerging CVD science and technology. Wiley, New York, p 419
78. Hayward IP (1991) Surf Coat Technol 49:554
79. Erdemir A, Halter M, Fenske GR, Zuiker GR, Csencsits C, Krauss R, Gruen DM (1997) Tribol Trans 40:667
80. Erdemir A, Fenske GR, Krauss R, Gruen DM, McCauley T, Csencsits C (1999) Surf Coat Technol 121:589
81. Gangopadhyay AK, Tamor MA (1993) Wear 169:221
82. Hayward IP, Singer IL, Seitzman LE (1992) Wear 157:215
83. Gardos MN, Ravi KV (1994) Diam Films Technol 4:139
84. Lim SC, Ashby MF, Brunton JH (1997) Acta Metall 35:1342

85. Hutchings IM (1992) Tribology. Edward Arnold, London
86. Rabinowicz E (1965) Friction and wear of materials. Wiley, New York
87. Liu Y, Erdemir A, Meletis EI (1996) Surf Coat Technol 82:48
88. Erdemir A, Bindal C, Pagan J, Wilbur P (1995) Surf Coat Technol 76–77:559
89. Strawbridge A, Evans HE (1995) Eng Fail Anal 22:85
90. Hayward IP, Singer IL, Seitzman LE (1991) Wear 157:215
91. Thorpe TP, Morrish AA, Hanssen LM, Butler JE, Snail KA (1990) In: Proceedings of the diamond optics-III, San Diego, July 1990, p 230
92. Blau PJ, Yust CS, Heatherly LJ, Clausing RE (1990) In: Dawson D, Taylor CM, Godets M (eds) Mechanics of coatings. Leeds Lyon symposium 16 (Tribology series 17). Elsevier, Amsterdam
93. Wong MS, Meilunas R, Ong TP, Chang RPH. In: Pope LE, Fehrenbacker LL, Winer WO (eds) New materials approaches to tribology. Proceedings of Material Research Society Symposium
94. Bühlmann FS, Blank E, Haubner R, Lux B (1999) Diamond Relat Mater 8(2–5):194
95. Fu Y, Yan B, Loha NL, Sun CQ, Hing P (2000) Mater Sci Eng A282:38
96. Haubner R, Lux B (2000) Diam Relat Mater 9:1154
97. Lindlbauer A (1992) Int J Refract Met Hard Mater 11:247
98. Michler J, Stiegler J, von Kaenel Y, Moeckli P, Dorsch W, Stenkamp D, Blank E (1997) J Cryst Growth 172:404
99. Joksh M, Wurzinger P, Pongratz P, Haubner R, Lux B (1994) Diam Relat Mater 3:681
100. Chromik RC, Winfrey AL, Lüning J, Nimanich RJ, Wahl KJ (2008) Wear 256:477
101. Erdemir E, Fenske GR, Krauss AR, Gruen DM, Macauley T, Csencsits RT (1999) Surf Coat Technol 120–121:565
102. Godet M (1984) Wear 100:437
103. Grillo SE, Field JE (1997) J Phys D Appl Phys 30:202
104. Bogus A, Gebeshuber IC, Pauschitz A, Roy M, Haubner R (1998) Diam Relat Mater 17 (2008)
105. Gardos MN, Ravi KV (1989) In: Proceedings of the international symposium on diamond and diamond like carbon, vol 89. Electrochemical Society, p 475
106. Gardos MN, Soriano BL (1990) J Mater Res 5:2599
107. Erdemir A, Fenske GR (1996) Tribol Trans 39:787
108. Erdemir A, Fenske GR, Wilbur P (1997) In: Pauleau Y, Barna B (eds) Protective coatings and thin film synthesis, characterisation and applications, Nato ASI High Technology Series, vol 21. Kluwer Academic, Dordrecht, p 667
109. Kohjaki M, Higuchi K, Noda S, Uchida K (1992) J Mater Res 7:1769
110. Hayward IP, Singer IL, Mowery R, Pehrsson PE, Colton RJ (1990) In: Proceedings of the AVS 37th annual symposium, Toronto, 8–12 October
111. Habig KH (1995) Surf Coat Technol 76–77:540
112. Mohrbacher H, Blanpain B, Celis JP, Ross JR (1993) Surf Coat Technol 62:583
113. Grillo SE, Field JE, Van Bouwelen FM (2000) J Phys D Appl Phys 33:985
114. Jahamir S, Deckman DE, Ives LK, Fieldman A, Farabaugh E (1989) Wear 133:73
115. Schmitt M, Paulmier D (2004) Tribol Int 37:317
116. Hayward IP, Singer IL (1991) In: Proceedings of the 2nd international conference on new diamond science and technology, Washington, DC, September 1990. Materials Research Society, Pittsburgh, PA
117. Wilks J, Wilks EM (1979) In: Field JE (ed) Properties of diamond. Academic, New York, p 351
118. Kohzaki M, Higuchi K, Noda S (1990) Mater Lett 9:80
119. Rickerby DS, Burnett PJ (1987) Surf Coat Technol 33:191
120. Doerner MF, Nix WD (1988) Crit Rev Solid State Mater Sci 14:232
121. Alahelisten A (1995) Wear 185:213
122. Alahelisten A, Olsson M, Hogmark S (1992) Tribologia 11(3):34

123. Alahelisten A, Olsson M, Hogmark S (1993) In: Proceedings of the 6th international congress on tribology, Eurorrib'93, Budapest, vol 3, p 280
124. Tolkowsky M (1920) D.Sc. Thesis, University of London
125. Gahlin R, Alahelisten A, Jaconson S (1996) Wear 196:226
126. Wilks EM, Wilks J (1959) Philos Mag 4:158
127. Schade A, Rosial SM, Singer RF (2006) Diam Relat Mater 15:1682
128. Wheeler DW, Wood RJK (1999) Wear 225–229:523
129. Kral MV, Davidson JL, Wert JJ (1993) Wear 166:7
130. Head WJ, Harr ME (1970) Wear 15:1
131. Feng Z, Tzeng Y, Field JE (1992) Thin Solid Films 212:35
132. Field JE, Nicholson E, Seward CR, Feng Z (1993) Philos Trans R Soc Lond A 342:261
133. Alahelisten A, Hollman P, Hogmarkm S (1994) Wear 177:159
134. Wheeler DW, Wood RJK (1999) Wear 233–235:306
135. Wheeler DW, Wood RJK (2001) Wear 250:795
136. Alzoubi MF, Ajayi OO, Woodford JB, Erdemir A, Fenske GR (2002) Lubric Eng 58:21
137. Mohrbacher H, Blanplain B, Celis JP, Roors JR (1993) Diam Relat Mater 2:879
138. Mohrbacher H, Blanplain B, Celis JP, Roors JR (1993) Surf Coat Technol 62:583
139. Liu E, Blanplain B, Celis JP, Roors JR, Alverez Verven G, Priem T (1996) Surf Coat Technol 80:264
140. Miyoshi K (1996) Mater Sci Eng A209:38
141. Skopp A, Klaffke D (1998) Surf Coat Technol 98:1027
142. Kraussa AR, Aucielloa O, Gruena DM, Jayatissaa A, Sumanta A, Tucek J, Mancini DC, Moldovan N, Erdemir A, Ersoy D, Gardos MN, Busmann HG, Meyer EM, Ding MQ (2001) Diam Relat Mater 10:1952
143. Hollaman P, Wänstrand O, Hogmark S (1998) Diam Relat Mater 7:1471
144. Abreu CS, Amaral M, Oliveira FJ, Fernandes AJS, Gomes JR, Silva RF (2006) Diam Relat Mater 15:2024
145. Belmonte M, Fernandes AJS, Costa FM, Oliveira FJ, Silva RF (2003) Diam Relat Mater 12:733
146. Erdemir A, Fenske GR, Krauss AR, Gruen DM, McCauley T, Csencsits RT (1999) Surf Coat Technol 120–121:565
147. Popov C, Kulisch W, Jelinek M, Bock A, Strnad J (2006) Thin Solid Films 494:92
148. Chromik RR, Winfrey AL, Luning J, Nimanich RJ, Wahl KJ (2008) Wear 265:477
149. Amral M, Abreu CS, Oliveira FJ, Gomes JR, Silva RF (2007) Biotribological performances of NCD coated Si3N4 bioglass composites. Diam Relat Mater 16:790
150. Abreu CS, Amaral M, Fernandes AJS, Oliveira FJ, Silva RF, Gomes JR (2006) Diam Relat Mater 15:739
151. Schneider A, Steinmueller-Nethl D, Roy M, Franek F (2010) Int J Refract Met Hard Mater 28:40
152. Bhushan B (1999) Principles and applications of tribology. Wiley, New York. ISBN 0-471-59407-5
153. Liang Q, Stanishevsky A, Vohra YK (2008) Thin Solid Films 517:800

Chapter 4
Tribology of Diffusion-Treated Surfaces

N. Krishnaraj and Manish Roy

4.1 Introduction

Diffusion treatments refer to surface modification processes in which interstitial or substitutional atoms from an external source are allowed to enter the surface of a component to modify the surface chemistry and properties. Since diffusion is strongly temperature dependent and diffusion is prominent at elevated temperature, these treatments are carried out usually at higher temperatures. In view of the use of elevated temperatures which alters the surface chemistry, these treatments are termed thermo-chemical diffusion processes. Though there is no intentional build-up of material, there may be marginal change in dimensions due to the density difference between the substrate and that of the modified surface. The rate of diffusion depends on the concentration gradient that exists between the diffusing species and the substrate. It also depends on the diffusivity of host material and the duration of treatment. The modification of the surface layers is primarily intended to enhance wear [1, 2] and corrosion resistance [3, 4] and hence the tribological properties of the treated components assume importance. Of all the diffusion treatments, carburising, nitriding or a combination of both processes are extensively employed in industry to improve the wear resistance of components. This chapter, therefore, attempts to review the various diffusion processes and the tribology of the treated components with special emphasis on the carburising and nitriding treatments. Other diffusion treatments which are essentially used for protection against corrosion but seldom used for improvement of wear resistance are also referred.

N. Krishnaraj (✉)
E.L. Forge Limited, Door No: 21C, ARK Colony, Eldams Road, Alwarpet, Chennai 600018, Tamil Nadu, India
e-mail: krishnarajdr2002@yahoo.co.in

M. Roy
Defence Metallurgical Research Laboratory, Kanchanbagh, Hyderabad 500058, Andhra Pradesh, India

M. Roy (ed.), *Surface Engineering for Enhanced Performance against Wear*,
DOI 10.1007/978-3-7091-0101-8_4,

4.2 Classification of Diffusion (Coatings) Treatments

Diffusion treatments may be classified as interstitial or substitutional depending on whether the diffusing element enters the surface and occupies the interstices of the crystal structure or its equilibrium site or whether diffusing elements substitute some existing elements of the treated substrate. Carburising [5], carbonitriding [6], nitriding [7] and nitrocarburising [8] are examples of interstitial processes. In contrast, aluminising [9], chromising [10], siliconising [11] are substitutional diffusion treatment processes. These treatments enhance corrosion resistance in addition to wear resistance. Detail discussion of these treatments is given below.

4.2.1 Carburising

Carburising is a surface modification treatment in which carbon is allowed to diffuse into a low carbon steel substrate at higher temperatures up to a predetermined depth, called the case depth. The higher temperature is chosen to ensure that the steel is the austenitic phase field of the iron–iron carbide system since this phase has a high solubility of carbon. The carbon content is not allowed to build to an extent that will exceed the solubility limit of the austenite at the processing temperatures. If the solubility limit is exceeded, carbide network develops around the austenite grain boundaries inducing brittleness. The network may also cause cracks during subsequent grinding operation.

The carburising treatment therefore involves the following steps:

Heat the component to austenitizing temperature, holding for sufficient length of time to allow penetration to the required depth and quench in oil or polymer quenching medium. The quenching may be followed by a low temperature tempering treatment, usually below 473 K to relieve the quenching stresses.

The rapid cooling suppresses the formation of ferrite–pearlite phases and converts the austenite to martensite. Unlike diffusion-based equilibrium structures like ferrite and pearlite, martensite transformation is diffusion-less and happens by lattice shear mechanism. Martensite is characterised by a high density of dislocations in the crystal structure and this accounts for its high hardness. As the hardness of martensite is dependent on the carbon content, the surface enriched with higher carbon acquires high hardness. The combination of high surface hardness and a relatively soft core makes the component suitable for applications requiring high wear resistance and toughness.

Hardening is accomplished by quenching the surface layer to form martensite. Although the process of carburising is limited to steel in particular, carburising is carried out for nonferrous materials as well. Ti is normally carburised using plasma carburising process as TiO_2 layer is formed on exposing Ti to elevated temperature and this layer prevents diffusion of carbon from the surface to core. Carburising of various Co base alloys is also attempted. The maximum carbon content in carburised steel is limited to 1 % carbon due to brittleness of martensite, complication for carbide formation and retain austenite. Carburising is carried out between temperatures 1,123 and 1,223 K.

4.2.1.1 Reactions in Carburising

The carburising reaction depends on three factors, viz. carbon potential [12], time and temperature.

Carbon Potential

Carbon potential is defined as the carbon content of pure iron that is in thermodynamic equilibrium with the carburising medium surrounding it. It follows, therefore, that the medium (atmosphere) providing the required carbon must be at a greater potential than that of the steel substrate for carbon transfer to take place. The difference in carbon potential is the driving force for carburising.

The reaction occurring at the surface of the component during carburising is primarily one of reduction. Irrespective of the carburising medium used, the essential reaction at the metal surface is [13]

$$2\mathrm{CO} \Leftrightarrow \mathrm{C} + \mathrm{CO_2} \tag{4.1}$$

The free carbon is absorbed at the surface of steel. The CO_2 formed reacts with the carbon of the medium to provide the CO needed for further carburising.

$$\mathrm{C} + \mathrm{CO_2} \Leftrightarrow 2\mathrm{CO} \tag{4.2}$$

For the reaction given in Eq. (4.1) to proceed forward, there must be an excess of CO available at any point of time.

At any temperature, the equilibrium carbon content in austenite based on Eq. (4.1) is given by

$$\left(K = \frac{{P_{\mathrm{CO}}}^2}{a_{\mathrm{c}} P_{\mathrm{CO_2}}}\right) \tag{4.3}$$

where P_{CO} and $P_{\mathrm{CO_2}}$ are partial pressure of CO and CO_2 respectively and a_{c} is activity of carbon. After a given carbon potential is established, the depth of case produced in a given carburising treatment is given by time-dependent diffusion of carbon from the surface to the interior of the part.

Time

The diffusion also depends on the duration of treatment, as per the relation given by Harris:

$$X(\text{case depth}) = K\sqrt{t} \tag{4.4}$$

where K is a constant and t is the time for carburising.

Temperature

The third and important factor in carburising is the austenitizing temperature. The strong dependency of temperature can be seen from the diffusion coefficient equation given by

$$D = 0.47\exp\left(-1.6C - \frac{37,000 - 6,600C}{RT}\right) \tag{4.5}$$

where D is in $\mathrm{cm}^2\ \mathrm{s}^{-1}$, C is the weight percent carbon, T is temperature in degrees Kelvin, and R is the gas constant.

Since carburising is faster at higher temperatures of austenitizing, the treatment is carried out at an optimum high temperature to ensure that there is no grain coarsening, but still be commercially economical. The relationship of diffusion with temperatures is shown graphically in Fig. 4.1 [14]. In this figure, the relationship between carburising time, temperature and case depth is shown.

4.2.1.2 Carburising Processes

Various carburising processes are

1. Pack carburising
2. Gas carburising
3. Liquid carburising
4. Vacuum carburising
5. Ion carburising

Pack Carburising

Pack carburising is the simplest and the oldest carburising process. The details of various carburising processes can be seen in other references [15–17]. This process cannot be employed for big parts. However, as the process is simple and requires

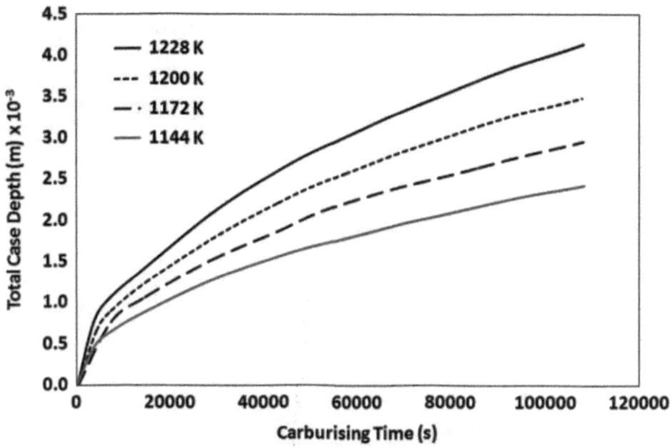

Fig. 4.1 Variation of total case depth with carburising time at different temperature [14]

minimal equipment, this process is still in practice. Hardening of the part can be achieved by quenching the pack or by reheating and quenching the part.

The process involves tightly packing the part in a solid carburising medium in a box. The medium is usually activated carbon and energisers like barium carbonate to accelerate the carburising process. The box is then heated in a furnace for a prolonged time to build up the case depth. The box is then unloaded and cooled to room temperature. The parts are removed, reheated to the hardening temperature and then quenched in oil or polymeric medium. The primary disadvantage of this process is the lack of control of carbon potential resulting in carbide network. Besides being labour intensive, it is time consuming. In view of these, presently pack carburizing is seldom employed in industry.

Gas Carburising

Gas carburising is widely practised in industry in view of the many advantages like economy, large volume processing per batch and good control on the case properties. The gaseous atmosphere used is mainly a saturated hydrocarbon gas like methane, propane or butane. The common practice is to use a mixture of carrier gas like endothermic gas and an enriching gas to provide the carbon potential. The main advantage of gas carburising is the accurate control of carbon potential by the use of oxygen probe connected to a controller that regulates the flow of the carrier gas and the carburising gas in real time. This minimises the chance of harmful carbide build up at the surface.

The feedstock may be either a mixture of hydrocarbon gas and air as stated above or a mixture of nitrogen and methanol. The furnaces used for gas carburising are either a pit-type carburising or an integral (sealed) quench furnace. The advantage of the latter is that the heated parts are transferred to the quenching medium in the controlled atmosphere itself thereby eliminating the scales on the surface.

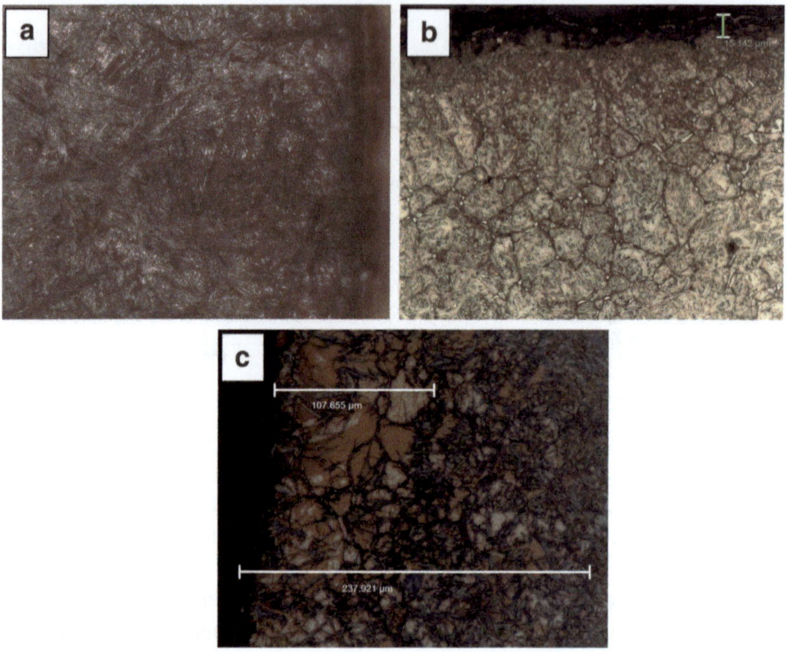

Fig. 4.2 Microstructure of gas-carburised parts. (**a**) Carburised case, (**b**) carbide network and (**c**) high amount of carbide network along with carbide network

Figure 4.2 shows the microstructures of gas-carburised parts. Figure 4.2a exhibits the microstructure of carburised case. Martensite can be recognised by the characteristic needle-like appearance. The white phase in between the needles is retained austenite. The base material is 17CrMoNi6. Figure 4.2b gives the carbide network around grain boundaries due to excess carbon potential. The gas carburising treatment has caused some inter-granular oxidation near surface to a depth of 0.015 mm. Base material is 17CrMoNi6. Microstructure of gas-carburised part showing detrimental features due to poor control of the carburising parameters is given in Fig. 4.2c. Very high percentage of retained austenite up to a depth of 0.108 mm along with carbide network has been caused by very high carbon potential, possibly beyond 1.2 %.

Liquid Carburising

In liquid carburising, treatment is carried out in salt bath [18]. Due to superior heat transfer characteristics of the salt bath solution, liquid carburising involves faster heat up resulting in shorter cycle time than gas carburising. Most liquid carburising bath contains cyanide. However, some bath contains only special grade of carbon. Cases produced by liquid carburising contain lower amount of nitrogen and higher amount of carbon. Further, liquid-carburised cases have significant higher depth (about 6.4 mm) as oppose to cyanide-treated case which has 0.25 mm case depth. Since most salt baths contain cyanide, disposal of the waste is a problem. As cyanide is toxic, it is to be disposed after chemical treatment which makes it a nontoxic waste.

Vacuum Carburising

The advantage of vacuum carburising is total freedom from scale formation and surface reaction. The carburised parts remain very clean and the composition distribution and case depth are very much uniform. The process control during vacuum carburising is also extremely precise. As the parts to be carburised are heated in oxygen-free atmosphere, substantially higher temperatures can be employed without the fear of surface or grain boundary oxidation. The higher temperatures increase the solid solubility of carbon in austenite and promote faster diffusion. As a result, the carburising time is substantially reduced. In view of the low pressures used, the requirement of carburising gas is also low. However, the reduced gas flow is disadvantageous in a way since deep recesses and blind holes do not get carburised due to the lower carbon potential. This results in nonuniformity of the case depth. If the gas pressure is increased to overcome this problem, free carbon or soot may form. Thus a delicate balance of the process condition must be dynamically adjusted to obtain a compromise of case uniformity and risk of sooting.

Plasma (Ion) Carburising

The ion carburising [19, 20] is more effective than other carburising processes as it bypasses several dissociation processes that produce active soluble carbon. The glow discharge dissociation of the hydrocarbon gas takes place at the steel surface directly due to the high negative potential of the work piece. The ready availability of the active carbon and the use of higher temperatures facilitated by vacuum accelerate the carburising process. Further, higher temperatures can be used as the process takes place in vacuum. The schematic representation of a plasma carburising is made in Fig. 4.3. Ion carburising or plasma carburising offers excellent uniformity of the case depth even for very complex shaped parts. In addition, presence of plasma which is responsible for active atomic species results in constant time for carburising independent of composition of the carburising parts. Thus the ion carburizing permits a wide range of steels to be carburised in a single batch. Further, any type of hydrocarbon can be used as source gas.

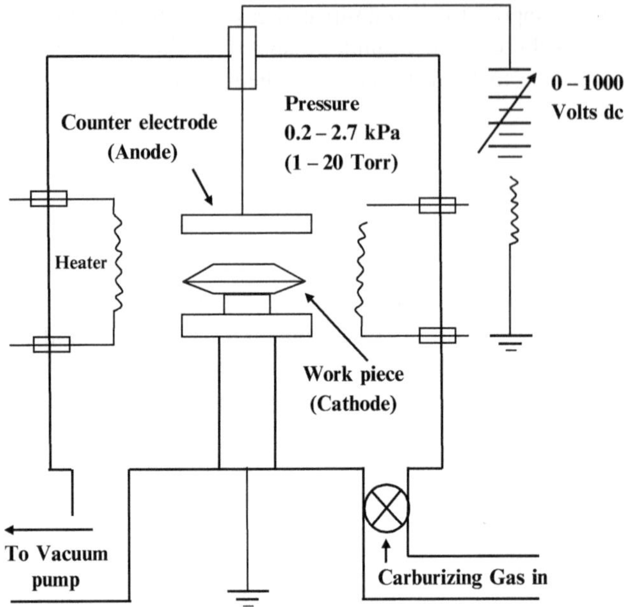

Fig. 4.3 Schematic representation of plasma carburising unit

4.2.2 Nitriding

Nitriding is a process of diffusing nitrogen into the surface of metals to produce a hard case. Although it is possible to get harder surface by nitriding than that by carburising, this process is slower than carburising and usually more complex. Hardness improvement by nitriding is not only by solid solution strengthening but also more importantly by the ability of nitrogen to form hard compound. At around 6 % nitrogen, nitriding of steel results in formation of gamma prime having composition Fe_4N. At nitrogen content greater than 8 %, the equilibrium reaction product is epsilon nitride Fe_3N. This hard layer is undesirable as this is quite brittle and may spall during use. In general, this layer is removed before use or nitriding process is modified to reduce this layer or to make this layer less brittle. Even though nitriding is carried out on a large variety of metals and alloys, it is significant for class of steel known as nitralloy which contains 1 % Al. On nitriding, Al forms discrete particles of AlN compound that strains ferrite lattice that strengthened and hardened. In addition to aluminium, chromium, titanium, molybdenum, vanadium and tungsten can form hard compound. A typical microstructure of austenitic plasma-nitrided AISI 4140 steel is presented in Fig. 4.4 [21]. In addition to compound layer, an intermediate transformed austenite zone can also be noted.

Nitriding is carried out on ferrite matrix i.e. it is a subcritical hardening process. The process does not involve quenching or subsequent heat treatment. Some steels are subjected to hardening and tempering treatment. The depth of nitriding layer

Fig. 4.4 SEM micrograph of plasma-nitrided AISI 4140 low alloyed steel [21]

can be increased by increasing the nitriding time. However, longer the nitriding time, greater is the tendency to form white layer and the thicker the white layer becomes. Another important characteristic of nitriding is formation of residual compressive stresses. This residual compressive stress is very much desirable as it improves fatigue resistance. Various nitriding processes are as follows.

Gas nitriding is the simplest nitriding process where NH_3 is introduced when the component attains the nitriding temperature. NH_3 dissociates into nitrogen and hydrogen [22]. Nitrogen diffuses into the substrate and hydrogen remains in the atmosphere of nitriding furnace. Gas nitriding is performed in the temperature range of 753–863 K. Nitriding can be done in salt bath known as liquid nitriding. Similar to carburising, salt nitriding is carried out using cyanide bath It is carried out in the temperature range of 783–893 K. The nitriding bath contains sodium cyanide (NaCN), sodium cyanate (NaCNO) and potassium cyanide (KCN) and some other salt. The advantages of salt nitriding over gas nitriding are easier handling of the work, no requirement for sealed retort, lower handling cost, greater work throughput. A variety of nitriding process known as Tufftride eliminates presence of white layer. The main problem of salt nitriding is disposal of cyanide salts. Consequently several varieties of non cyanide nitriding processes have been developed. These processes are slightly different from conventional salt nitriding processes.

Glow discharge plasma nitriding is one of the most important plasma surface engineering techniques. Plasma nitriding set up is shown schematically in Fig. 4.5. It has been widely used in industry over the past decades to improve the hardness, wear and corrosion resistance and fatigue strength of various metallic materials such as low alloy steels, titanium alloys and hot working tool steels [23–26]. The basic mechanism of plasma nitriding is the formation of Fe–N precipitates as the result of a reaction between plasma and the surface of the metal. In addition, the plasma mass transfer plays an important role on the formation of nitride precipitates [27]. Nitride precipitates formed on the surface consist of different phases such as γ-Fe_4N, ε-$Fe_{2-3}N$ and other M_xN_y precipitates. The morphology of these precipitates is effectively influenced by gas composition, temperature and time of plasma-nitriding process [28]. Ion nitriding has several advantages over other nitriding process. In ion nitriding, simple mixture of hydrogen and nitrogen is used and thus has fewer

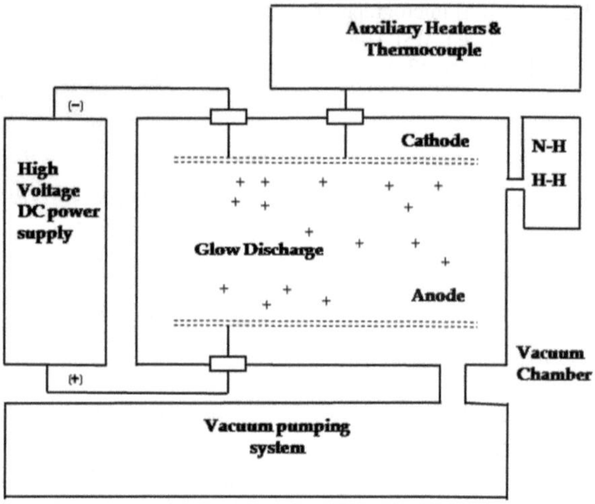

Fig. 4.5 Schematic diagram of plasma nitriding set up

problems with environment and disposal. Broader and lower temperature can be used in ion nitriding eliminating many post-nitriding operation. Duration of nitriding can be reduced considerably. Further masking of the parts during ion nitriding is substantially simple [29].

4.2.3 Carbonitriding

Carbonitriding is a modified form of carburising in which nitrogen is also diffused into the case along with carbon [30]. The temperatures of treatment ensure that the steel is taken to the austenitic phase field. Ammonia introduced along with the carbon providing species dissociates to hydrogen and nitrogen. The ammonia catalytically dissociates on the heated steel surface allowing the nascent nitrogen to enter the surface by diffusion. Ammonia is introduced usually towards the end of the carburising cycle and the temperature is lower than the carburising temperatures. At the end of the cycle, the component is quenched in oil to produce a martensite case. Carbonitriding case depths are normally thinner than carburised cases [31]. Carbonitriding is different from nitrocarburising in that the latter is carried out in the ferritic range of iron–nitrogen–carbon system at temperatures lower than 863 K. Nitrocarburising does not require quenching as the increase in surface hardness is not on account of martensite formation, but by the development of a compound layer [32].

Nitrogen increases the hardenability of the steel and hence allows even plain low carbon steels to form martensite although it is the enriched surface carbon that gives the hardness of martensite. The nitrogen forms carbonitrides in the case increasing the wear and scuffing resistance and in this respect the carbonitrided case is superior to carburised case. Besides, the presence of nitrogen gives resistance to softening at higher temperatures and it is customary to temper at moderately higher temperatures up to typically 493 K. It is for this reason that gears are often carbonitrided when the application demands superior wear and scuffing resistance.

However, nitrogen being an austenite stabiliser increases the retained austenite of the case. Retained austenite is a soft phase and hence adversely affects the wear resistance of martensite. Furthermore, increased nitrogen level has a tendency to form porosity which again affects the wear resistance. It is, therefore, customary to restrict the nitrogen percentage to 0.40 % to minimise the retained austenite and to contain the level of pores. However, the presence of some voids does not significantly reduce the wear resistance but even helps to retain the oil in lubricated wear applications. If the level of retained austenite is considered detrimental to the wear application, a sub-zero treatment can be carried out after hardening to convert the retained austenite to martensite.

The case depth aimed in carbonitriding is generally much less, typically not exceeding 0.75 mm. This is because the treatment temperatures are lower, at around 1,143 K, and shorter than in carburising; longer durations allow uncontrolled dissociation of ammonia increasing the nitrogen availability and consequent undesirable retained austenite percentage of the case. Higher temperatures are not employed since the nascent nitrogen recombines to form nitrogen molecules and dilutes the carburising potential.

Low carbon and low carbon alloy steels are suitable for carbonitriding. The increased hardenability imparted by nitrogen is due to its lowering the critical cooling rate needed to form martensite. The case hardness is also higher and more uniform than achievable from carburizing treatment alone, particularly in plain low carbon steels. In some special cases, medium carbon and medium carbon alloy steels are also carbonitrided. Carbonitriding imparts adhesive wear resistance in addition to abrasive wear resistance. This property is used advantageously in components like gears, cams, shafts that are involved in metal to metal sliding and rolling.

Carbonitriding is ideal for case hardening powder metallurgy components such as electrolytic iron compacts [33]. Carburising of these components poses many difficulties due to lower hardenability, high martensite transformation temperature and the high levels of porosity causing deep penetration into the material. The low temperature treatment possible with carbonitriding at around 1,153 K enables low rate of diffusion and control of case depth with adequate levels of carbon in the case for hardening. Nitrogen also helps to suppress pearlite formation during oil quenching.

Combining carburising and carbonitriding treatments offers several advantages. Initially a deeper case is built during carburising at higher temperatures up to 1,220 K for a long duration and after dropping the temperature to less than 1,173 K, ammonia is added to the carburising atmosphere to permit nitrogen diffusion. The duration during carbonitriding phase is much shorter than the carburising stage. The case now consists of a high wear resistant near surface layer backed by a deeper carburised layer for enhanced fatigue resistance. Case depths up to 3 mm are possible with the combined treatment.

4.2.4 Nitrocarburising

Another type of surface hardening is introduction of carbon and nitrogen into ferritic steel and the process is known as nitrocarburising. In contrast to carbonitriding, in this process carbon and nitrogen are added to ferrite.

Compared to conventional gas nitriding, the liquid nitrocarburising technique (Tufftriding) is performed at a higher temperature (50 K) and the necessary process duration is two times less for the same case depth. Also, it is suggested for the surface treatment of intricate shapes because of the lack of the distortion of the work piece after treatment. Additionally, liquid nitrocarburising produces nitride layers of higher wear and corrosion resistance, enhanced fatigue strength and toughness and is less expensive and simpler than other non-conventional nitriding techniques, such as plasma or ion-nitriding and fluidised-bed nitriding. Thus, it is well-suited for industrial applications. During liquid nitrocarburising, the temperature is kept under the eutectoid point (591 K) of the Fe–N phase diagram, allowing simultaneous diffusion of nitrogen and carbon atoms into the ferrite lattice, a fact that results in the formation of the so-called nitrocarburised case. This case consists of two distinct layers. The first layer, on the top of the treated surface, is the compound or 'white layer', composed mainly of ε-carbonitride $Fe_{2-3}(C,N)$. It exhibits higher toughness, compared to that of conventional ε-nitride phase formed during gas nitriding and also possesses high wear resistance under dry or lubricating friction due to its hexagonal-layered structure and its microporosity [34]. The layer underneath is the diffusion zone, mainly composed of α-(Fe,N) solid solution, in which the N and C atoms are dissolved interstitially in the ferritic matrix, leading to improved fatigue resistance [35, 36]. A comprehensive study using plain carbon steel, low alloy medium carbon steel and a tool steel indicated that increasing nitrocarburising treatment temperature from ferritic phase to austenitic phase increases the thickness and hardness of the compound layer due exclusive formation of ε carbonitride. It also increases the porosity due to increase of etching action of molten salt and nitriding action. Sub-zero treatment followed by ageing increases the hardness of the compound zone and the zone beneath the compound layer by transforming austenite to martensite and bainite. A typical microstructure of nitrocarburised low alloy medium carbon steel (En19) after treatment at 843 K (ferritic nitrocarburising) is given in Fig. 4.6. A compound layer of 10 μm thickness at the surface backed up by a diffusion zone with a nitride networks close to the compound layer can be noticed. These networks white in appearance seem to follow prior ferrite grain boundaries. In addition, alloy nitrides and carbonitrides such as γ-Fe_4N and, rarely, ζ-Fe_2N can be precipitated in both the compound and the diffusion zone. Considering that the diffusion coefficient of carbon in a-Fe is approximately 100 times lower than that of nitrogen at the temperature range used, it can be safely assumed that only the compound layer is enriched in carbon. Figure 4.6b, c illustrates the microstructures of specimens treated in austenitic region at temperature 923 K and 973 K respectively. Treatment at higher temperature results in increase of thickness of compound layer. Further there appears a zone

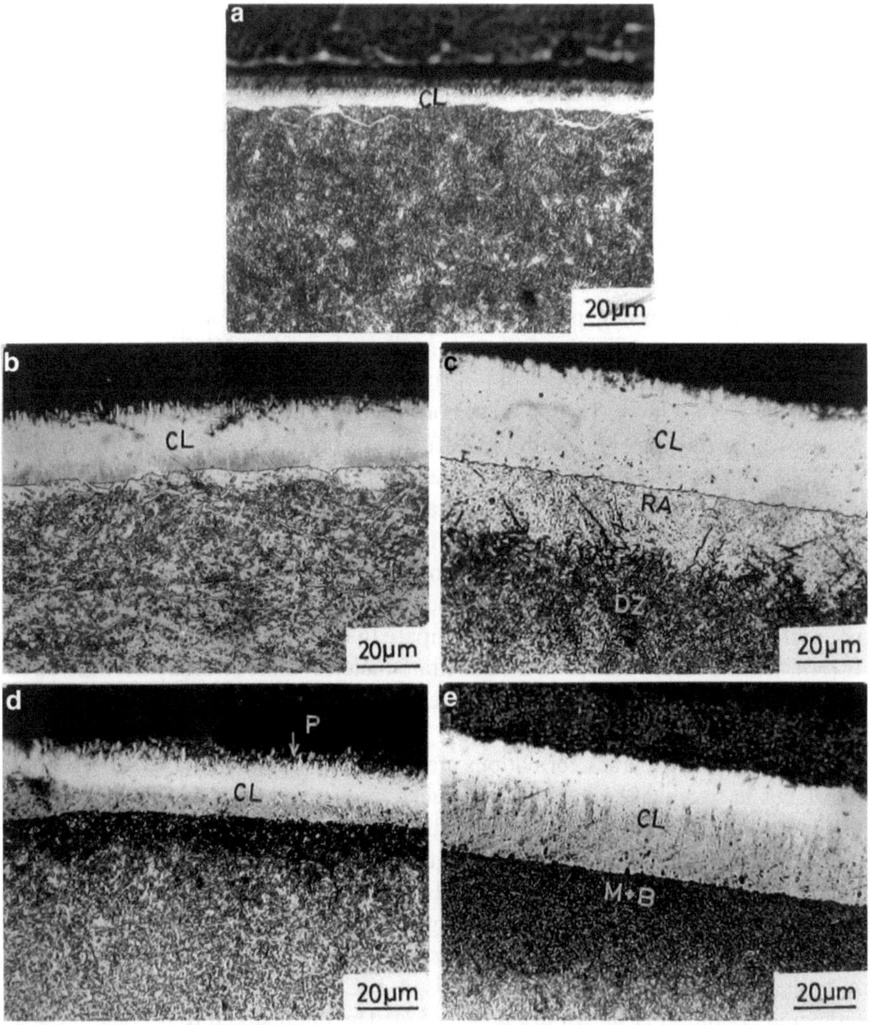

Fig. 4.6 Optical micrograph of nitrocarburised low alloy medium carbon En19 steel. Specimen treated at (**a**) 843 K, (**b**) 923 K, (**c**) 973 K, (**d**) subzero treated and aged (843 K) and (**e**) subzero treated and aged (973 K). *CL* compound layer, *RA* retained austenite, *DZ* diffusion zone, *M* + *B* martensite and bainite

termed as retained austenite layer which is etched white. When these specimens are subjected to sub-zero temperature treatment and subsequently aged, retained austenite layer is converted into martensite and bainite layer as shown in Fig. 4.6d, e. Nitrocarburising can be carried out in plasma. Recently, nitrocarburising/quenching [37] and nitrocarburising/oxidising [38] duplex techniques were proposed for the achievement of even higher wear resistance of the metallic component.

4.2.5 Boronising

Boronising is a process to form borides with the substrate by diffusing boron into metal surface to produce hard layer that enhances wear resistance. Boron behaves more like carbon in strengthening metals. Boronising can be carried out using several processes such as metalliding, pack cementation, CVD process, gas boronising, ion boronising, etc. [39–43]. In CVD process, components are put inside a retort and the retort is heated. Subsequently boron containing gas is introduced into the retort. This process can be carried out at 673 K. Pack boronising is similar to carburising process and it is carried out at around 873 K. Pack boronising mixtures comprise 5 % B_4C as the source, 5 % KBF_4 as activator and 90 % SiC as the diluting agent [44, 45]. Boronising does not need quenching to harden the surface layer. The nature of boronised layer depends on the substrate and process of boronising [46]. With the diffusion of boron into the steel, iron borides (FeB, Fe_2B) form and the thickness of the boride layer is determined by the process temperature and time [47, 48]. The crystal structure of both layers exhibits columnar shapes oriented along the diffusion axis [49]. Generally, saw-tooth-shaped single phase (Fe_2B) layer formation is preferred over a layer with both FeB and Fe_2B [39, 45]. The boron-rich FeB phase that has approximately 16.23 wt.% B is not preferred because it is less brittle compare to the Fe_2B phase that has 8.83 wt.% B [50]. Boride layers adhere to the substrate material more strongly because of the saw-tooth morphology. Brittleness of the boronised layer increases with the layer thickness [39, 51]. Also, since the FeB and Fe_2B phases have different thermal expansion constants, crack formations are frequently found on the FeB/Fe_2B phase interfaces in two phase layers. These cracks frequently cause spalling and scaling under a mechanical load [50]. By controlling the boronising parameters, that is the boronising powder composition, temperature and process time, the pack boronising method can reliably produce the Fe_2B phase [39]. Usually compound layer containing various hard borides having hardness 1,500–2,000 HV is formed. Boronising can be performed on various ferrous and nonferrous metals [41, 51]. A typical microstructure of cast-boronised 13Cr–4Ni steel in as-boronised and tempered condition is shown in Fig. 4.7 [53]. The microstructure becomes courser after boronising.

For salt bath boronising, appropriate molten salts which includes borax ($Na_2B_4O_7$) and sodium borofluoride ($NaBF_7$) are employed. Salt bath boronising enjoys the same simplicities and advantages as salt bath nitriding. However, the range of operating temperature is somewhat restricted by the melting point and stability of boron containing salts. For ion boriding, boron halides such as boron trichloride (BCl_3), boron trifluoride (BF_3) or boron tribromide (BBr_3) are commonly used as source gas along with hydrogen for glow discharge. Halides are easily dissociated in presence of hydrogen although halide dissociate products are corrosive. Ion boronising produces single phase Fe_2B layer as oppose to double phase FeB and Fe_2B is produced in powder pack method. With ion boronising a single phase boride layer can be obtained at as low temperature as 823 K [52, 54].

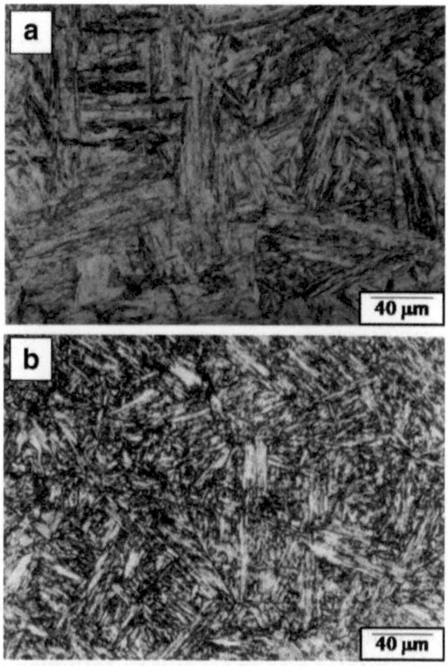

Fig. 4.7 Microstructure of boronised cast iron, (**a**) as boronised, (**b**) tempered [52]

Boronising can be carried out on non-ferrous materials such as nickel, cobalt, titanium, refractory metals and their alloys. Boride layers of tantalum, niobium, tungsten, molybdenum, etc. do not exhibit tooth-shaped morphology. Addition of alloying elements generally retards the growth of borided layer. In recent times, fluidised bed boriding has become quite popular as it offers several advantages. Metal boride coating or deposition of TiB_2, ZrB_2, rare earth boride, etc. by chemical vapour deposition is also finding several applications due to their inherent advantages. Among various boride coating developed so far, TiB_2 layer has been investigated most extensively.

4.2.6 Other Diffusion Process

The pack cementation process originally involved pack carburising. This process is capable of diffusing many different elements in many different substrates to change the surface characteristics of engineering materials. These processes have given rise to other pack diffusion processes such as aluminising [55, 56], siliconising [57, 58], chromising [59], etc.

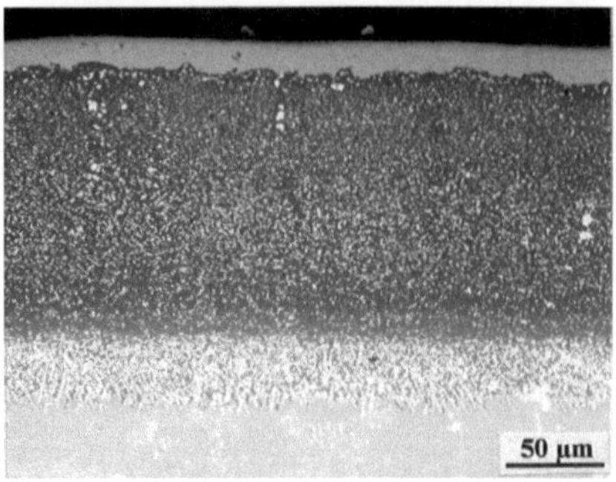

Fig. 4.8 Typical microstructure of aluminide coating on CM-247LC(DS) alloy [56]

4.2.6.1 Aluminising

Aluminising is a pack cementation process commercially practised in all range of alloys including nickel base, cobalt base, low alloy and stainless steels. This coating offers resistance against atmospheric corrosion, elevated temperature oxidation and environment. SEM micrograph showing the microstructure of aluminide coating is presented in Fig. 4.8 [56]. An Al-rich single phase NiAl layer containing various precipitates rich in Cr and W followed by an interdiffusion layer can be noted as the characteristics of the aluminised layer. However, this coating may spall at elevated temperature due to thermal cycling. Interdiffusion may cause further degradation of the coating. Two different types of aluminide coating can be produced by pack cementation. One type involves Al as dominant diffusing species into the substrate and it is referred as 'inward diffusion' coating. This coating requires high activity pack. The second type of coating involves outward diffusion of Ni as dominant diffusing species and requires a low activity pack.

Diffusional transport occurs within the coating to supply Al to the oxide scale, whereas diffusional transport between coating and substrate results in Al transport out of the coating away from the oxide scale. The concentration gradient of all elements in the coating and the substrate must satisfy Fick's second law of diffusion and this can be stated as

$$\frac{\partial C_i}{\partial t} = \tilde{D}_i \frac{\partial^2 C_i}{\partial X^2} \tag{4.6}$$

where D_i is the diffusion coefficient for the ith element in the alloy. In the above equation it is assumed that there is no interaction between the elements in the alloy. The Al concentration at the coating/oxide interface is determined by the rate at which Al is consumed by oxidation. Assuming no interaction between the elements, Al flux, J_{Al} is given by

$$J_{\mathrm{Al}} = -\tilde{D}_{\mathrm{Al}} \frac{\partial C_{\mathrm{Al}}}{\partial X} \tag{4.7}$$

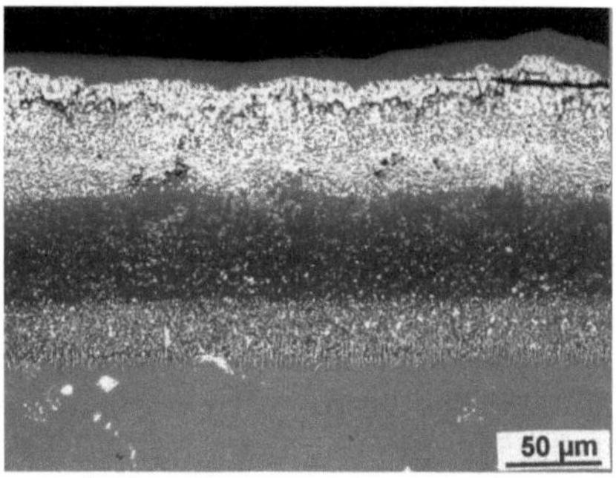

Fig. 4.9 Typical microstructure of Pt aluminide coating on CM-247LC(DS) alloy [60]

Two types of modified aluminide coatings have received considerable attention in recent times. The first type is Pt-modified aluminide which are produced by electroplating 5–10 μm Pt on the substrate prior to aluminising treatment. During aluminising $PtAl_2$, Pt_2Al_3 phases are formed. A typical microstructure of Pt aluminide coating is given in Fig. 4.9 [60]. The microstructure of the coated layer contains three different zones. The first zone is platinum-rich layer containing $PtAl_2$ and NiAl. The second layer is single phase NiAl layer and finally interdiffusion zone. The details of microstructure and process of Pt-modified aluminide coatings can be found in literature [61, 62]. Other variety is Cr-modified aluminide coating. Cr can either be co deposited during pack aluminising or it can be deposited prior to aluminising. The Cr-modified aluminide coating has been reviewed by Streiff and Boone [63].

4.2.6.2 Siliconising

Siliconising is a process similar to aluminising. There are packs and retort processes in which components are subjected to silicon containing gas at elevated temperature and this process produces nascent silicon that diffuses into the substrate to produce a silicon-rich surface layer. In the first process, pure silicon pack is activated with NH_4Cl, $SiCl_4$. $SiHCl_3$ gases, which are reduced by hydrogen gas to deposit elemental silicon on the substrate. The other process involves tumbling part in retort filled with SiC. When the temperature reaches 1,283 K, $SiCl_4$ gas is introduced. This gas reacts with the substrate and Si to produce silicon-rich surface layer. High Si content on the surface produces SiO_2 layer which is stable in oxidising environment. Various modification techniques such as aluminide silicide coating [64] or Si-modified overlay coating [65, 66] are also examined to take advantage of beneficial effects of Si in corrosive environment.

4.2.6.3 Chromising

Chromising is an established process conducted mainly in a salt bath [67] or by pack powder processing [68–71]. In general, chromising of steel with less than 0.1 wt.% carbon is termed 'soft' and provides high temperature oxidation resistance, while chromising of steels with more than 0.3 wt.% carbon is termed 'hard' and is used to produce wear and corrosion-resistant layers [68, 69]. However, a limited number of studies exist explicitly on the chromising of steels via FBR-CVD. Chen et al. [72] have chromised tool steels, Perez et al. have published [73–75] results on chromised AISI 304 austenitic stainless steel, and Priyantha et al. [76] have chromised AISI 409 stainless steel and carbon steel. Perez et al. produced non-uniform layers with a porous appearance, while both Perez et al. and Priyantha et al. created layers with relatively low chromium content (20–30 wt.%). To date, the production of satisfactory chromised layers on steels from FBR-CVD has been reported in the literature [77].

4.3 Tribology of Diffusion Coatings

4.3.1 Sliding Wear

An extensive work is done on characterising the sliding wear behaviour of various diffusion-treated surfaces. Most of these investigations are related to the effect of various degrading condition of wear and friction performances. Friction coefficient is found to be reduced with diffusion surface treatment. According to adhesive wear theory, friction coefficient is given as

$$\mu = \frac{\tau_f}{\sigma_y} \tag{4.8}$$

where τ_f is shear strength and σ_y is yield strength of the material undergoing wear. With surface treatment there is an increase in strength of the treated layer. This resulted in decrease in shear strength and increase in yield strength. Consequently friction coefficient decreases.

Friction coefficients of diffusion-treated surfaces depend on the wearing conditions such as applied load, sliding velocity, etc. It also depends on the diffusion treatment conditions. The influence of applied load on friction coefficient of liquid nitrocarburised cold worked steel as obtained by Pantazopoulos et al. [78] is presented in Fig. 4.10. The friction coefficient decreases with increasing load and this may be attributed to contact geometry in combination with relative sticking friction mechanism. This trend follows the relation presented in Eq. (1.10). Thus, in this case adhesion governs the friction behaviour for diffusion-treated surfaces. Similarly, influence of sliding velocity on friction coefficient is investigated by

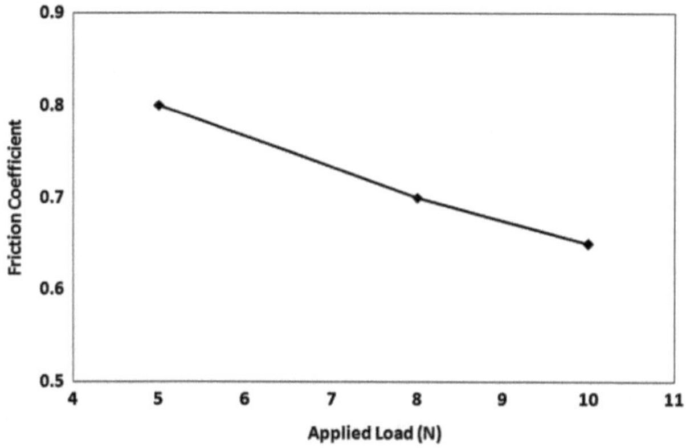

Fig. 4.10 Variation of friction coefficient with applied load [78]

Nicoletto et al. [79] as shown in Fig. 4.11. Friction coefficient tends to increase with increase in sliding velocity. The friction coefficient of carbonitrided lamellar cast iron initially increases then decreases with increase of sliding velocity.

The influence of processing condition on the friction coefficient is examined by Qiang et al. [35]. They studied friction behaviour of salt bath-nitrocarburised 1045 steel. The steel was diffusion treated for three different conditions consisting of one single stage process and two different two stage processes. Friction coefficient decreases considerably as compared to untreated steel as per Eq. (4.8). Their observation is presented in Fig. 4.12. It is clear from Fig. 4.12 that single stage salt bath nitrocarburised steel exhibited minimum friction coefficient.

Karamis et al. [80] investigated the influence of temperature on the friction coefficient of plasma-nitrided 722M24 steels. Their observation is shown in Fig. 4.13. In Fig. 4.13, the friction coefficient as function of test temperature is shown for 722M24 steels nitrided by single-stage and two-stage processes. Friction coefficient tends to increase although marginally with temperature. Such an increase in friction coefficient is noted in spite of the fact that roughness decreases with temperature and the wear mechanisms are microploughing, microcutting and microcracking. Such response is attributed to decrease in hardness of the nitrided layer with increase in temperature.

Plasticity dominated wear prevails at relatively low load and low sliding velocity. Wear in this regime is governed by the deformation behaviour of the material undergoing wear. Material loss is primarily by delamination. The wear rate (W) in this regime is given by [81, 82]

$$W = k\frac{F}{H} \tag{4.9}$$

where k is known as wear coefficient, F is applied load and H is hardness of wearing material.

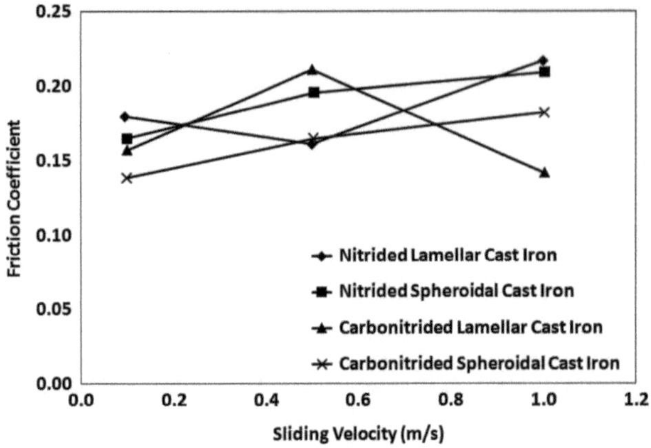

Fig. 4.11 Influence of sliding velocity on friction coefficient of various nitrided cast iron [79]

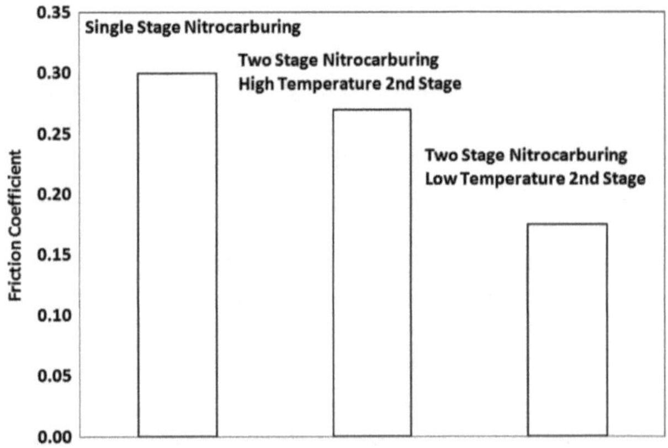

Fig. 4.12 Bar diagram showing the influence of various processing condition of the friction coefficient of salt bath nitrocarburised 1045 steel [35]

The wear rate in this regime can be reduced by modifying the surface of a component with a layer having

1. Low wear coefficient *k* or
2. High hardness *H*

The wear coefficient values estimated for liquid nitrocarburised three different types of steels after 30,000 sliding cycles are presented in Fig. 4.14 [83], as the ratio of wear coefficient of annealed steel to the wear coefficient of nitrocarburised steel for all the normal loads. The nitrocarburised tool steels exhibited wear coefficient values much lower than those of the non-treated materials. For the nitrocarburised AISI D6, the wear coefficient exhibited a mean value of 100 times lower than that of the as-annealed tool steel. The wear coefficient ratio decreased with the increase of

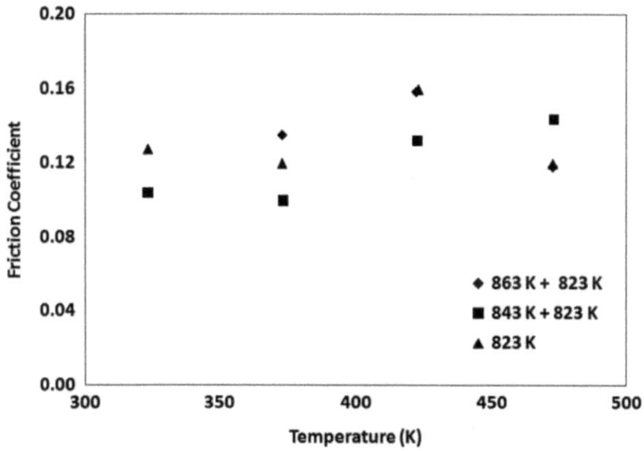

Fig. 4.13 The influence of temperature on the friction coefficient of plasma-nitrided 722M24 steel [80]

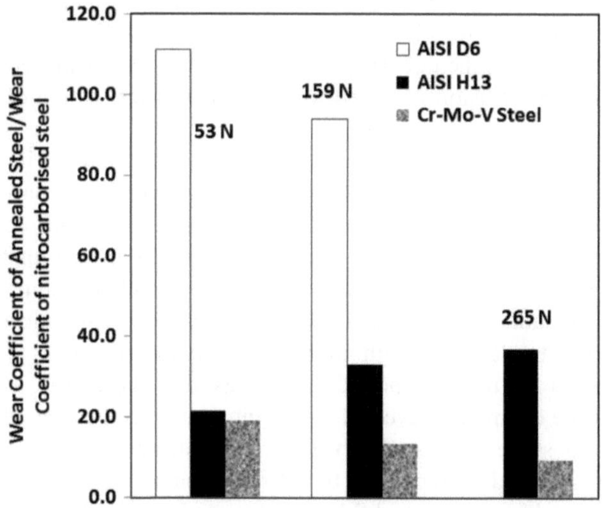

Fig. 4.14 The wear coefficient values estimated for liquid nitrocarburised three different types of steels after 30,000 sliding cycles [83]

the normal load (Fig. 4.14), indicating that the formation of elongated nitrocarbides and carbide bands led also to the impairment of the mechanical behaviour against compressive loading. For the nitrocarburised AISI H13, the mean value of the wear coefficient occurred 30 times lower than that of the as-annealed tool steel. Contrary to the previous case, the ratio increased (Fig. 4.14) with increasing normal load, indicating that nitrocarburising enhanced not only the wear resistance of the material but also the mechanical behaviour under compressive loads. This is due to the presence of finely dispersed microstructural constituents (mainly carbides) in

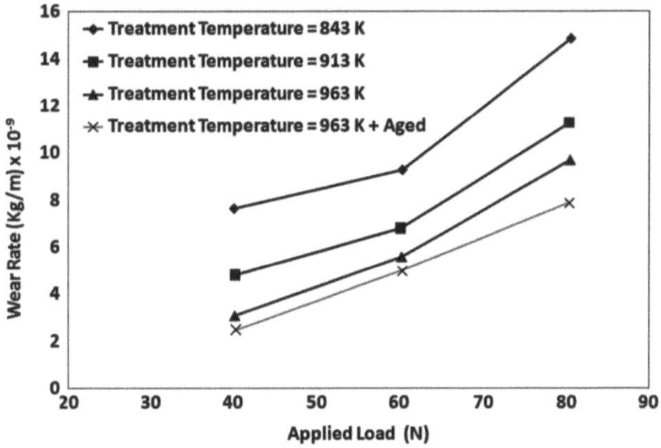

Fig. 4.15 The effect of applied load on the wear rate of nitrocarburised H11 steel [34]

the diffusion zone and in the substrate. For the nitrocarburised Cr–Mo–V tool steel, the mean value of wear coefficient was 14 times lower than that of the as-annealed material. As in the case of AISI D6 tool steel, increasing the applied normal load resulted in a decrease of the ratio (Fig. 4.14), indicating that the presence of lamellar precipitates in the diffusion zone led to subsequent deterioration of compressive behaviour.

According to Eq. (4.9), the wear rate increases with increase in applied load. This trend has been observed by Krishnaraj et al. [34] for nitrocarburised H11 steel as shown in Fig. 4.15. In the regime of applied load, no obvious change in wear mechanism was observed. The wear rate of nitrocarburised steel improved significantly with respect to bare hot working dies steel and this improvement was attributed to formation of a surfacial compound layer of ε carbonitride. Wear rate was also found to decrease with increase in compound layer thickness. One important aspect of this compound layer was that there was no spalling or debonding of the compound layer due to jagging interface between the compound layer and the substrate. Similar observation of increasing wear rate with increasing applied load was also reported by Pantazopoulos et al. [78] for liquid nitrocarburised cold working steel. Microcutting and micropolishing are the principal wear mechanisms. In addition, localised adhesion phenomena also took place during sliding. Fine wear debris and relatively smooth wear tracks observed for 5–10 N applied load indicated the occurrence of mild wear regime. The relative wear resistance of nitrocarburised steel was found 34–112 times higher than that of the annealed steels.

The influence of sliding velocity on the wear rate can be noted from the work of Nicoletto et al. [79]. They nitrided and nitrocarburised spheroidal and lamellar cast iron. These surface-treated cast irons were subjected to sliding wear at different sliding velocities. Figure 4.16 shows variation of wear rate as function of sliding velocities. Wear rates of all tested materials exhibited decrease of wear rate with sliding velocity. It can be noted that for all sliding velocity spheroidal cast iron

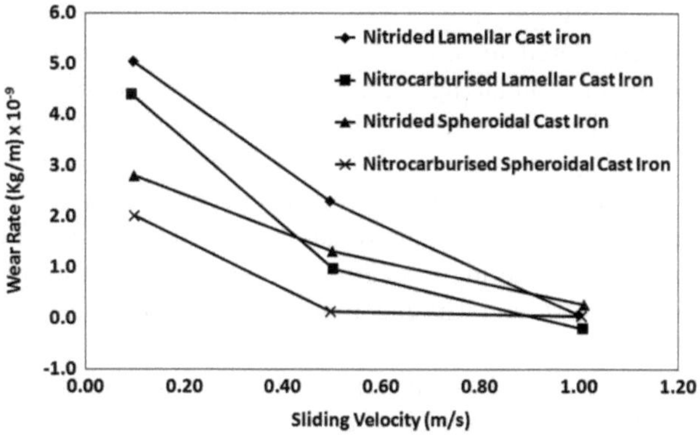

Fig. 4.16 Influence of sliding velocity of the wear rate of various nitride cast iron [79]

exhibited lower wear rate than lamellar cast iron. For both materials, nitriding performed better at low sliding velocity whereas nitrocarburising was superior at high sliding velocity.

While engineering the surface of a material for improved performances, attention should be paid in the bulk temperature of the wearing surface. This bulk temperature (T_b) is given as [81, 82]

$$T_b = T_0 + \frac{\alpha \mu F v l_b}{A_n K_m} \tag{4.10}$$

where T_0 is room temperature, α is a constant known as heat distribution coefficient, μ is the friction coefficient, F is applied load, v is sliding velocity, l_b is mean heat diffusion distance, A_n is nominal area of contact and K_m is thermal conductivity of the metal. If T_b is high, the hardness of the hardened layer tends to come down. One way of reducing the temperature is to lower the friction coefficient. For a given sliding velocity, the friction coefficient depends on the crystal structure of the surface. The crystal structure will determine the number of slip system available. If the number of slip system is high, μ and T_b will be low. This, in turn will reduce the tendency of the surface to soften.

To illustrate an example where such principle can be applied advantageously is carburising and nitriding. In both the cases, the wear resistance is improved by hardening the surface of the component. In carburising, hardening is achieved by transforming the carbon-rich austenite into martensite, which has BCT structure. In contrast, nitriding is done on ferritic matrix having BCC structure. The available slip system in BCC-nitrided layer is higher than for the BCT-carburised layer giving rise to low friction and less increase of temperature. Consequently the nitrided layer does not get soften by increasing in temperature. The work due to Krishnaraj et al. [84] clearly confirms the above statement. They noted superior performances of nitrocarburising to martensite hardening for hot working die application. On the contrary, Rizvi and Khan [85] found opposite results with

H13 tool steel which was gas alloyed by tungsten arc. The higher wear resistance of carburised layer compared to nitrided layer was attributed to formation of fine dendrite microstructure containing chromium–vanadium carbides.

The thermochemical treatments with the rare earth (RE) addition were investigated at the beginning of the 1980s by Wei et al. [86]. In the following many years, a number of advantages of this new technique have been ascribed to addition of RE [87–90]. They suggested that the RE elements could diffuse into a significant depth of steels during gas carburising, gas carbonitriding, gas nitrocarburising and plasma nitriding. The incorporated RE elements can help interstitial species (such as carbon and nitrogen) diffuse into a significant depth of the steel and enhance mass transfer as well as kinetics and then improve the microstructure and the property of the modified layer. Liu et al. [91] found that the wear rates of the un-nitrocarburised, nitrocarburised and RE plasma nitrocarburised PH stain less steel were about 4.3×10^{-12}, 2.1×10^{-12} and 0.05×10^{-12} $m^3 m^{-1}$ respectively. The wear mechanisms of steel which is plasma nitrocarburised with and without RE addition are different mainly because of the differences in the microstructure, the phase proportion and the hardness of the modified layer.

Forati Rad et al. [92] plasma-nitrided AISI H11 steel. The samples were austenitized at 1,293 K for 30 min, quenched in oil media and then tempered at 833 K for 2 h in an electric furnace. They were subsequently plasma nitrided. The wear rate of the nitrided samples is summarised in Fig. 4.17. The untreated specimen exhibited poor wear resistance with respect to plasma-nitrided samples. It is to be stated that the wear resistance of the nitrided specimens increases with increasing the surface hardness which is in agreement with Eq. (4.9). The lowest wear rate is related to the sample that has the highest hardness and the greatest amount of ε phase. Normally, hcp crystalline structures in comparison with bcc and/or fcc structure show better tribological properties due to slip of basal planes. It was mentioned that fracture properties of surface layer also played an important role on the wear behaviour of specimen [93]. Although the wear rate of various nitride samples is lower than untreated sample, they exhibit greater wear weight loss with respect to other nitrided samples. This difference can be attributed to the fact that these samples have thin compound layer that can be broken in small sliding distance and produce abrasive particles resulting in severe scrape wear and increase the wear rate of the samples.

The wear resistance of the nitrided layer increases with distance from the surface. The wear resistance of the porous compound layer is low. The wear resistance of the compact compound layer is higher than the wear resistance of the diffusion layer. The structure and composition of the compound also influence the wear resistance. The wear resistance of the compound layer increases with increase in the volume of ε-nitride. For a given volume fraction of ε nitride, carbinitrided layer shows higher wear resistance than nitrided layer. If the nitrided alloy is hardened by precipitation or solid solution hardening, the wear resistance improves. Post nitriding treatment such as ageing after nitriding results in improved wear resistance. Higher the ageing time higher is the wear resistance. The wear rate is sensitive to the type and distribution of nitride precipitation. A matrix with finely

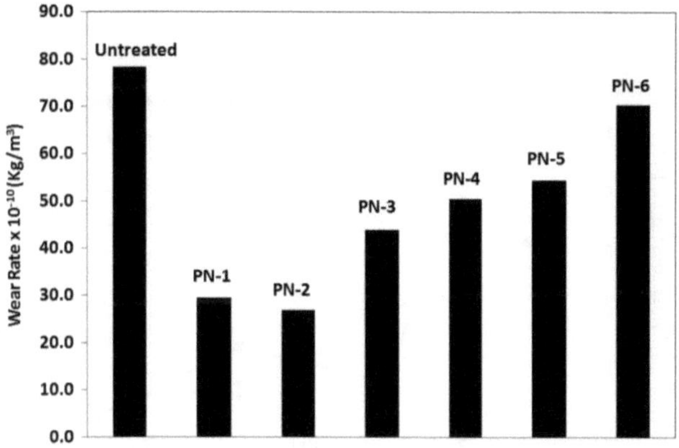

Fig. 4.17 The wear rate of the plasma-nitrided AISI H11 steel [92]

distributed $Fe_{16}N_2$ nitride has a lower wear resistance than one with course distribution of F_4N nitride.

In general, tribological properties of nitriding layer can be improved by (1) increasing the ε-nitride content instead of γ-nitride content by using higher nitriding potential, (2) changing nitride to carbinitride by adding carbon donors such as Endogas and CO_2 or by using material with higher carbon content and (3) by minimising porous zone by polishing.

Plasma chromising and carburisation following plasma chromising (duplex treatment) were carried out on the Ti_2AlNb-based (O phase) alloys by the double glow plasma process by Wu et al. [94]. Dry sliding friction tests on the samples against silicon nitride (Si_3N_4) balls were completed at high temperature by a ball-on-disc tribometer. The results indicated that the duplex-treated layer was mainly composed of $Cr_{23}C_6$ and Cr_2Nb, while the chromised layer was consisted of Al_8Cr_5 and Cr_2Nb phases. The ultramicro hardness of the duplex-treated layer was higher than that of the chromised layer, whereas the elastic modulus of the duplex-treated layer was lower than that of the chromised layer. The wear rates of the duplex-treated layer and the chromised layer were lower than that of substrate. However, the duplex-treated layer exhibited excellent high temperature wear resistance. Aluminised and chromised high carbon steel also exhibited promising wear resistance in physiological solution especially at high and intermediate load [95].

The effect of temperature on wear rate of plasma-nitrided 722M24 steels is shown in Fig. 4.18 [80]. The wear rate decreases with increase of temperature. It is clear that the higher the test temperature, the lower the weight loss for plasma-nitrided cases under lubricated abrasive conditions. In other words, at higher temperatures, the softer the nitrided case, the more abrasive particles and wear debris penetrate the worn surfaces. The oil layers affected by temperature also decrease the weight loss of sample by lower friction. It is also to be stated that as the temperature increases, the wear rate and the surface roughness both decrease.

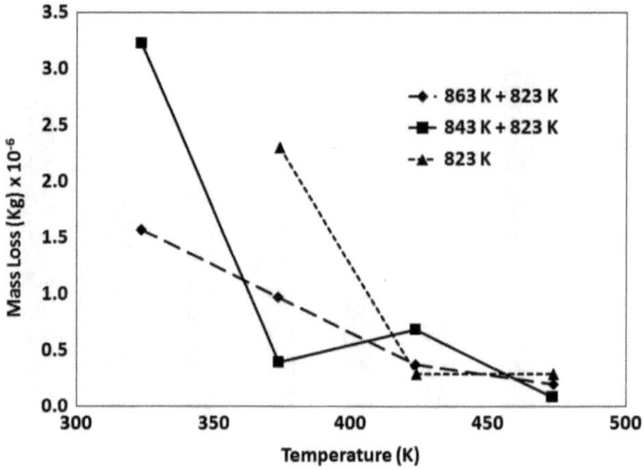

Fig. 4.18 The influence of temperature on the wear rate of plasma-nitrided steel [80]

SEM images showing the morphology of worn surfaces of ferritic nitrocarburised low carbon steel En 8 is shown in Fig. 4.19 by Krishnaraj [96]. Figure 4.19a presents worn surfaces of as-nitrocarburised En 8 steel and Fig. 4.19b represents the transverse section of the worn surface of same steel. The worn surface exhibits severe plastic deformation with patched of mild oxidation marks. In contrast, subsurface region indicates presence of delamination cracks.

4.3.2 Solid Particle Erosion

Degradation of diffusion-treated surface due to solid particle erosion is also studied to a considerable extent. Solid particle erosion rate as per localisation model [97] is given as

$$E \infty \left(\frac{L}{r}\right)^3 \left(\frac{\Delta\varepsilon_{\mathrm{m}}}{\varepsilon_{\mathrm{c}}}\right) \tag{4.11}$$

where E is the erosion rate, L is the depth to which the deformation extends beneath the eroded surface, r is the radius of erodent, $\Delta\varepsilon_{\mathrm{m}}$ is the mean strain increment for each impact and ε_{c} is the critical strain for initiation of localisation of plastic deformation. If it is possible to increase the hardness of a material by diffusion treatment, the depth of deformation decreases leading to increase in erosion resistance provided other parameter remains unchanged. The validity of this statement can be found below.

Wen [98] characterised the erosion response of liquid nitrocarburised tool steel. Figure 4.20 shows the cumulative weight losses versus erosion time for the untreated and the nitrocarburised specimens at normal impact. It is clear from the

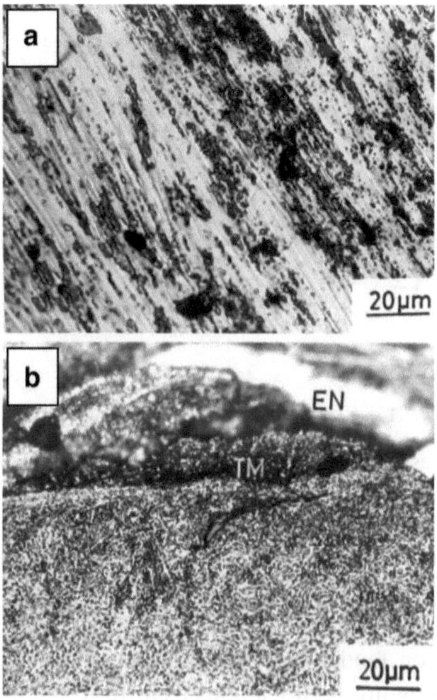

Fig. 4.19 SEM image showing morphology of worn surfaces of (**a**) as received tempered martensite tool steel and (**b**) salt bath nitrocarburised tempered martensite tool steel [96]

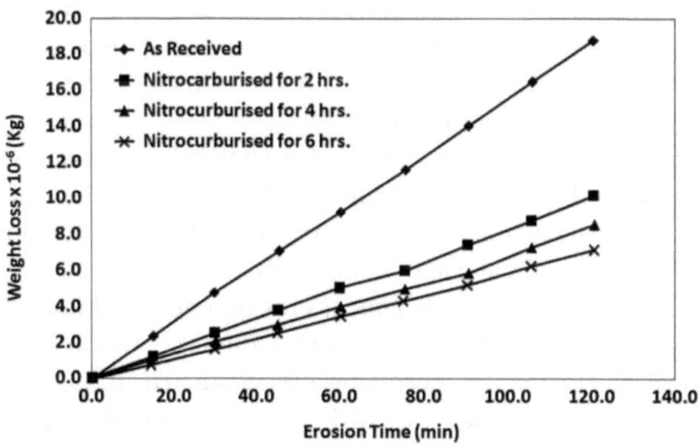

Fig. 4.20 The cumulative weight losses versus erosion time for the untreated and the liquid nitrocarburised tool steel at normal impact [98]

results that the erosion rate of the nitrocarburised specimens decreases with increasing nitrocarburising holding time: i.e. the erosion rate decreases with increase of the case depth and surface hardness. The highest weight loss is obtained in the untreated specimen. The weight losses are found to increase linearly with increasing erosion time for these four specimens. This indicates that the erosion mechanism does not change noticeably, implying a steady erosion damage during the

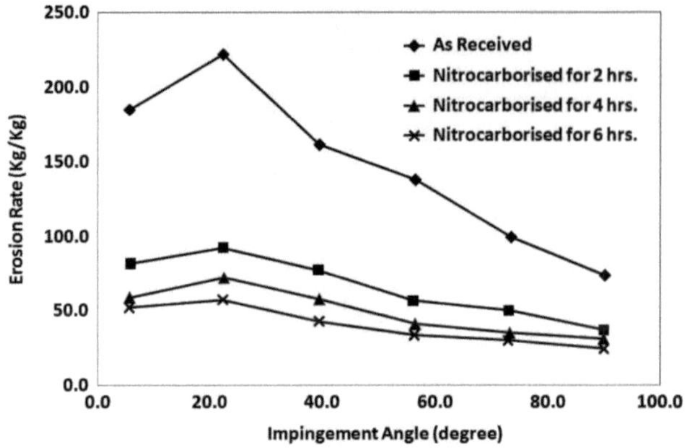

Fig. 4.21 The erosion rates versus the impinging angle for the untreated and the nitrocarburised specimens [98]

impingement processes, regardless of the variation in the impinging angle. Figure 4.21 shows the erosion rates versus the impinging angle for the untreated and the nitrocarburised specimens after a 2-h test. The erosion rates for all specimens show a similar variation tendency, in which the erosion rate increases and then decreases with increasing impinging angles from 15 to 90° and reaches a maximum around 30°. It shows a typical ductile erosion behaviour [99]. For ductile materials, material loss is by formation of lips and their subsequent fracture [100]. The erosion rate of the nitrocarburised specimens decreases with increasing nitrocarburising holding time: i.e. the erosion rate decreases with increase of the case depth and surface hardness. This improvement is effective even at high impinging angles because increasing surface hardness after nitrocarburising is not accompanied by loss in ductility of the substrate material.

Dong et al. [101] compared the erosion–corrosion behaviour of low-temperature plasma surface-alloyed AISI 316 austenitic stainless steel with untreated material with the help of a slurry pot erosion–corrosion apparatus containing a slurry comprising 20 wt.% silica sand and 3.5% NaCl at 40 °C. The total erosion–corrosion wastage, the mechanical erosion under cathodic protection and the electrochemical corrosion were measured directly. Based on the data, the synergistic effect of erosion and corrosion was estimated. It has been shown that the erosion–corrosion resistance of AISI 316 austenitic stainless steel can be effectively improved by low-temperature plasma alloying with carbon (carburising) and nitrogen (nitriding) by 50 % and 70 %, respectively. The degradation process of the untreated steel is dominated by erosion whilst that of the low-temperature plasma-carburised material is by an erosion–corrosion mechanism and that of low-temperature plasma-nitrided AISI 316 is mainly by corrosion–erosion. The synergy between erosion and corrosion is only 1.7 % in case of the untreated material. This increases to 30.0 % for the plasma-carburised material and to 69.4 % in case of the plasma-nitrided steel. Boronising is also found to improve

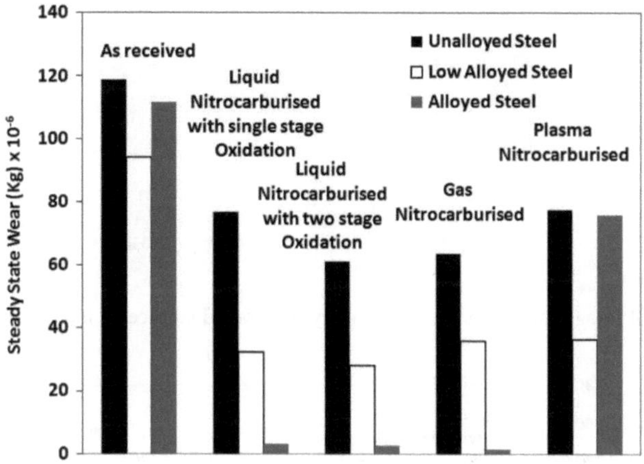

Fig. 4.22 The two body abrasive wear resistance of a large number of steel by several nitrocarburising process [103]

erosion resistance. Improved erosion resistance of various boronised steel is reported by Biddulph [102] and Singhal [40].

4.3.3 Abrasion

Significant literature is available on the abrasive wear behaviour of diffusion-treated surfaces. Kassim et al. [103] compiled the two body abrasive wear resistance of a large number of steel by several nitrocarburising processes. Part of their result is provided in Fig. 4.22. In general, it is noted that abrasive wear rate decreases with increase in hardness of the treated layer in accordance with Eq. (1.16) as discussed in Chap. 1. There is significant improvement in abrasive wear rate by nitrocarburising. The improvement is maximum for alloyed steel although such improvement is marginal for alloyed steel subjected to plasma nitrocarburising. Abrasive wear rates of unalloyed steel and low alloyed steel are nearly independent of nitrocarburising processes. Further, nitrocarburised low alloyed steel performs better than unalloyed steel under two body abrasion irrespective of the nitrocarburising process. The abrasive wear resistance is found to be governed by the hardness, thickness, composition and porosity of the compound layer. In general, thicker and harder compound layer results in lower wear rate. SEM images showing the morphology of abraded surface of low alloyed steel is illustrated in Fig. 4.23. Microcutting and microcracking appear to be the material removal mechanism. Krishnaraj et al. [104, 105] also noted that scuffing resistance of compound layer increases because of presence of ε carbonitride having hexagonal crystal structure. For alloyed steel scuffing resistance of the compound layer is

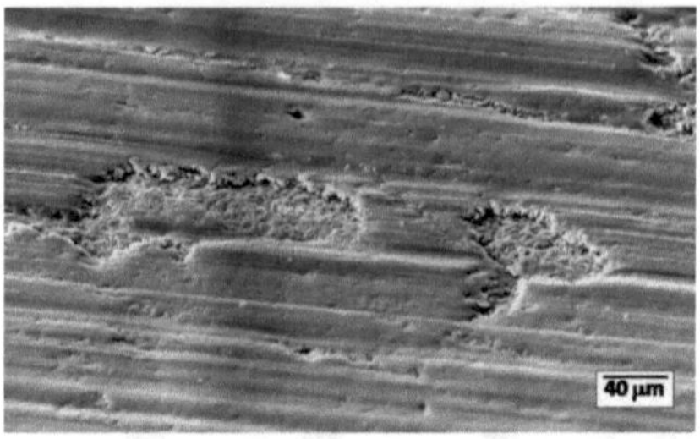

Fig. 4.23 SEM images showing the morphology of abraded surface of low alloyed steel

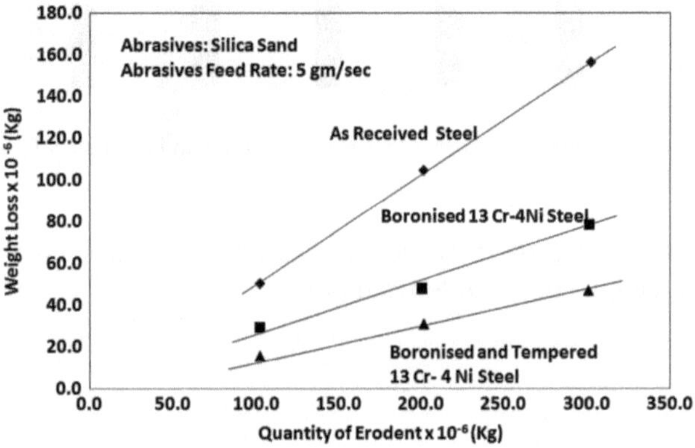

Fig. 4.24 The abrasive wear rate of borided, borided and tempered, and as-received 13Cr–4Ni steel [52]

superior to that of plain carbon steel as the hardness and compactness of the compound layer is better for alloyed steel than plain carbon steel.

Mann [52] studied abrasion resistance of (1) as-received, (2) borided and (3) borided and tempered 13Cr–4Ni steel. His observation which is an average of five samples is presented in Fig. 4.24. Borided steel has exhibited the highest abrasion resistance which is about 300 % improvement compared to the as-received 13Cr–4Ni steel. The abrasion rates are based on weight loss per unit silica sand used (kg kg^{-1}). The values for borided, borided and tempered, and as-received 13Cr–4Ni steel are 0.15, 0.27 and 0.5×10^{-6} kg kg^{-1}, respectively. This is mainly due to its hardness, although after boronising the ductility was reduced significantly. A substantial reduction in abrasion rate is observed when tempered at 873 K. After tempering, the hardness of borided layers was reduced slightly.

The time of boronising also influences the abrasive wear rate. The study of Tabur et al. [106] shown in Fig. 4.25 essentially validates such observation. Boronising was conducted for AISI 8620 cementation steel using the solid-state pack boronising method containing a mixture of 5 % B_4C, 5 % KBF_4 and 90 % SiC. Samples were boronised by packing inside stainless steel boxes with at least 10 mm

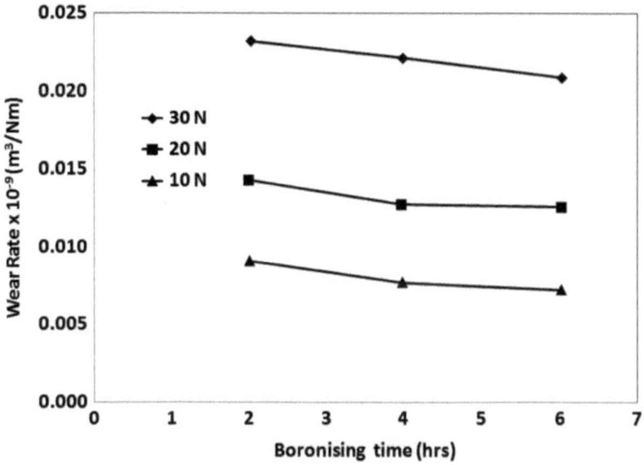

Fig. 4.25 The influence of the time of boronising on the abrasive wear rate of AISI 8620 cementation steel [104]

of powder mix packing. After boronising, the samples were air cooled. The wear resistance of specimens increased with increasing boronising time, at the same boronising temperature. Grooves were much finer in sample boronised for 6 h and worn surface appeared smoother. The wear resistance of specimens decreased with increasing boronising temperature at the same boronising time. Grooves were much wider and deeper for higher boronising temperature. The wear rate was found to be increasing with increasing load. It is also observed that wear scars on the sample surfaces become deeper and wider with increasing load and abrasive wear increased with increasing abrasive grain size. Finally, it is observed that the FeB layer that forms on the sample surface exhibited a crisp and brittle structure.

Meric et al. [50] studied the abrasive wear rate of pack-boronised cast irons. The thickness of boronised layer increased with increasing boronising time. Among three different cast iron namely grey cast iron, compacted graphite cast iron and ductile iron, boronised compacted graphite cast iron exhibited best performances as shown in Fig. 4.26. It is also noted that the wear rate decreases with increase in boronising time which in turn increases the thickness of the boronised layer.

Boron-aluminised layers were formed on plain carbon steels Q235 (1020), 45 (1045), and alloyed steel GCr15 (E52100) by dipping them in a molten aluminium bath at 993 K and then diffused. Aluminised samples were subsequently boronised at 1,223 K for 6 h by Luo et al. [107]. The XRD analysis results indicated presence of Fe_2B, Fe_2AlB_2 and Fe_2Al_5 in the layer. Under dry abrasive conditions, wear resistance of the complex boronised layer was found to be superior to the aluminium hot-dipped layer.

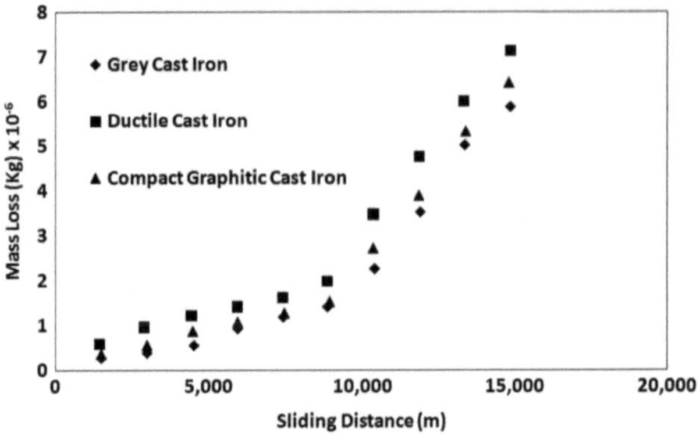

Fig. 4.26 Variation of abrasive wear rate of various cast iron with sliding distance [105]

4.4 Application of Diffusion-Coated Components

Diffusion coated components particularly steels are used extensively in various engineering applications.

Various machine parts such as gear, gear teeth, spline, bearing, pinions, etc. are gas carburised. Parts like cramshaft, stamped are normally liquid carburised. Induction carburising is used to create unusual surface layers having high carbon content. These diffusion coatings consist of ledeburite structures or graphite particles. Ledeburite structures have structural properties very similar to those of high-alloy materials, which are now widely used in harsh, abrasive conditions such as agricultural machinery, mining, oil and gas equipment, excavators and road machinery. In these applications, very expensive and complicated flame, welding, and induction surfacing technologies are currently used. Structures that contain graphite particles can be used for self-lubricating parts in friction pairs and in many other applications. Life-cycle costs can be reduced by creating ledeburite surface layers on parts that now have short life spans. Figure 4.27 displays gas-carburised clutch pin and clutch plates (Maheswari Carburising Engineering Private Limited, Gujarat, India) [108].

Nitriding is used extensively in aircraft engine parts such as cams, cylinder liners, valve stems, shafts, piston rods, etc. Among various automotive components, nitrided layer is found in ABS hardware, drive plates, rocker arm shafts, clutch springs, engine valves, rocker arms, ball joints, engine valve guides, brake pistons, fuel injectors, seat rails, etc. Bushings, parts of fuel injection mechanisms, parts of landing gear, parts of gearbox assembly, engine components, actuators of auxillary power units, actuators, wing flaps, etc. are some precision components that are nitrided. Various mechanical components such as gears, gun barrels, pinions and shafts, etc. are also used in nitrided condition. Aluminium die casting dies, aluminium die casting sleeves, aluminium extrusion dies, ejector pins, forging dies,

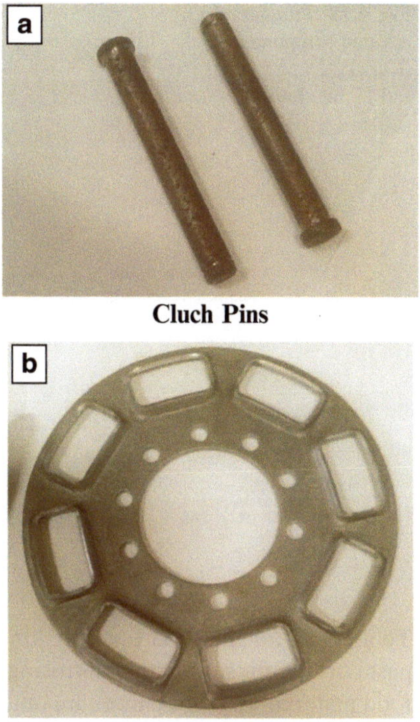

Cluch Plate

Fig. 4.27 Photographs of gas-carburised clutch pin and clutch plates (Maheswari Carburising Engineering Private Limited, Gujarat, India)

forming dies, etc. are examples of nitrided tools and dies. Some nitride components are shown in Fig. 4.28 (Kunakeshwar Springs Pvt. Ltd., Pune, India).

For small screw machine parts and stampings where a hard wear-resistant surface is required, along with core strength, carbonitriding is typically used. This produces a file hard surface with case depths from 0.13 to 0.89 mm. Front axle bearing, washer, distributer, drive shaft, bushing brake operating cam, hinge pin, mandrel spring pin, helical gear, hub, pinion gear, ring gear constitute several parts that are carbonitrided.

Boronising has not become as popular as carburising and nitriding and its usage is still very less as compared to other diffusion treatments for tribological applications. However, for high temperature applications boronising finds extensive applications. It is used in steel rings, steel rope, steel thread guide bushings, grooved cast iron drums of textile machinery, drive worm and helically toothed steel gear of various high performances vehicle and stationery engines [109]. Nozzles of bag filling equipment, bend and baffle plates of conveying equipments, steel mould for manufacturing ceramic bricks, extruder barrels. Boronised permalloy is used for magnetic head applications.

Aluminising is used primarily for gas turbine blades. Platinum aluminising is one method which finds application in aero engines. But such application is for resistance against high temperature exposure rather than protection against wear.

Fig. 4.28 Photographs of nitrided components (Kunakeshwar Springs Pvt. Ltd., Pune, India)

Tools used for the food industry for the closing of cans filled with, for example pasta or ravioli, suffer from corrosion, probably due to the high salt content of the food material. CVD of chromium onto forming tools for the food Industry resulted chromium carbide coating up to 15 μm thick and has a hardness of 1,500 HV (0.1 kg). The tools are heat treated after coating and this is carried out in a vacuum furnace at 1,303 K with a rapid quench after 30 min at temperature. The chromium carbide coating shows no distress as a result of this treatment. The resulting coated tool was able to withstand the arduous condition in the food canning plant and acceptable lives are routinely achieved.

4.5 Summary

In this chapter an attempt is made to take stock of the state of art of diffusion treatment and the tribological performances of diffusion treated surfaces. Research on diffusion surface treatment has taken place for last several decades. Although most diffusion treatments for improved tribological performances are carried out on ferrous materials, nonferrous materials are receiving increased attention in recent times especially after advent of plasma diffusion treatment process. Although the equipment and the initial investment required for plasma processing is high, there are several advantages which are responsible for increasing popularity of this process. The cost of equipment has come down considerably due to widespread application of such process and improved technology. Further reduction of cost and further application is expected in near future. One key issue for glow discharge process is 'hallow cathode effect'. With stringent and improved process control this effect has been minimised in recent time. Plasma diffusion treatment offers more flexibility and promises to provide large spectrum of high performance surface engineered components. Various hybrid processing or duplex treatment requires extensive investigation to realise their advantages in applications involving synergistic effect of wear and corrosion.

Multicomponent boronising such as boroaluminising, borosiliconnising, borochromising, etc. involves consecutive diffusion of boron or one or two metallic components [110]. This multicomponent boronising although used for improved tribological application, is expected to draw attention of research community. Vacuum carburising is another area where significant development is expected to realise in recent future. Rapid development in the field of gas pressure quenching leaves for tremendous opportunity for development of high performance low distortion gearing and bearing systems.

References

1. Sun Y, Bell T (2002) Wear 253:689
2. Leyland A, Lewis DB, Stevenson PR (1993) Surf Coat Technol 62:608
3. Esfandiari M, Dong H (2007) Surf Coat Technol 202:466
4. Goward GW, Boone DH (1971) Oxid Met 3:475
5. Genel K, Demirkol M (1999) Int J Fatig 21:207
6. Hombeck F, Bell T (1991) Surf Eng 7:45
7. Karamis MB (1991) Wear 147:385
8. Nie X, Tsotsos C, Wilson A, Yerokhin AL, Leyland A, Matthews A (2001) Surf Coat Technol 139:135
9. Hickl J, Heckel RW (1975) Metall Trans A 6A:431
10. Zeng D, Yang S, Xiang ZD (2012) Appl Surf Sci 258:517
11. Ivanov VE, Smovo AI (1964) Vacuum 14:247
12. Batz W, Mehl RF, Wells C (1950) Trans AIME 188:553
13. Hurvey FJ (1978) Met Trans 9A:1507
14. Metals handbook, heat treating, vol 4. ASM International, Materials Park, OH
15. Krauss G (1990) Steel: heat treatment and processing principles. ASM, Materials Park, OH, p 286
16. Stickels CA, Mack CM (1989) Overview of carburising processes and modelling. In: Krauss G (ed) Carburising processes and performances. ASM, Materials Park, OH, p 1
17. St. Pierre J (1991) Vacuum carburising, heat treating, vol 4, ASM handbook. ASM, Materials Park, OH, p 348
18. Selcuk B, Ipek R, Karamis MB, Kuzucu V (2000) J Mater Process Technol 103:310
19. Suh B-S, Lee W-J (1997) Thin Solid Films 295:185
20. Buhagiar J, Qian LM, Dong H (2010) Surf Coat Technol 205:388
21. Fattah M, Mahboubi F (2010) Mater Des 31:3915
22. Kochmański P, Nowacki J (2005) Surf Coat Technol 200:6558
23. Oliveira SD, Tschiptschin AP, Pinedo CE (2007) Mater Des 28:1714
24. Habig KH (1980) Mater Des 2:83
25. Yang J, Liu Y, Ye Z, Yang D, He S (2011) Mater Des 32:808
26. Karakan M, Alsaran A, Çelik A (2004) Mater Des 25:349
27. Karamis MB (1993) Wear 161:194
28. Priest JM, Baldwin MJ, Fewell MP (2001) Surf Coat Technol 145:152

29. Edenhofer B (1974) Heat Treat Met 2:59
30. Liu J, Dong H, Buhagiar J, Song CF, Yu BJ, Qian LM, Zhou ZR (2011) Wear 271:1490
31. El-Hossary FM, Negm NZ, Khalil SM, Abed Elrahman AM, McIlroy DN (2001) Surf Coat Technol 141:194
32. Howsow HJ, Pistorius PGH (1999) Surf Eng 15:476
33. Charizanova S, Dimitrov G, Rousseva E, Marcov V (1998) J Mater Process Technol 77:73
34. Krishnaraj N, Bala Srinivasan P, Iyer KJL, Sundaresan S (1998) Wear 215:123
35. Qiang YH, Ge SR, Xue QJ (1998) Wear 218:232
36. Torchane L, Bilger P, Dulcy J, Gantois M (1996) Metall Mater Trans A 27:1823
37. Qiang YH, Ge SR, Xue QJ (1999) Tribol Int 32:131
38. Hoppe S (1998) Surf Coat Technol 98:1199
39. Sahin S (2009) J Mater Process Technol 209(4):1736
40. Singhal SC (1977) Thin Solid Films 45:321
41. Bejar MA, Moreno E (2006) J Mater Process Technol 173:352
42. Takeuchi E (1979) Wear 55:121
43. Habig KH, Chatterjee-Fischer R (1981) Tribol Int 14:209
44. Knotek O, Lugscheider E, Leuschen K (1977) Thin Solid Films 45:331
45. Ipek R, Selcuk B, Karamis MB, Kuzucu V, Yucel A (2000) J Mater Process Technol 105:73
46. Ozdemir O, Usta M, Bindal C, Ucisik AH (2006) Vacuum 80:1391
47. Ozbek I, Bindal C (2002) Surf Coat Technol 154:14
48. Dybkov VI, Lengauer W, Barmak K (2005) J Alloys Compd 398:113
49. Galibois A, Boutenko O, Voyzelle B (1980) Acta Metall 28:1753
50. Meric C, Sahin S, Backir B, Koksal NS (2006) Mater Des 27:751
51. Sen S, Ozbek I, Sen U, Bindal C (2001) Surf Coat Technol 135:173
52. Mann BS (1997) Wear 208:125
53. Sinha AK (1991) ASM handbook, vol 4. ASM, Materials Park, OH, p 437
54. Buckley DH, Spelvin T (1985) Mater Lett 3:181
55. Parzuchowski RS (1980) Thin Solid Films 55:146
56. Das DK, Singh V, Joshi SV (1998) Metall Mater Trans A 29A:2173
57. Nicoll AR, Hilderbrandt UW, Wahl G (1979) Thin Solid Films 64:321
58. Grunling HW, Bauer R (1982) Thin Solid Films 3:95
59. Nicoll AR, Hildebrandt UW, Wahl G (1979) Thin Solid Films 64:321
60. Das DK, Singh V, Joshi SV (2002) Oxid Met 57:245
61. Das DK, Roy M, Singh AK, Sundararajan G (1996) Mater Sci Technol 12:295
62. Mervel R, Duret C, Pichoir R (1986) Mater Sci Technol 2:201
63. Streiff R, Boone DH (1985) In: Sisson RD (ed) Coatings and bimetallics for aggressive environments. ASTM, Materials Park, OH, p 159
64. Smialek JL (1974) NASA TM X-3001. National Aeronautics and Space Administration, Washington, DC
65. Daimer J, Fitzer E, Schlichting J (1991) Thin Solid Films 84:119
66. Nicoll AR, Wahl G (1982) Thin Solid Films 95:21
67. Arai T (1990) In: Sudarshan TS (ed) Surface modification technologies III. The Minerals, Metals and Materials Society, Warrendale, PA, p 587
68. Lee J-W, Duh J-G, Tsai S-Y (2002) Surf Coat Technol 153:59
69. Lee J-W, Duh J-G (2004) Surf Coat Technol 177–178:525
70. Meier GH, Cheng C, Perkins RA, Bakker W (1989) Surf Coat Technol 39–40:53
71. Lee SB, Cho KH, Lee WG, Jang H (2009) J Power Sources 187:318
72. Chen F-S, Lee P-Y, Yeh M-C (1998) Mater Chem Phys 53:19
73. Perez FJ, Pedraza F, Hierro MP, Carpintero MC, Gomez C (2004) Surf Coat Technol 184:47
74. Perez FJ, Hierro MP, Pedraza F, Gomez C, Carpintero MC (1999) Surf Coat Technol 120–121:151
75. Perez FJ, Hierro MP, Pedraza F, Gomez C, Carpintero MC, Trilleros JA (1999) Surf Coat Technol 122:281

76. Priyantha N, Jayaweera P, Sanjurjo A, Lau K, Lu F, Krist K (2003) Surf Coat Technol 163–164:31
77. Ralston KD, Fabijanic D, Jones RT, Birbilis N (2011) Corros Sci 53:2835
78. Pantazopoulos G, Psyllaki P, Kanakis D, Antoniou S, Papadimitriou K, Sideris J (2006) Surf Coat Technol 200:5889
79. Nicoletto G, Tucci A, Esposito L (1996) Wear 197:38
80. Karamis MB, Gercekcioglu E (2000) Wear 243:76
81. Lim SC, Ashby MF (1987) Acta Metall 35:11
82. Roy M (2009) Trans Indian Inst Met 62:197
83. Psyllaki P, Kefalonikas G, Pantazopoulos G, Antoniou S, Sideris J (2002) Surf Coat Technol 162:67
84. Krishnaraj N, Iyer KJL, Sundaresan S (1997) Wear 210:237
85. Rizvi SA, Khan TI (1999) Tribol Int 32:567
86. Wei YD, Liu ZR, Wang CY (1983) Acta Metall Sin 19:B197
87. Liu ZR, Zh FY, Cui YX, Sh YX, Wang CG (1983) J Rare Earths 11:196
88. Bell T, Sun Y, Liu ZR, Yan MF (2000) Heat Treat Met 27:1
89. Yan MF, Pan W, Bell T, Liu ZR (2001) Appl Surf Sci 91:173
90. Yan MF, Sun Y, Bell T, Liu ZR, Xia LF (2002) J Rare Earths 20:330
91. Liu RL, Yan MF, Wu DL (2010) J Mater Process Technol 210:784
92. Forati Rad H, Amadeh A, Moradi H (2011) Mater Des 32:2635
93. Habib KA, Saura JJ, Ferrer C, Damra MS, Gimenez E, Cabedo L (2006) Surf Coat Technol 201:1436
94. Wu YH, Zhang P, Xu Z (2008) Surf Eng 24:464
95. Gulhane UD, Roy M, Sapate SG, Mishra SB, Mishra PK (2009) In: Proceedings of ASME/STLE international joint tribology conference, IJTC 2009, Memphis, TN, 19–21 October
96. Krishnaraj N (1997) Ph.D. thesis, Indian Institute of Technology, Madras
97. Roy M, Tirupataiah Y, Sundararajan G (1993) Mater Sci Eng A165:51
98. Wen DC (2010) Wear 268:629
99. Sundararajan G, Roy M (2010) Tribol Int 268:629
100. Bellman R Jr, Levy A (1981) Wear 70:1
101. Dong H, Qi P-Y, Li XY, Llewellyn RJ (2006) Mater Sci Eng A 431:137
102. Biddulph RH (1977) Thin Solid Films 45:341
103. Kassim A, Al-Rubaie S, Steinmeier F, Pohl M (2000) Wear 243:112
104. Krishnaraj N, Iyer KJL, Sundaresan S (1997) Mater Lett 32(5–6):355
105. Krishnaraj N (1998) Ph.D. Thesis, Indian Institute of Technology, Madras
106. Tabur M, Izciler M, Gul F, Karacan I (2009) Wear 266:1106
107. Luo X, Li D, Chen K (2005) Int J Microstruct Mater Prop 1:88
108. http://www.heattreatmentservice.com
109. Dearnley P, Bell T (1985) Surf Eng 1:248
110. Chaterjee-Fischer R (1986) Met Prog 129(5):24

Chapter 5
Hardfacing for Wear, Erosion and Abrasion

E. Badisch and Manish Roy

5.1 Introduction

Weld overlay coatings also known as hardfacing is a method which involves solidification for applied coatings. This method is as old as welding process and was first applied by J.W. Spencer in 1896. This technique offers unique advantages over other processes in that the overlay/substrate weld provides a metallurgical bond which is not susceptible to spallation and can easily be applied free of porosity and other defects. Weld deposits can have a thickness as high as 10 mm. Hardfacing process can be carried out easily even in on site. The process is also extremely versatile as a large variety of materials can be deposited for protection against degradation.

Weld overlay coatings can be classified under four different categories. These are (1) surface cladding where a thick layer of materials is applied and the overlay layer can have completely different composition from the substrate, (2) hardfacing where a materials more resistant to degradation is applied with good metallurgical bonding and the composition can be close in terms of composition to that of the substrate, (3) build up where a weld deposit is applied to restore the original dimension of the component and the composition of the build layer is exactly similar to that of the substrate and (4) buttering where an intermediate layer is deposited before applying the final coating.

The materials used for hardfacing should have melting point close to or lower than the substrate materials. During hardfacing, the temperature of the coating

E. Badisch (✉)
Austrian Center of Competence for Tribology, Viktor Kaplan-Strasse 2,
2700 Wiener Neustadt, Austria
e-mail: badisch@ac2t.at

M. Roy
Defence Metallurgical Research Laboratory, Kanchanbagh, Hyderabad 500058,
Andhra Pradesh, India

M. Roy (ed.), *Surface Engineering for Enhanced Performance against Wear*,
DOI 10.1007/978-3-7091-0101-8_5,

material is increased to the melting point and then allowed to solidify on the substrate. The effectiveness of hardfacing depends on the process of application of the hardfaced layer and the composition of the said layer. The process should be optimised to have high deposition rate, high thermal efficiency, excellent dilution, excellent control of composition and coating thickness.

The present chapter is concerned with various hardfacing processes, their advantages and disadvantages and finally the tribological performances of hardfaced surfaces, along with their high temperature properties.

5.2 Analytical Model for Heat Flow Equations

In order to balance the heat flow in an element, it is to be assumed that

Heat in–heat out = heat accumulated + heat generated. This can be represented mathematically as

$$-\frac{U}{\alpha}\Delta T + \Delta^2 T - \frac{1}{\alpha}\frac{\partial T}{\partial t} = -\frac{H}{k} \tag{5.1}$$

For one-dimensional heat flow model, it is assumed that heat flows in one direction and there is no convection or heat generation leading to the equation

$$\frac{\partial^2 T}{\partial z^2} = \frac{1}{\alpha}\frac{\partial T}{\partial t} \tag{5.2}$$

If it is assumed that the heat input is constant, thermal properties are independent of temperature, no heat loss due to convection and no latent heat effect, then the boundary conditions are

at $z = 0$, surface power density F_0 is given by

$$F_0 = \left(\frac{P_{\mathrm{tot}}(1 - r_f)}{A}\right) = -k\left[\frac{\partial T}{\partial z}\right]_{\mathrm{surf}}$$

at $z =$ infinity

$$\frac{\partial T}{\partial z} = 0$$

At $t = 0, T = T_0$

Hence the solution is as given below:

$$T_{z,t} = \frac{2F_0}{k}\left\{(\alpha t)^{1/2}\mathrm{ierfc}\left[\frac{z}{2(\alpha t)^{1/2}}\right]\right\} \tag{5.3}$$

where ierfc is "integral of the complementary error function".

When there is no heat source, the material will cool for $t > t_1$ according to the relationship given by

$$T_{z,t} = \frac{2F_0}{k}\alpha^{1/2}\left\{t^{1/2}\mathrm{ierfc}\left[\frac{z}{2(\alpha t)^{1/2}}\right] - (t-t_1)^{1/2}\mathrm{ierfc}\left[\frac{z}{2(\alpha(t-t_1))^{1/2}}\right]\right\} \tag{5.4}$$

where T is temperature, z is depth, t is time, t_0 is the start time of heat source, t_1 is the end time of heat source, k is thermal conductivity, α is thermal diffusivity and F_0 is absorbed power density. This one-dimensional heat flow equation is applicable when the heat source is large compared to the depth considered. This solution does not take care any variation in beam size structure, speed or substrate thickness. Still solution of this model is useful, Brien and Kear [1] developed graphs of cooling rates, thermal gradient, solidification rates expected from this model.

The governing equation for the conduction of heat during hardfacing assuming no heat loss due to convection or radiation is [2]

$$\left(\frac{\partial^2 T}{\partial x^2} + \frac{\partial^2 T}{\partial y^2} + \frac{\partial^2 T}{\partial z^2}\right) = \frac{1}{\alpha}\frac{\partial T}{\partial t} \tag{5.5}$$

For a stationary instantaneous point source of energy $Q\rho C$, the solution for the above equation is

$$T = \frac{Q}{8(\pi\alpha t)^{2/3}}\exp\left\{-\left[(x-x')^2 + (y-y')^2 + (z-z')^2\right]/4\alpha t\right\} \tag{5.6}$$

For a continuous point source, if heat is liberated at a rate of $\phi(t)\rho C$ from $t = 0$ to $t = t'$, then temperature distribution can be found by integrating the Eq. (5.5) over the time period as

$$T = \frac{1}{(8\pi\alpha)^{3/2}}\int \phi(t')\exp\left[\frac{-r^2}{4\alpha(t-t')}\right]\frac{\mathrm{d}t'}{(t-t')^{3/2}} \tag{5.7}$$

where $r^2 = (x-x')^2 + (y-y')^2 + (z-z')^2$.

When $\phi(t)$ is constant and equal to q

$$T = \frac{q}{4\pi\alpha r}\mathrm{erfc}\left[\frac{r}{(4\alpha t)^{1/2}}\right] \tag{5.8}$$

More often than not the heat source during welding operation is a finite source with Gaussian heat distribution. In that case for stationary source temperature distribution is obtained as

$$T_{\text{cont-gauss}} = \frac{2P(1 - r_f)D}{\pi D 2kT^{1/2}} \tan^{-1}\left[\frac{2(\alpha t)^{1/2}}{D}\right] \tag{5.9}$$

As the heat source moves during hardfacing at a velocity of v m/s the solution for temperature distribution becomes

$$T - T_0 = \frac{Q}{2\pi k} \mathrm{e}^{-vx/2\alpha} \frac{\mathrm{e}^{-vR/2\alpha}}{R} \tag{5.10}$$

The above equation appears to be most applicable analytical solution for hardfacing to find temperature distribution, cooling rate and heat affected zone, etc.

5.3 Hardfacing Processes

Most welding processes can be used for hardfacing. These processes can be grouped as torch processes, arc welding process and high energy beam processes.

5.3.1 Torch Process

Oxy fuel welding (OFW) and oxyacetylene welding (OAW) are the oldest and simplest torch hardfacing process which involves simply heating the surface with the flame until the surface is glassy and on the verge of melting and then melting the filler rod or powder to get the hardfacing to the melt. The filler rods or powders are heated until droplets from them wet the surface and form a continuous deposit. These processes are most widely used for cobalt, nickel and copper base hardfacing alloys. Although fuel gases other than acetylene can be used, most recommended gas is acetylene. These processes are one of the slowest processes but one which requires minimum alloy dilution. Since the process is slow, it is used for small jobs or where field welding torch is only available equipment. Usage of these processes requires welders with high skill. These processes have drawback that the base metal surface picks up carbon from the reducing flame.

5.3.2 Arc Welding Process

In the arc welding process, heat is generated by striking an arc between the electrode and the work piece [3]. Arc welding process can be grouped as consumable electrode process and non-consumable electrode process. In the consumable arc welding process, arc is maintained between a consumable electrode and the work piece. In shielded metal arc welding (SMAW), weld overlay is deposited by melting a consumable wire covered by flux in an arc. The flux forms a liquid slag upon melting and the gas protects the molten metal pool. The electrode is made of an alloy or coating materials that forms wear-resistant deposit. The main advantage of this process is ready availability of the equipments and consumables even in small quantities. A large variety of ferrous and non-ferrous materials can be deposited by this process. This process is very slow and hence small area can be covered. This process is also dependent on the manual skill of the operator.

In flux cored arc welding (FCAW) which is widely used since 1975, a metallic tubular electrode is filled with flux materials. The flux material volatilises and provides protective shielding to the weld pool. The coiled tubular electrode wire is fed through the gun and an arc is established between the wire and the substrate. Sometimes metal powder can also be added to the core materials to create an alloy weld deposit. Gas shielding can also be provided to reinforce shielding effect. This process can be used as semiautomatic or automatic process. This process is very fast. However, few alloys are available in the form of tube and these alloys are mostly ferrous. This process is mainly suitable for large area ferrous materials.

During gas metal arc welding (GMAW), the consumable wire electrode and the substrate are protected from the atmosphere by a gas fed axially with wire through welding gun nozzle. Deposition of the weld overlay can be accomplished by several methods. The wire can short against the wire and this process is called short-circuiting transfer or short arc. It is also possible to get the wire melt off without touching the work. This process is called globular transfer. Deposition can also be done by a process known as spray transfer which is preferred mode for out-of-position welding. This process is very fast, can be applied to large area and can be used to clad inside a large tank; a wide variety of materials can be deposited. Sometimes it is possible to carry out this process without the gas shielding known as open arc process.

Submerged arc welding (SAW) process is a fast process which uses consumable wire flux applied by ancillary equipment [4]. The arc is struck between the work and the filler wire. During welding, the arc is covered with granular flux fed through a hopper that follows the welding head. Using additives with the flux necessary alloys can be deposited. This process can be easily automated. It is a fast process. It is suitable for cladding a large area but it is used mostly for ferrous materials. This can be used for anything that can be positioned for flat welding process.

Among non-consumable welding process, gas tungsten arc welding (GTAW) and plasma arc welding (PAW) process are used extensively. Both processes involve a tungsten arc electrode and introduction of filler materials in the form of wire in case of GTAW and powder in case of PAW. In GTAW, arc is generated by striking a non-consumable W electrode with the work piece. Filler material is supplied by putting a bare rod manually in the arc or by using other devices which introduces bare wire in the arc. Any materials can be deposited by GTAW although the process is very slow especially in manual mode and only small area can be covered. This process produces a neat deposit which is an advantage on finished machine parts.

In PAW process, an inert gas is flowed through an electric arc in the welding torch. The filler material is melt by plasma flame for hardfacing. The temperature of the flame can be as high as 50,000 K giving the flame high penetrating capability. When the plasma flame is obtained by an arc between the work piece and the plasma gun, the process is called plasma transfer arc (PTA). In case the arc is maintained in the torch between the tungsten electrode and the water cooled nozzle, the process is called non-transferred arc plasma. In this process filled material is added by spool feeding of bare wire or powder into the arc zone. This process is very fast although the equipment cost is very high. Hardfacing by PAW requires minimum machining. There is no requirement of flux removal. Thin or thick layer can be deposited by this process. Low melting substrate can also be coated by this process. The PTA process exhibits enormous potential because the PTA overlays have a lower production cost and a higher productivity compared to thermal sprayed coatings, as well as easy operation and no need for any special surface treatment [5–7]. Furthermore, the PTA technique allows the production of high quality coatings (good metallurgical bonding and low level of porosity) consisting of metal matrix and carbide hard phases

5.3.3 High Energy Beam Process

Among high energy beam processes, laser hardfacing [8–10] and electron beam hardfacing are important [11–13]. Surface modification by laser is discussed in details in Chap. 7. Hardfacing by electron energy beam has been developed for many years and increasingly implemented for industrial applications [14–17]. In this process, electrons are generated by heating negatively charged filament (cathode) to its thermoionic emission temperature range and accelerating the electron by passing them through an electric field. The kinetic energies of these electrons are converted into heat which is used for depositing the surface layer on the substrate. Accurately controllable energy density, small beam size, accurate beam alignment, low heat input, pure environment are important features of electron beam surfacing which makes this process superior to many other hardfacing processes for several application. High cost of equipment, high running cost and several other shortcomings prevent extensive usage of this process. However, rapid development for the equipment of electron beam hardfacing, requirement of demanding quality of surfaces, economic benefit of the process lead to widespread use of this process in recent time.

5.3.4 Other Hardfacing Process

Friction surfacing is a recent hardfacing process [18] for depositing thick coating where the consumable is forced to move parallel to the surface and made to rotate simultaneously under applied load. As a result a hot plasticized layer typically 1–2 mm thick, depending on the consumable material, is formed. The temperature of the layer is nearly 40 °C below its melting point. A high contact stress is required to avoid the formation of oxide layer between the substrate and the consumable. Aluminium, titanium, mild steel, tool steel, stainless steel, satellite etc can be deposited on a wide range of materials employing this process [19–21].

Electroslag surfacing is another hardfacing process which is similar to SAW [22, 23]. This process uses equipment similar to SAW primarily for strip cladding. Slag from particulate flux shields the process. This process has high rate of deposition and this process can be applied on flat as well as curved surfaces. Low heat input, small and uniform penetration of the base materials, low dilution, extra low carbon in the deposited layer are the main characteristics of this process. The regular deposit profile obtained by this process reduces the machining cost and sometimes no machining is required. There is no ultraviolet radiation in this process. It is difficult to surface the internal surfaces of tubes, etc., employing this process. There is a limitation of the strip size and the thickness of the substrate.

Pulsed electrode surfacing (PES) and its variants pulsed air arc deposition processes are high energy density microwelding processes [24–26]. These processes use short duration high current electrical pulses via discharge capacitance. There is very little distortion or metallurgical changes of the substrate materials. The surface layer is rapidly solidified forming a fine grain structure. The equipment is compact, simple and portable. It can be operated manually or can be integrated to machine tools. It is mainly used for treatment of cutting tools.

5.4 Hardfacing Materials

From metallurgical point of view, hardfacing materials can be classified in to several types [27, 28] although many other points should be considered during selection of hardfacing materials [29]

- Iron-based alloys
- Cobalt-based alloys
- Nickel-based alloys
- Copper-based alloys
- Carbides
- Composite materials
- Materials deposited by friction surfacing

5.4.1 Iron-Based Materials

Iron-based hardfacing materials constitute the bulk of the hardfacing materials [30, 31]. Due to low cost and moderate wear resistance, these materials are ideal for large area coating of crushing, grinding and earth moving equipment. They are often used for restoration of surfaces. Following types of iron-based hardfacing materials are available for deposition of coatings.

- Pearlitic steel
- Austenitic steel
- Martensitic steel
- Cast iron

5.4.1.1 Pearlitic Steel

Pearlitic steel variety of hardfacing materials is actually low carbon weldable steel containing limited alloy addition. Normally submerged arc welding is used for depositing pearlitic steel primarily to surface the components on routine basis. The microstructure and mechanical properties of the weld deposit are controlled by controlling composition of the feeding wire and flux. In general, a finer microstructure giving better wire resistance is preferred [32]. Sometimes higher addition of alloying element can be made. But such situation requires faster cooling rate in order to get desired microstructure. However, it is also necessary to preheat the substrate in order to avoid cracking due to generation of thermal stress during cooling.

5.4.1.2 Austenitic Steel

Austenitic steel, Hadfield manganese steel is the most widely used austenitic variety of hardfacing materials. This steel contains 12–16 % Mn and up to 1 % C. The steel can retain completely austenitic microstructure on cooling, but this microstructure is metastable. This material typified EFeMn–C is extremely tough, wear and shock resistance. It is sensitive to plastic deformation. During abrasion, as the material deforms, austenite transforms to martensite due to plasticity induced transformation [33, 34]. This increases the work hardening rate of the steel and this in turn increases the wear resistance. This material with high work hardening capability and moderate yield strength is capable of responding plastically to abrasion and impact loading. This plasticity helps in dissipating energy, and in the process cracking and spalling of the coating is avoided [35].

However, there are some inherent problems associated with this material. Annealing or slow heating followed by cooling results in embrittlement of the material due to precipitation of carbides in the grain boundary. In order to avoid carbide precipitation, high interpass temperature is avoided. Another way of avoiding this embrittlement is to reduce impurities such as Sb, Sn, P, As, N, etc. High purity steel does not undergo embrittlement. This embrittlement can further be reduced by fast cooling through the critical embrittlement temperature. Si is also added to avoid embrittlement problem. During hardfacing, if the deposited layer is diluted highly, it results in cracking as the diluted layer may not be stable austenite resulting in peeling of the layer. High dilution is avoided by selecting highly alloyed steel. Highly alloyed composition allowed the austenite to be stable even with high dilution. Austenitic manganese steel is not corrosion resistance. Because of embrittlement problem, they are not used at elevated temperature. These steels are difficult to machine. Hence they should preferably be used in as-deposited condition.

5.4.1.3 Martensitic Steel

The composition of martensitic steel is so chosen that the microstructure becomes martensitic after air cooling. Martensitic surface has generally less impact resistance compared to pearlitic or austenitic structure although it has higher hardness and abrasive wear resistance. However, the toughness of martensitic structure can be improved by appropriate tempering treatment. Martensitic coating is deposited by putting two layers in order to avoid dilution. If machining is required, three layers are deposited. Martensitic layers are susceptible to cracking due to thermal stress as it is cooled in air relatively fast to avoid austenitic formation. To avoid such cracking preheating of the substrate is carried out. Care should be taken during preheating. If heated beyond some temperature, the microstructure may not remain fully martensitic. Part of it may be transformed to austenitic.

When protection is required against wear in corrosive environment, martensitic stainless steel is used as protective layer. A typical composition for such layer is 0.3 (wt%) C, 12 % Cr, 0.8 % Mo, 0.6 % Mn, 0.6 % Ni and 0.5 % Si. This steel is austenitic at elevated temperature and transforms completely to martensite on slow cooling. Before application, this steel is usually tempered to 773–873 K. A low carbon variety of this steel which contains higher amount of Cr in addition to Mo, V and W, could improve the life of continuous caster rolls by as high as 50 % [36]. If the application is concerned with wear protection at elevated temperature, martensitic variety of tool steels is used. The coated components are cooled to ambient condition to ensure martensitic structure and subsequently tempered to 923 K. During tempering, various alloyed carbides precipitate resulting in increase in hardness. This phenomenon is known as secondary hardening and this helps retaining the strength at elevated temperature.

Six different hardfacing alloys produced as flux-cored wires on iron basis were selected and welded onto 1.0038 mild steel plates. The chemical compositions of these alloys are listed in Table 5.1 [37]. Table 5.2 shows the welding parameters, which are optimised related to the welding behaviour and the composition of the different flux cored wires. The welding was carried out in flat position in two layers. Typical microstructures of the welded deposits are shown in Fig. 5.1 [37]. Alloy A is mainly martensitic with some islands of austenite. Fine primary Niobium carbides are well distributed throughout the micro-section. The hardness of the martensite is about 800 HV0.1. The content of Nb carbides is given in Table 5.3 to 7 % at a size of <3 μm in a spattered shape. Alloy B consists of primary Fe/Cr

Table 5.1 Chemical composition of Fe base alloys

	Chemical composition					
Alloys	Fe	C	Cr	Nb	B	Others
A	Bal	<1.0	6.0	3.0	–	1.5
B	Bal	5.5	21.0	7.0	–	1.0
C	Bal	2.5	7.0	–	<1.0	<1.0
D	Bal	<1.0	–	–	4.0	1.5
E	Bal	1.3	15.4	4.2	4.2	11.5
F	Bal	–	<1.0	–	–	1.8

Table 5.2 Welding parameter of Fe based alloys

Alloys	Current (A)	Voltage (V)	Interpass temperature (°C)	Wire speed (mm/s)
A	175	22.9	168	4.4
B	247	27.5	197	8.4
C	246	21.7	205	5.4
D	188	23.3	182	4.7
E	244	23.6	175	9.3
F	113	21.0	151	3.0

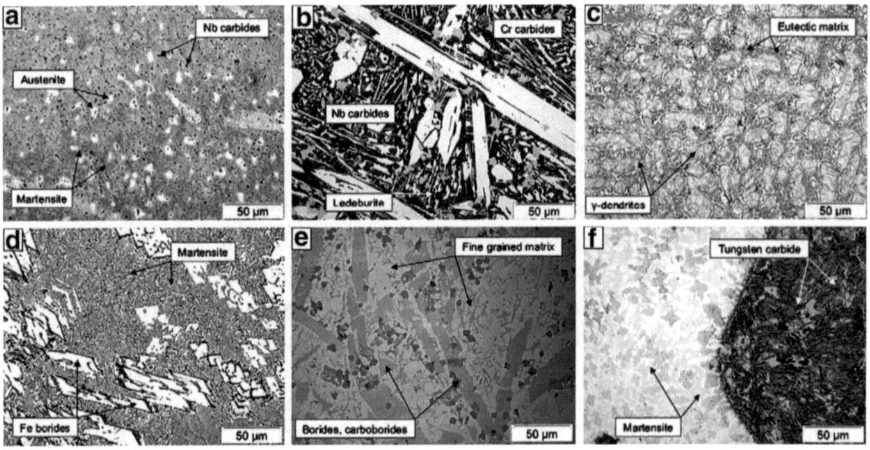

Fig. 5.1 SEM images showing the microstructures of various iron-based alloys [37]

carbides with a micro-hardness of roughly 1,600 HV0.1 in a ledeburitic matrix. The content of Fe/Cr carbides is listed in Table 5.3 to 57.1 % at a size of 30–200 μm. The chemistry of the Fe/Cr carbides is reported for hypereutectic FeCrC alloys in literature to M_7C_3 structure [38–41]. The hardness values of the ledeburitic matrix determined to be about 800 HV0.1, are close to previous investigations from Fischer [42] and Buytoz [43]. Besides, small and evenly distributed primary Nb carbides (light grey in Fig. 5.1b) at a volume content of approximately 5 % can be detected. These are supposed to be of major ingredient for increasing the resistance against erosion and abrasion due to their high hardness. Alloy C has solidified in

Table 5.3 Details of the hard phases (HP) of the Fe-based alloys

Alloys	Type of HP	Content of HP (vol.%)	Size of HP (μm)	Shape of HP
A	Nb carbides	7.0	<3.0	Spattered
B	Fe/Cr carbides	57.1	30–200	Columnar
	Nb carbides	5.4	<7.0	Granular
C	Fe/Cr carbides	48.3	–	Closed skeleton
D	Fe carbo-borides	18.5	20–80	Columnar
E	Fe/Cr carbo-borides	52.0	10–100	Columnar
	Nb carbides	4.6	5–10	Blocky
	Mo/W carbo-borides	5.3	10–25	Blocky
F	Synthetic WC	38.7	65–250	Rounded

hypoeutectic γ-dendrites with about 920 HV0.1, which are embedded in an eutectic matrix of about 1,000 HV0.1 (see Fig. 5.1c). A closed net or skeleton of brittle Fe/Cr carbides at a volume content of 48.3 % (see Table 5.3), is clearly surrounding the primary dendrites. This appearance comes close to a solidification of the finally solidifying eutectic melt described as N-type [42]. Alloy D is built up of Fe carbo-borides in columnar structure with a hardness of about 1,500 HV0.1 in a hard eutectic matrix of about 1,000 HV0.1 (see Fig. 5.1d). The distribution of hard phases in this alloy is quite uniform. Volume content and size of the Fe carbo-borides are listed in Table 5.3 to 18.5 % and 20–80 μm, respectively. The complex alloy E which contains an amount of boron similar to alloy D, but a much higher level of other elements like W, Mo, Nb and Cr and shows a dense and uniform distribution of very hard complex carbides and carbo-borides (see Fig. 5.3) with hardness values between 1,200 and 1,900 HV0.1. Type and content of the hard phases were determined to be Fe/Cr carbo-borides at a volume content of 52 % at a size of 10–100 μm, Nb carbides and Mo/W carbo-borides at a volume content of approximately 5 % in blocky shape. In ref. [44] hard phases of a very similar alloy are described as $M_{23}(BC)_6$ and $M_7(CB)_3$ carbo-borides phases in a matrix exhibiting high fracture toughness up to 73.3 MPa $m^{1/2}$ due to an effective distribution of fine carbide and boride phases in ductile dendrites/cells. The synthetic multiphase alloy F shows the original fused and crushed tungsten carbides (2,500–2,700 HV0.1) which are extensively dissolved in the Fe-based matrix and lead to well-distributed reprecipitated carbides with a decreased hardness of 1,200–1,600 HV0.1. Content and size of the synthetically added tungsten carbides is determined to be 38.7 % and 65–250 μm, respectively. The matrix has a hardness between 800 and 1,100 HV0.1. Higher welding amperage increases the rate of tungsten carbide dissolution. Overall the tungsten carbides are irregularly distributed. There is a higher density of original carbides close to the fusion line, whereas at the surface only rests of carbides are visible.

Wang et al. [45] produced a Fe-based alloy hardfacing coating reinforced by TiC particles. TiC particles were formed by metallurgical reaction of ferrotitanium (Fe–Ti) and graphite during the arc welding rather than the TiC particles being directly added into the welding pool. Figure 5.2 exhibits TEM image of lath martensite morphology of Fe-based hardfaced layer. TEM observation indicates

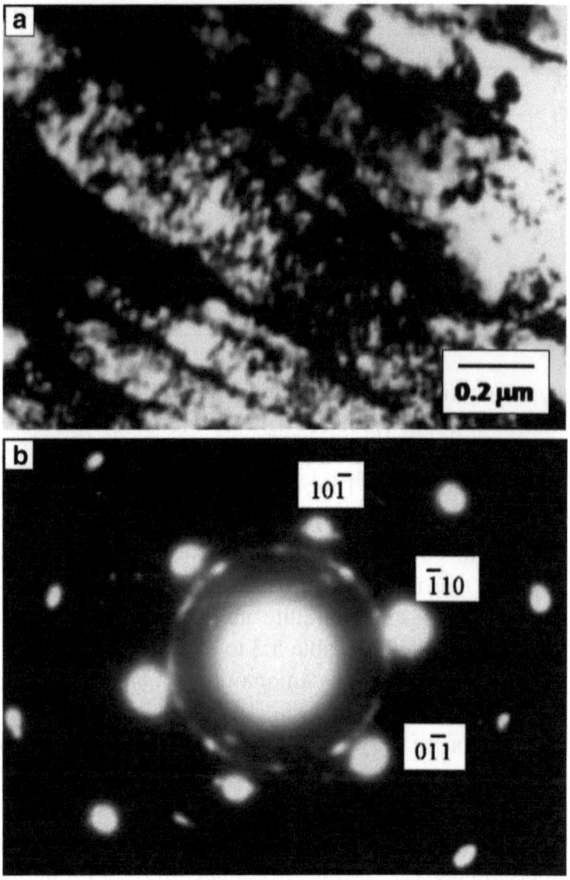

Fig. 5.2 The TEM micrograph showing martensite lath and the corresponding diffraction pattern of an iron-based hardfacing alloy reinforced with TiC [45]

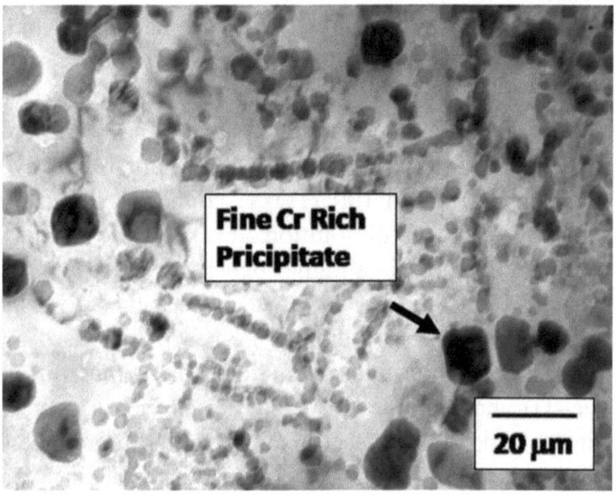

Fig. 5.3 TEM image shows the microstructure of Ni base alloy Colmonoy 5 [48]

that there is considerable dislocation in the subcrystal structure inside the lath martensite, but no twin martensite can be seen. The low carbon lath martensite has the features of auto-tempering. When low carbon lath martensite is formed, the microstructure of the hardfacing coating has a higher strength and the toughness is still largely maintained.

5.4.1.4 Cast Irons

Cast irons are essentially high carbon (usually more than 2 % C) containing iron. There are several varieties of cast irons available and these are obtained by changing the cooling rate, heat treatment and additives during cooling. Among these cast irons, white cast iron is used against wear protection. One variety of white cast iron which contains 12 % Cr is popular as hardfacing cast iron. Depending on the amount of carbon and Cr present, the hardfaced layer can be classified as hypoeutectic (2–3 % C, 5–29 % Cr), near eutectic (3–4 % C, 12–29 % Cr) and hypereutectic (4–7 % C, 15–36 % Cr) alloys. In general, hypereutectic alloy has higher abrasion resistance and lower impact resistance than hypoeutectic alloy. This high wear resistance is due to presence of high volume fraction of hard high chromium carbide [46]. These alloys contain various types of carbides mainly M_7C_3 types although other variety of carbides such as M_6C and M_3C are also available. The sizes of primary carbides are between 50 and 100 μm whereas the sizes of eutectic carbides are generally less than 10 μm [47]. Matrix surrounding these carbides can be pearlitic, austenitic or martensitic. Martensitic matrix is good for high stress abrasion resistance whereas austenitic matrix is preferred if wear resistance along with high fracture toughness is required [27]. Sometimes Ti, Nb and Mo are added to produce fine dispersed hard carbides. Such fine dispersion enhances the abrasion resistance along with toughness. This also leads to improvement of low stress and high stress abrasion resistance [38].

These alloys can be deposited by GTAW, FCAW, SAW and SMAW, etc. these alloys are one of the lowest cost hardfacing materials. Their overall wear performance is excellent and they can be used under variety of tribological degradation conditions. They have good availability and good applicability.

5.4.2 Nickel Base Alloys

These families of hardfacing alloys are used primarily for wear-resistant application. Boron is the main alloying element in these alloys. These alloys cannot be hardened by heat treatment. Further, these alloys are generally used in as-deposited condition. For similar hardness range, these alloys are more wear resistant than iron-based alloys. Physical and corrosion characteristics of these alloys are similar to Ni base alloys. These alloys are not ferromagnetic. They can retain their hardness up to 800 K. They cannot be formed at ambient condition and can be considered to be brittle. These materials can be applied very easily and they can wet ferrous-based materials and other substrate better than most nonferrous hardfacing alloys. They can be machined more easily than other hardfacing materials with comparable hardness. These alloys can be classified under three heads.

- Boride-containing alloys
- Carbide-containing alloys
- Laves phase-containing alloys

Boride-containing alloys are available as bare cast rod, tubular wires and powders. The microstructure of this coating consists of borides and nickel borides in Ni-rich matrix. Nickel boride (Ni_3B) is the main hard phase at low chromium content. With increase in chromium content, nickel boride is gradually replaced by chromium boride. At low chromium content, chromium boride is CrB and at higher chromium content it is Cr_5B_3. Si is an important matrix element which functions mainly as self-fluxing element and remains as an intermetallic compound Ni_3Si. The matrix also contains complex carbides such as $M_{23}C_6$ and M_7C_3. These materials have least corrosion resistance compared to all non-ferrous hardfacing materials. They, however, have good high temperature hardness. A typical TEM bright field image of Ni base-hardfaced alloy Colmonoy 5 is shown in Fig. 5.3 [48]. The TEM bright field images of the coating showed needle (5–20 μm), fine spherical (20–200 nm), cuboidal (50–200 nm), blocky (2–3 μm) and floret (15 μm)-type precipitates TEM-EDS analysis confirms that the needles, floret blocky and fine spherical precipitates are basically rich in Cr with Ni and Fe as minor elements. Row of fine spherical precipitates of about 100 nm in size were observed (of about 1 μm length) as seen in Fig. 5.3. It has been reported that the Colmonoy 5 deposit can withstand its as-deposited hardness to the large range of temperatures [49]. These precipitates are beneficial to withstand high temperature hardness of the coating.

Carbide-containing alloys have excellent resistance against abrasion, galling, impact and corrosion. In these respects, these alloys are better than boride-containing Ni base materials. These alloys also contain complex carbides such as M_7C_3 and M_6C. Co-based hardfacing alloys were found to be sources for radioactive co isotopes [50], and hence carbide-containing nickel base alloys were developed.

Laves phases are essentially intermetallic compounds with general formula AB_2. Although these phases improve the wear resistance, they generally bring down the toughness. Laves phase-containing Ni base alloys have less wear resistance than laves phase-containing Co base alloys. These laves phases are NiMoSi and Ni_3Mo_2Si. Generally laves phases constitute around 50 % of volume of the hardfacing layer. The structures of laves phases are hexagonal and dihexagonal [51, 52]. These alloys contain Cr for oxidation resistance. These alloys have good high temperature corrosion, oxidation and hot corrosion resistance. It has good metal to metal wear resistance especially at elevated temperature [53, 54]. They can be easily deposited by GTAW and PTA process. Since these alloys are devoid of Co, these alloys find application in valve and valve seat of nuclear reactor.

5.4.3 Cobalt Base Alloys

Cobalt base hardfacing alloys are well established for a considerable time. The main advantage of these alloys is there excellent corrosion and oxidation resistance in addition to wear resistance and that is why they are successful in replacing iron-based alloys. These alloys are however more expensive than other variety of alloys. The first Co-based hardfacing alloy was developed as Stellite.

These alloys are also known as carbide-containing alloys. Subsequently other families of alloys were developed and they are known as Triboaloy. These alloys can be classified as laves phase-containing alloys. These alloys cannot be hardened by heat treatment and they do not respond to annealing. Co base hardfacing alloys are available as bare rod for GTAW and OAW deposition, as coated electrode for SMAW and as wire for GMAW, FCAW, SAW and PTAW.

Stellite alloys contain varying amount of carbon and tungsten or carbon and molybdenum to change hardness and other properties. The matrix is essentially cobalt strengthened by chromium and tungsten or cobalt strengthened by chromium and molybdenum. The wear resistance is governed by formation of carbide, their volume fraction and size and distribution. High chromium content leads primarily to chromium carbide formation. However, volume fraction, size, shape and distribution of carbides are controlled by the composition of the alloy and their deposition technique. Depending on the amount of carbon present, these alloys can be divided as hypoeutectic, eutectic and hypereutectic alloys. In hypoeutectic alloys, especially low carbon content variety, ductile chromium carbide in the form $M_{23}C_6$ is normally present. In some of the alloys, a network of eutectic Co–M_7C_3 is present. These alloys have limited ductility but exhibit improved wear resistance. The size, nature and distribution of the carbides can powerfully be influenced by addition of molybdenum [55]. Addition of molybdenum results in formation of two types of carbides, M_6C carbide primarily at the Co-rich dendrite interface and eutectic chromium-rich M_7C_3 and $M_{23}C_6$ in the interdendritic region. Further, molybdenum addition decreases the arm spacing of cobalt-rich dendrite arm spacing and the size of the chromium-rich carbides. Molybdenum addition also helps in improving the hot hardness of the layer. Figure 5.4a–d [55] shows optical micrographs of the etched cross section of the Mo-free Stellite 6 and Mo-modified Stellite 6 hardfacing alloys deposited by the PTA. The first phase to form during cooling from the liquid state of the Stellite 6 alloy produced by PTA process is the primary Co-rich dendrite and the remaining liquid eventually solidifies by a eutectic reaction into an interdendritic, intimate lamellar mixture of Co-rich phase and Cr-rich carbides.

It is mentioned earlier that laves phases are intermetallic compound. In case of cobalt base alloys, these phases are generally CoMoSi or Co_3Mo_2Si. Wear behaviour is primarily governed by volume fraction of laves phases. These alloys have superior high temperature strength than carbide-containing alloys. Accordingly these alloys exhibit superior high temperature tribological properties. These alloys are not machinable.

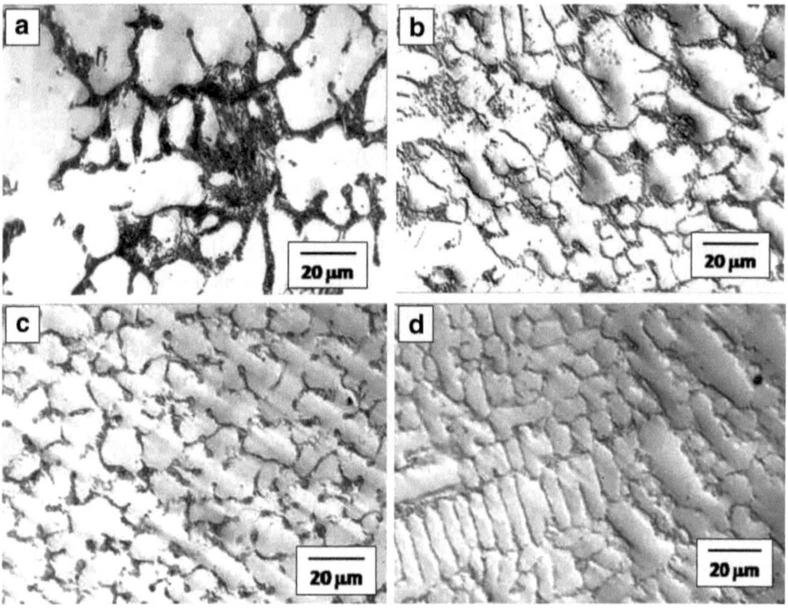

Fig. 5.4 Optical micrographs of cobalt base hardfacing alloys Stellite 6 containing increasing amount of Mo. (**a**) Mo free, (**b**) 1.5 % Mo, (**c**) 3 % Mo and (**d**) 6 % Mo [55]

5.4.4 Copper Base Alloys

Copper base alloys as various bearing materials are used in the industry for quite some time. Although these materials are well known as tribological materials, they are good for specific tribological condition. They are especially poor against abrasion. Copper base alloys can be classified under four categories as given below.

- Brasses, Cu–Zn alloys
- Aluminium bronzes, Cu–Al–Fe alloys
- Phosphor bronze, Cu–Sn–P alloys
- Silicon bronze, Cu–Sn–Si alloys

Brasses are single phase Cu–Zn solid solution alloy. This alloy cannot be hardened by heat treatment. This alloy is utilised in metal to metal contact wear applications. On the contrary, aluminium bronze can be hardened by forming martensite although hardfaced aluminium bronze is normally used as deposited condition. Other two varieties of alloys, namely phosphor bronze and silicon bronze cannot be heat treated. None of these alloys can be used for temperature in excess of 473 K. These alloys are available in the form of rod, small diameter wire and coated electrode and these alloys can be deposited by almost all available welding process for hardfacing. These alloys can be applied to ferrous, nickel and copper substrates.

5.4.5 Composite Materials

This category material contains carbides, nitrides or borides of tungsten, titanium, vanadium or chromium in very high proportion. Although tribological properties such as abrasion resistance or erosion resistance are found to improve by order of magnitude [56, 57], hardfaced layers produced by such materials suffer from the propensity of thermal fatigue and crack initiation at the interface of incoherent carbide and poor impact resistance. In general, wear performances improve with increasing volume fraction of hard phases. However, a detrimental effect of increasing the proportion of hard phase in the layer is decrease in fracture toughness at high volume fraction of carbides [58, 59]. Another limitation on the use of carbide-reinforced metal matrix composites has been dissolution of carbide in the matrix phase during deposition [60]. Intermetallic compounds such as Ni_2W_4C formed by dissolution of WC were identified as a potential site for subsequent crack initiation [61]. During deposition of these materials especially WC-containing materials, care should be taken to prevent excess dissolution of WC. If dissolved in Co base or Ni base alloys, these carbides can reprecipitate as W_2C or η phase. W_2C and η phases are very brittle and they have inferior wear and corrosion resistance [62]. During processing of these materials, care should be taken to avoid excess welding current. If the excess welding current is employed during deposition, some of the wear resistance carbides may be dissolved. Some of the important varieties of composite materials are Stelcar® 1215 Stelcar® 60, Super Stelcar®, etc. In recent years, a new type of hardfacing alloy with a very fine dispersion of carbide phases has been introduced in order to increase both fracture toughness and abrasion resistance [63, 64]. This range of hardfacing alloys has been described as nanostructured materials [64] and complex high alloy type of hardfacing deposits [64]. The high level of alloying additions (Nb, Cr, W and Mo) has enabled a significant undercooling of the melt prior to nucleation and solidification, resulting in a very strong refinement of the carbide phases and matrix. In addition, the alloying elements have acted to enhance the hardness of the carbide phases. The structure of these highly alloyed deposits has been identified as containing complex type precipitates in a matrix of either α-Fe or γ-Fe [64].

The XRD patterns of the Cr_3C_2 and Cr_3C_2–Ni reinforced alloys and NiCrBSi matrix alloy are shown in Fig. 5.5 [65]. XRD analysis of the original NiCrBSi matrix alloy indicated a high amount of $FeNi_3$ solid solution in Ni-based matrix with presence of small amount of Ni_3B and Cr_7C_3 hard phases (Fig. 5.3a). When 40 vol.% Cr_3C_2 particles were added to the NiCrBSi alloy, Cr_3C_2, Cr_7C_3 and $M_{23}C_6$ phases could be detected by XRD analysis in the hard phased layer as shown in Fig. 5.3b. Intensive dissolution of carbides results in an increased content of free Cr and C in the liquid alloy, which influences the matrix composition and leads to the formation of an austenite $Ni_{2.9}Cr_{0.7}Fe_{0.36}$ phase. The complex phase Cr_7C_3 is formed in the alloy as reprecipitated carbides during the rapid solidification process. It should be outlined that XRD peaks for many chromium carbides and borides are very closely placed and with significant alloying in these phases, confirmation of

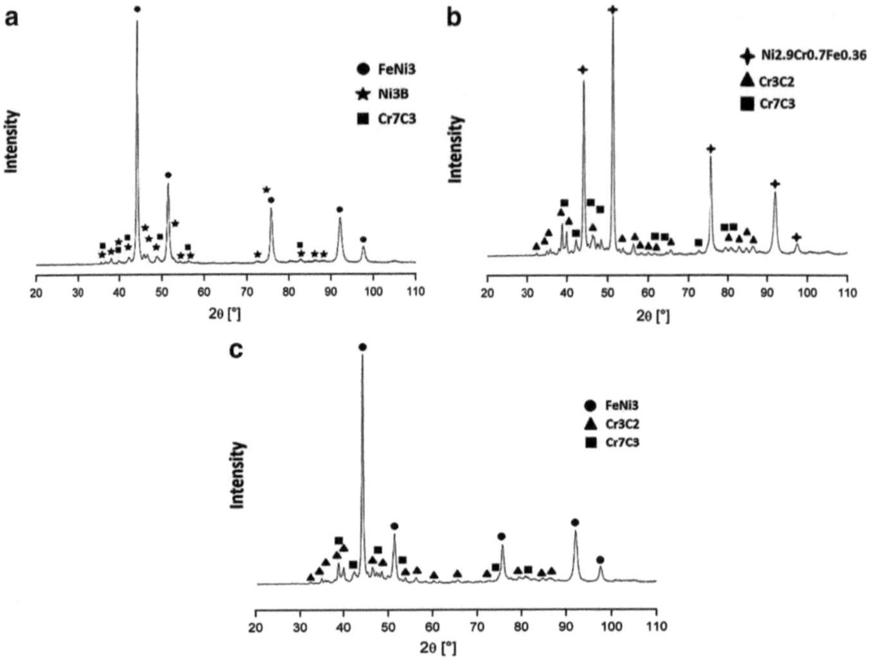

Fig. 5.5 The XRD patterns of the Cr_3C_2 and Cr_3C_2–Ni-reinforced alloys and NiCrBSi matrix alloy [65]

the type of carbides from XRD alone is difficult. Therefore the presence of $M_{23}C_6$ phase is very doubtful. Almost all peaks of $M_{23}C_6$ match with Cr_7C_3 and Cr_3C_2 peaks. Furthermore, $M_{23}C_6$ phase is usually not formed by high free carbon levels and the microstructural analysis hasn't revealed any $M_{23}C_6$ carbides in the hardfacing matrix. Most probably those peaks correspond to some mixed Cr–B–Ni phase presented as precipitations in the matrix. Figure 5.5c presents the XRD analysis of the NiCrBSi alloy with addition of 40 vol.% Cr_3C_2–Ni powder. In this case, due to the Ni binder in the cermet particles, the dissolution of chromium carbides is negligible and the matrix phase was identified as $FeNi_3$ with a high content of Cr_3C_2 and Cr_7C_3 hard phases. The presence of Cr_7C_3 can be explained by the structure of the precursor Cr_3C_2–Ni bulk cermet, where a certain amount of Cr_7C_3 was formed during sintering.

Figure 5.6 illustrates the typical cross-sectional OM micrographs of composite coatings. The reference material (NiCrBSi self-fluxing matrix alloy) exhibits a nickel-rich dendritic matrix containing borides and carbides (Fig. 5.4a) that is a typical phase composition for alloys with low chromium content. According to phase distribution analysis, the content of Cr_7C_3 phase is 2 ± 1 wt%, whereas the content of Ni_3B phase is 19 ± 1 wt%. Figure 5.6b shows the microstructure of the Cr_3C_2 (Sulzer Metco 70C-NS) reinforced hardfacing layer. The micrograph reveals the presence of a high amount of re-precipitated spine-like carbides (Cr_7C_3)

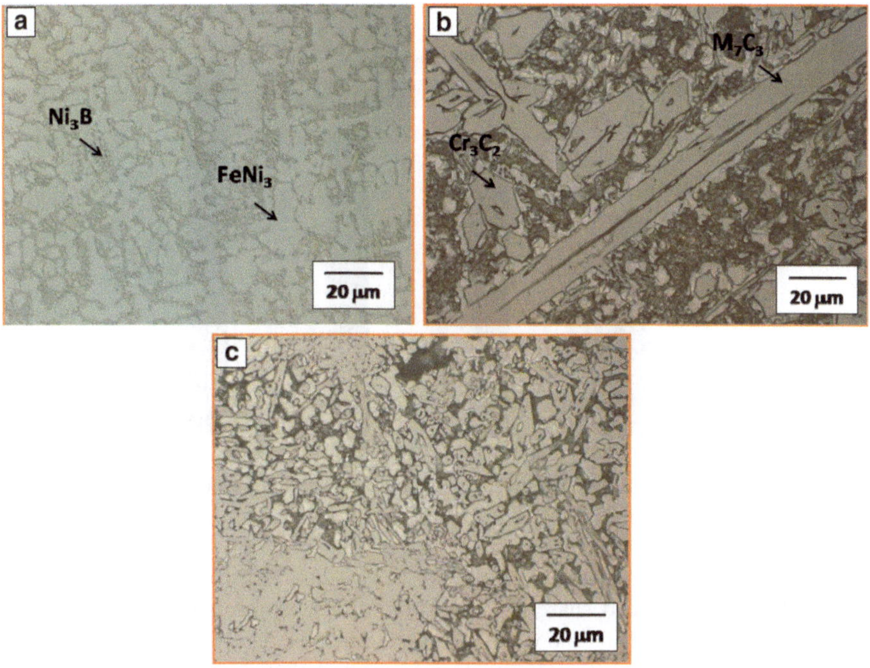

Fig. 5.6 OM cross-sectional images of composite materials: (**a**) NiCrBSi matrix; (**b**) Cr_3C_2 reinforced coating; (**c**) Cr_3C_2–Ni reinforced coating [65]

resulting from the extensive carbide dissolution in the matrix. Figure 5.6c illustrates the microstructure of the coating produced from cermet powder in combination with NiCrBSi matrix. The microstructure of this coating can be characterised by three apparent phases: hard phases-rich hypereutectic matrix material, some amount of dissolved and re-precipitated Cr_7C_3 carbides and a certain content of particles of the initial cermet particles homogeneously distributed throughout the matrix. It is assumed that the nickel binder in the cermet particles protects the primary carbides from dissolution and helps keeping the particles in the original composition.

5.5 Tribology of Hardfaced Surfaces

5.5.1 Sliding Wear

Sliding wear behaviour of hardfaced surfaces is studied extensively. Comparative wear tests on a regular share and three kinds of hardfacing with electrodes were conducted by Bayhan [66]. The influence of different hardfacing electrodes on the

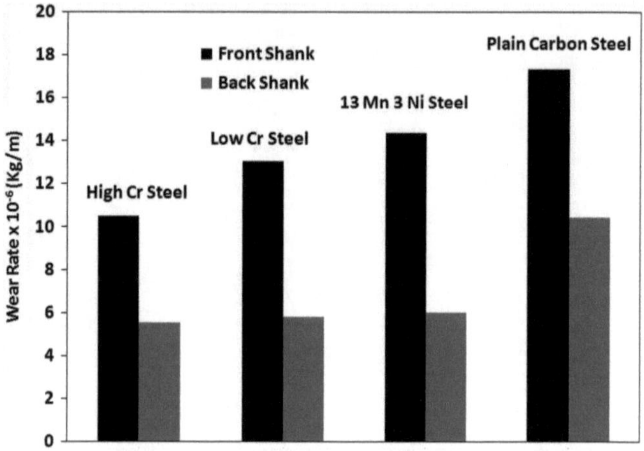

Fig. 5.7 The influence of different hardfacing electrodes on the wear rate of the share cutting edge of a chisel plough [66]

wear rate of the share cutting edge of a chisel plough is depicted in Fig. 5.7. It was found that the hardfaced share had a positive effect, as the hardness of the share cutting edge increased and the wear decreased. Su and Chen [67] deposited Ni base hardfacing layer on AISI 1045 steel and reported that addition of Nb reduced friction and wear loss. Addition of Mo and C resulted in formation of carbides with excellent wear resistance.

In a study by Wang et al. [68, 69], different hardfacing layers were produced by shield manual arc welding (SMAW) process in which a bare electrode of H08A was coated with fluxes, to which different measures of ferrotitanium (Fe–Ti), ferrovanadium (Fe–V), ferromolybdenum (Fe–Mo) and graphite had been added. The influence of added alloy elements on the microstructure and wear properties of the Fe-based hardfacing layers was investigated. The results showed that complex carbides of TiC–VC–Mo_2C were synthesised via metallurgical reaction during hardfacing. Carbides are uniformly dispersed in the matrix. The addition of graphite and ferromolybdenum can enhance macro-hardness and wear resistance of the hardfacing layer significantly. However, it increases the crack sensitivity of the hardfacing layer. With the increasing of the additions of Fe–Ti and Fe–V, the macro-hardness and wear resistance of the hardfacing layers are increased. A good resistance against cracking and wear resistance of the hardfacing layer could be obtained, when the amounts of graphite, Fe–Ti, Fe–V and Fe–Mo were controlled within a range of 8–10 %, 12–15 %, 10–12 % and 2–4 %, respectively as shown in Fig. 5.8 [68]. The deposited layer surfaced by Fe–Ti–V–Mo–C hardfacing alloy possesses a higher wear resistance and less friction coefficient than that of the deposited layer surfaced by EDRCrMoWV-A3-15 hardfacing alloy.

Wang et al. [68] also recorded the features of worn surface of hardfacing layer containing increasing amount of Mo. Figure 5.9 [69] shows the SEM micrographs of the worn surface of the specimens containing increasing amount of Mo after

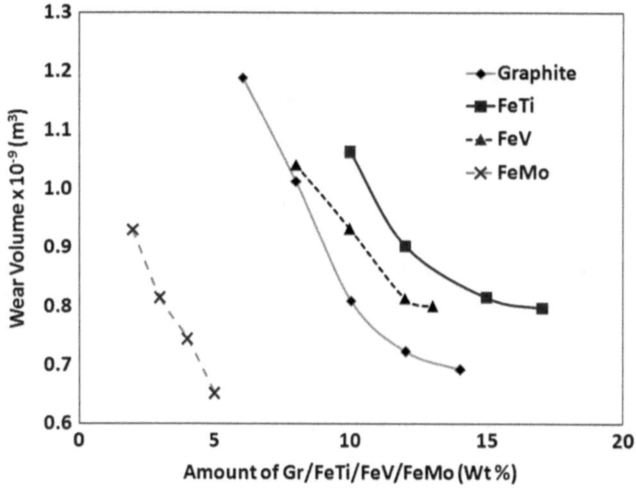

Fig. 5.8 Wear rate of the iron-based hardfacing layer when the amounts of graphite, Fe–Ti, Fe–V and Fe–Mo in the flux were controlled within a range of 8–10 %, 12–15 %, 10–12 % and 2–4 % [68]

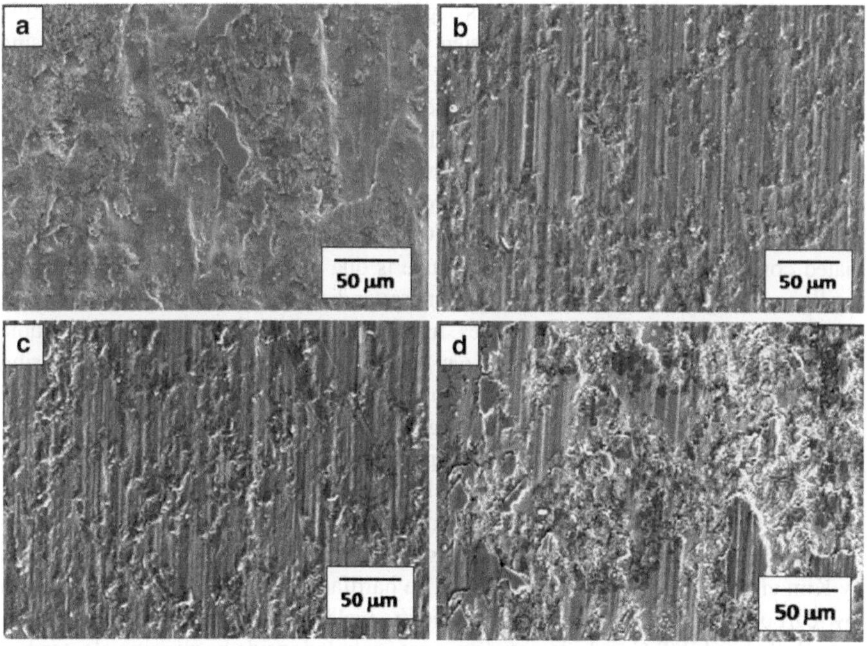

Fig. 5.9 SEM micrographs of the worn surface of the specimens containing increasing amount of Mo after wear test in air at normal load of 98 N and sliding distance 1,260 m [69]

wear test in air at normal load of 98 N and sliding distance of 1,260 m. Compared with the specimens containing higher amount of Mo (Fig. 5.9b–d), the worn surface of the Mo-free hardfacing layer (Fig. 5.9a) shows serious plastic deformation and deep groove. This results from the lower hardness and less volume fraction of carbide in the Mo-free specimen. However, for specimens containing Mo, the hardfacing layers exhibit relatively smooth worn surface, and there is no indication of brittle failure or loose debris formation of carbides. It indicates that hardfacing layers with Mo alloy possess a much higher resistance to plastic deformation and removal of the edges of grooves. In addition, it is found that some void formation at carbide/matrix interfaces and the fall-off of carbides in the specimen containing highest amount of Mo. Compared to wear scar of specimen containing highest amount of Mo, fall-offs of carbide of specimens containing lower amount of Mo are lower. It reveals that specimens with intermediate amount of Mo offer the best overall properties of wear and crack resistance.

Figure 5.10 shows the wear damage of a series of Co base and Ni base alloys deposited by gas tungsten alloy welding (Fig. 5.10a) and a series of cobalt base alloy Triboaloy deposited by plasma transfer arc welding (Fig. 5.10b) rubbed against a series of counter-body materials [28]. Wear rate is low when rubbed against Hastelloy C-276. It is generally high when the counter-body is 304 SS. The lowest wear rate is exhibited by ERCoCr-C alloy against Hastelloy although it has the highest wear rate against 304 SS. Wear rates of the alloys deposited by plasma transfer arc are generally lower than that deposited by gas tungsten welding. These data are obtained by using pin on block rig at an applied load of 2,722 kgf. Triboaloy T-400 has excellent galling resistance and it is particularly suitable when lubrication is not possible. In fact, under self-mated condition Triboaloy T-400 exhibits superior wear resistance than satellite 6. Triboaloy T-800 contains high amount of Cr and hence suitable for high temperature wear resistance. The wear properties of a series of satellite base alloys have been investigated by Foroulis et al. [70]. An excellent property especially in self-mated condition is noted. Such performance is due to low stacking fault energy of satellite coatings. The low stacking fault energy indicates less tendency for cross slip and this tendency, in turn, facilitates transformation from body cantered cubic structure to hexagonal close packed (h.c.p.) [71] structure during deformation. During sliding, these hcp structures reorient themselves to have basal planes parallel to the sliding surfaces [72]. Basel plane sliding reduces the friction force to minimum. Further, there is significant strain hardening during deformation. Thus satellite creates a layer which combines low friction with high load bearing capacity.

A correlation was made among microstructure, high-temperature wear resistance and surface roughness in hardfacing alloys reinforced with complex carbides by Kim et al. [73] and Choo et al. [74]. The hardfacing alloys were deposited on a low-carbon steel substrate by a submerged arc welding method. Different fractions of FeWTiC and WTiC carbide powders were included inside hardfacing electrodes. Microstructural analysis indicated that cuboidal and rod-type complex carbides were homogeneously distributed in the bainitic matrix. As volume fraction of these complex carbides increased, hardness and wear resistance increased as shown in Fig. 5.11 [73]. The alloy reinforced with FeWTiC carbides contained

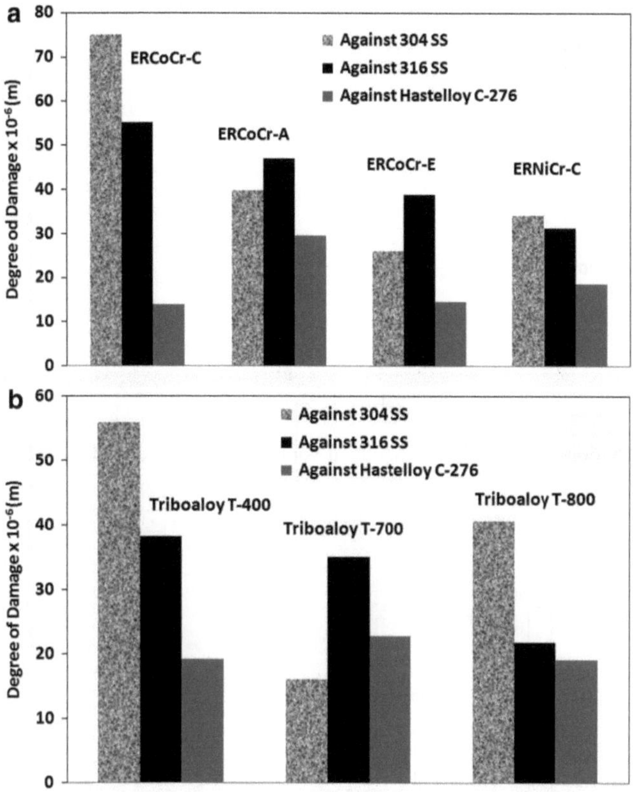

Fig. 5.10 The degree of damage of various (**a**) Co base alloys and (**b**) Triboaloy sliding against various counter bodies [28]

more complex carbides in the harder matrix than those reinforced with WTiC carbides because of efficient melting and solidification during hardfacing, and thus showed the best wear resistance and excellent surface roughness. Hardness, wear resistance and surface roughness of the hardfacing alloys reinforced with complex carbides were better than high speed steel rolls. In-situ observation of the fracture process showed that microcracks were initiated at complex carbides and that shear bands were formed between them, leading to ductile fracture. The hardness, wear resistance and fracture toughness of the hardfacing alloys reinforced with complex carbides were improved in comparison with high-chromium white-iron hardfacing alloys, because of the homogeneous distribution of hard and fine complex carbides in the bainitic matrix [75].

Relative wear resistance of a series of Ni base alloys is presented in Fig. 5.12. It is clear that the wear resistance of boride-containing alloys is lower than carbide-containing alloys. Among boride-containing alloys, colmonoy 6 has best wear resistance. Carbide-containing alloys have excellent and comparable wear resistance. Interestingly alloys containing tungsten also exhibit similar wear resistance.

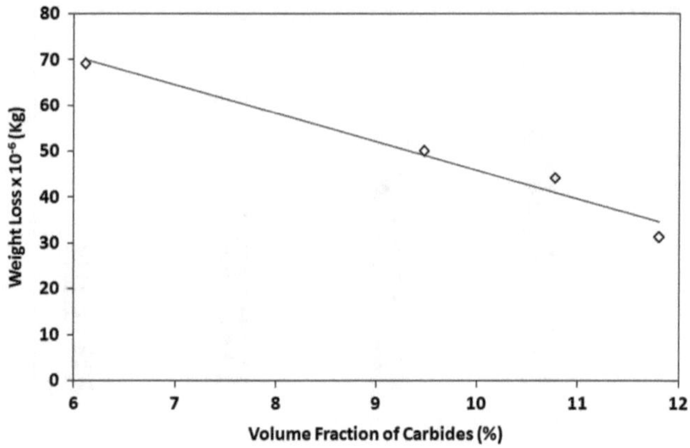

Fig. 5.11 Variation of wear rate as function of volume fraction of carbides of various iron-based composite alloys [73]

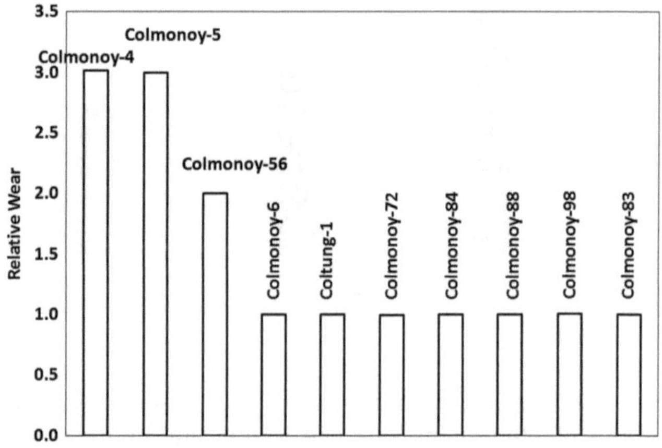

Fig. 5.12 Bar diagram showing the relative wear of a series of Ni base alloys

In another study by Wu and Wu [76], it is reported that Co based alloys deposited using plasma transfer arc are not suitable for low alloy steel. However, Ni base alloy deposited by PTA showed good performance.

Wear at elevated temperature for hardfaced materials has received considerable attention. Nichols [54] investigated a large number of hardfacing alloys for high temperature wear and noted that Triboaloy 700, 800 and Stellite 6, 12 and 20 are excellent coatings. In a study by Lee et al. [77], sliding wear behaviour of newly developed Fe-base Co-free hardfacing alloy (Fe–Cr–C–Si) was investigated and compared to that of Stellite 6 and Fe-base NOREM 02 in the temperatures ranging from 300 to 575 K under a contact stress of 103 MPa (15 ksi) using a sliding wear test apparatus installed in an autoclave filled with distilled water. The weight loss of Fe–Cr–C–Si was equivalent to that of Stellite 6 over all temperatures range in 100-

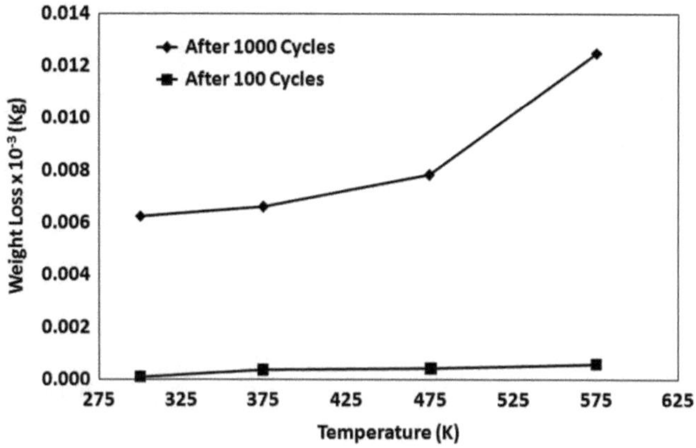

Fig. 5.13 Influence of temperature on the wear rate of newly developed Fe-base Co-free hardfacing alloy (Fe–Cr–C–Si) [77]

cycle wear test. The weight loss of Fe–Cr–C–Si 1000-cycle wear test increased almost linearly with increasing temperature up to 575 K as presented in Fig. 5.13. It is noted that for Fe–Cr–C–Si alloy, the tendency for oxidative wear resulting in the formation of glaze layers increased with increasing sliding distance and test temperature. Lee et al. [78] also noted that under pressurised water and up to 573 K, best wear resistance was exhibited by Fe–20Cr–1.7C–1Si alloy. This behaviour was primarily due to strain induced martensite transformation. Figure 5.14 gives the TEM images showing the transformation of the matrix to martensitic due to strain induced transformation [78]. The wear resistance of this alloy was comparable to that of satellite 6. The martensite was observed at the worn surfaces tested at the room temperature. TEM micrograph of the specimen before wear test showed no strain-induced martensitic transformation (Fig. 5.14). Therefore, external stress induced martensitic transformation and affected the wear behaviour. Kesavan and Kamaraj [48] examined dry sliding wear behaviour of nickel base hardfaced of nearly 4–5 mm thickness coating deposited using plasma transferred arc (PTA) welding process on 316 L (N) stainless steel substrate without any defects. Results showed that the wear resistance of coating improved significantly with increase in test temperature, and the high wear resistance was observed at 823 K.

5.5.2 *Abrasive Wear*

Maximum tribological work of hardfaced surfaces is done under abrasive wear condition [79–81]. The abrasive wear resistance is related to the hardness of the hardfaced layer. Interestingly microstructure also plays important role. Plain carbon-hardfaced steel alloys have different microstructures. In non-hardenable steel, ferrite is the matrix and hard phases can be carbide or cementite. In

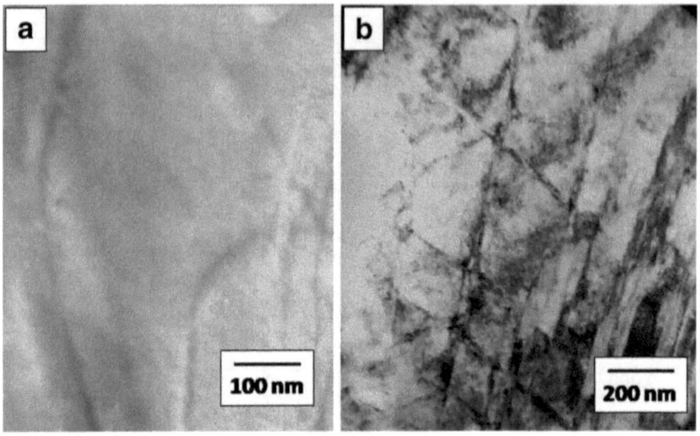

Fig. 5.14 TEM image showing martensite free and strain induced martensite in the matrix of Fe–Cr–C–Si alloy [78]

hardenable steel, matrix can be austenite with other hard constituent. In hardened condition it is martensite or bainite matrix. Figure 5.15 shows steel with similar hardness and different microstructure have different abrasive wear rates [81]. Clearly martensitic microstructure is the best for wear resistance application followed by bainitic microstructure. Interestingly they also have high hardness.

Hardfacing is one of the most useful and economical ways to improve the performance of components subjected to severe abrasive wear conditions. A study was made by Buchely et al. [82] to compare the microstructure and abrasion resistance of hardfacing alloys reinforced with primary chromium carbides, complex carbides or tungsten carbides. The hardfacing alloys were deposited onto ASTM A36 carbon steel plates by a shielded metal arc welding (SMAW) method. Three different commercial hardfacing electrodes were employed to investigate the effect of the microstructure. The abrasion tests were carried out in a dry sand–rubber wheel abrasion machine according to the procedure A of ASTM G65 standard. The results showed that the wear resistance is determined by the size, shape, distribution and chemical composition of the carbides, as well as by the matrix microstructure. The best abrasion resistance was obtained in microstructures composed of eutectic matrix and primaryM_7C_3 or MC carbides, while the higher mass losses were measured in completely eutectic deposits. The main wear mechanisms observed at the surfaces included microcutting of the matrix and brittle fracture of the carbides.

When a contacting abrasive particle forms a groove by plastic deformation, volume of material removal can be represented as

$$\delta V = k_1 \times k_2 \times A \times S \tag{5.11}$$

where k_1 is the probability of materials removal, k_2 is the proportion of the groove volume that forms wear debris, A is the cross-sectional area and S is the sliding distance. The cross-sectional area of the groove is related to the depth of indentation of the particle p^2 as

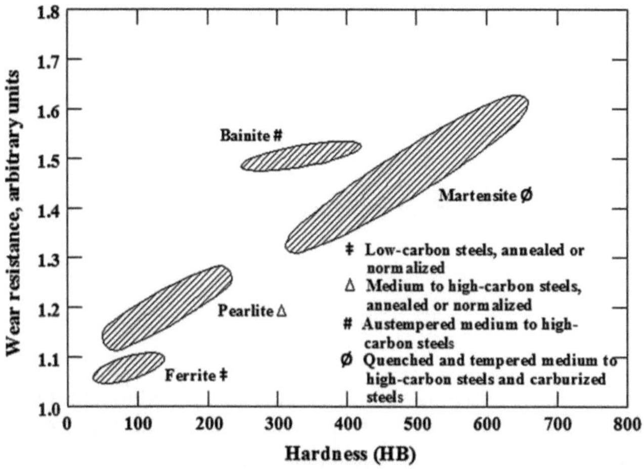

Fig. 5.15 Effect of microstructure and hardness on abrasive wear resistance of various steels [81]

$$A = k_3 \times p^2 \tag{5.12}$$

where k_3 is particle shape dependent constant. According to Brookes et al. [83] the depth of indentation is given by

$$p^2 = \frac{k_4 \times L}{H} \tag{5.13}$$

where k_4 is another constant which depends on the shape of the particle, L is the load on the particle and H is the hardness of the contacting surface. Now the mean load on the particle is

$$\bar{L} = \frac{\sigma}{N} \tag{5.14}$$

where σ is the applied load per unit area and N is the number of contacting particles per unit area and is given as

$$N = \frac{k_5}{d^2} \tag{5.15}$$

where k_5 is a constant and d is mean particle diameter. From Eqs. (5.13)–(5.15) the following equation can be obtained:

$$p \propto \frac{d}{\sqrt{\sigma H}} \tag{5.16}$$

Combining Eqs. (5.12) and (5.16) cross-sectional area of the groove is obtained as

$$A = \frac{k_6 d^2}{\sigma H} \tag{5.17}$$

Thus, volume wear (W) per unit area per unit sliding distance can be obtained as [84]

$$W = \frac{k_1 \times k_2 \times k_6 \times S}{\sigma H} \tag{5.18}$$

where k_6 is a constant. Abrasion wear resistance was found to be related to hardness of the hardfaced layer by Buchely et al. [82] as given in Fig. 5.16. This behaviour is in line with Eq. (5.18). Hardness dependence of abrasive wear rate also complies with Eq. (1.16) of Chap. 1. Similar observation is also noted by Chang et al. [85]. They deposited a series of Fe–Cr–C layers by gas tungsten arc welding (GTAW) on ASTM A-36 steel. They also used Cr and graphite filler rod. Depending on amount of graphite addition, various hypereutectic microstructure containing $(CrFe)_7C_3$ carbides and eutectic colonies of [Cr–Fe + $(CrFe)_7C_3$] were obtained. With increase of carbon content, volume fraction of primary carbides $(CrFe)_7C_3$ increased, their size decreased, hardness and abrasive wear resistance increased.

Similarly, Buchanan et al. [86] studied the abrasive wear behaviour of commercially available Fe–Cr–C hardfacing alloys, a hypereutectic type and a hypoeutectic type. Their result under dry sand abrasive test shown in Fig. 5.17 indicates increase in abrasive wear rate with increasing applied load in accordance with Eq. (5.18). Although there was significant difference in their hardnesses, there was no clear superiority in wear resistance under the conditions examined. However, the wear rate was not always directly proportional to load. The hypereutectic coating exhibited a transition from low to high wear coefficient when the higher loads were applied. This was due to severe fracturing of the primary carbides. The abrasion resistance of the hardfacing is determined primarily by the microstructure. On the one hand, the hypoeutectic material attained its wear resistance through the dispersion of the carbides within the matrix and its capacity to undergo strain hardening, with subsequent enhancement in its resistance to microploughing and microcutting. On the other hand, the hypereutectic resisted wear by virtue of its primary and eutectic carbides. However, the electrode coating was too enriched with carbon and chromium, producing a weld with excess carbide volume that eventually reduced wear resistance.

Bern and Fischer [87, 88] investigated 15 different commercial Fe–Cr–C hardfacing alloys containing B, Ti and Mn. They noted that the microstructure primarily consisted of solidified coarse hard phases embedded in fine eutectic of hard lamellas and metallic constituent. Depending on the $B/(B + C)$ ratio, M_3C, M_7C_3 and MC carbide-containing boron or M_2B, M_3B_2 and MB_2 boride-containing carbon were formed. MC and MB_2 hard phases contained mainly Ti. Other

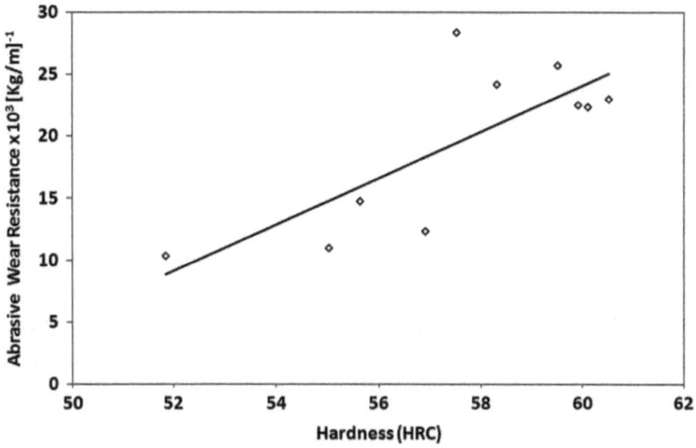

Fig. 5.16 Influence of hardness on abrasive wear resistance [82]

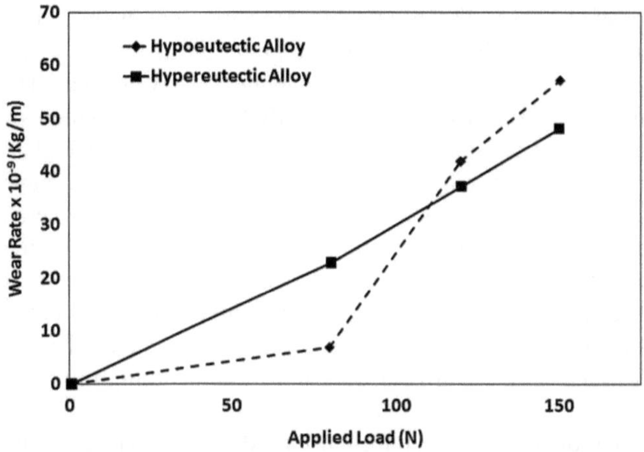

Fig. 5.17 The increase in abrasive wear rate with increasing applied load for Fe–Cr–C hardfacing alloy [86]

elements like, Cr, Fe and Mn were equally distributed in all phases and tried to increase retained austenite. Both the hardness and abrasive wear resistance increased with increase in hard phase volume fraction up to 50 % as long as hard phases were harder than the abrasives and this trend was more pronounced harder the eutectic is.

Kirchgaßner et al. [37] developed a new complex Fe–Cr–W–Mo–Nb hardfacing alloy with high boron content and compared the abrasive wear resistance of the alloy with lower alloyed hardfacing materials on base Fe–Cr–B–C, a synthetic multiphase hardfacing alloy on iron base with around 50 wt% tungsten carbides, a crack-free martensitic Fe–Cr–C hardfacing alloy containing finely precipitated niobium carbides and a conventional hypereutectic Fe–Cr–Nb–C hardfacing alloy. Either synthetic materials using tungsten carbides or complex alloys providing both hard phases and a hard and tough nanostructured matrix at the same time performed best and their abrasive wear properties were comparable.

Similarly Dwivedi [89] found the influence of the composition and heat treatment of overlays on the abrasive wear resistance of iron base hardfacing overlays alloy. Overlays were deposited using a shielded metal arc (SMA) welding process on structural steel using two commercial hardfacing electrodes, i.e. Fe—6 % Cr—0.7 % C and Fe—32 % Cr—4.5 % C. It was found that the wear resistance of the high Cr–C coating is better than the low Cr–C hardfacing under identical conditions. Heat treatment of overlays affected wear resistance. Soaking at low temperature (503 K, 673 K) enhanced wear resistance but high temperature exposure (773 K, 873 K) adversely affected it.

The influence of welding process on the abrasive wear resistance can be obtained from the work of Coronado et al. [90]. In their research, four kinds of welding deposits applied through two different welding processes: flux cored arc welding (FCAW) and shielded metal arc welding (SMAW). The other variable of the tests was the deposited layers. The hardfacing deposits were evaluated using the dry sand–rubber wheel test rig. Their result is provided in Fig. 5.18. FCAW welds presented higher abrasive wear resistance than the SMAW deposits. The hardfacing deposit formed by uniformly distributed carbides rich in titanium presented the highest abrasive wear resistance. Abrasive wear resistance was higher when three layers were applied. The results showed that the most important variable to improve abrasion resistance is the microstructure of hardfacing deposits, where the carbides act as barriers to abrasive particle cutting. Hardfacing using FCAW presents higher abrasive wear resistance than hardfacing by SMAW process because it presents a microstructure composed of an eutectic matrix with carbides rich in titanium. Such microstructure acts as barriers to cutting by abrasive particles. The abrasive wear mechanisms found on the worn surfaces were microcutting, microploughing and wedge formation.

Metal matrix dispersed with tungsten carbide has provided hardfacing layers with the highest available resistance to abrasive wear. The main role of the carbide particles in these coatings has been to act as a wear-resistant barrier to abrasives. Accordingly, the resistance to abrasive wear of metal matrix composites has been shown to increase with the proportion of carbide phase in the metal matrix [91–93]. Badisch and Kirchgaßner [58] found that welding process parameters profoundly influence the abrasive wear rate. Weld deposit of a typical NiCrBSi hardfacing alloy reinforced with tungsten carbide obtained by plasma transfer arc at high current and high voltage by double layer deposition resulting in a high content of uniformly distributed tungsten carbides in a metal matrix gives the maximum resistance to wear in pure abrasion. In contrast, Katsich and Badisch [94] deposited a WC/W_2C-reinforced Ni-based hardfacing by plasma transferred arc (PTA) welding using different welding currents. Their investigation showed significant carbide degradation with increasing welding current, resulting in a significantly reduced primary carbide content and carbide diameter. Reduced carbide content indicated a significant wear rate increase under pure three-body abrasion

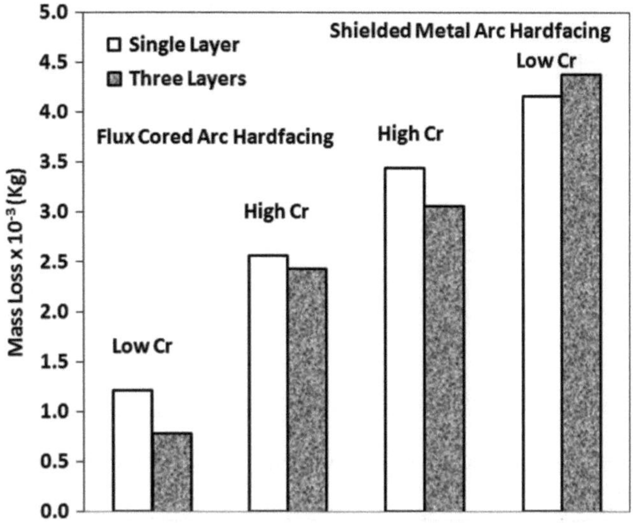

Fig. 5.18 The influence of welding process on the abrasive wear resistance [90]

conditions. The abrasive wear rate of metal matrix composites in pure abrasion has been shown to increase strongly with a rise in the level of degradation of the primary carbide during the deposition process [95, 96]. A complex high alloy deposit has shown a higher resistance to abrasive wear than either a WC or CrBC type of reinforced composite [97]. But very little data exists on the abrasive wear of hardfacing deposits of the complex high alloy type.

Carbide-containing cobalt base alloys are quite popular hardfaced layers for protection against abrasive wear. The abrasive wear rate of several standard Co-based alloys is shown in Fig. 5.19 [28]. These alloys were deposited by gas tungsten arc welding (GTAW). ERCoCr-C alloy with 26–33 wt% Cr and 11–14 wt% W exhibits the lowest abrasive wear rate. The carbon content of this alloy is over 2 % and it contains large primary carbide and Co–M_7C_3 eutectic. Alloys containing low carbon is reasonably ductile but abrasive wear resistance is low (alloy ERCoCr-E). When carbon content is increased to 1 % as in alloy ERCoCr-A and alloy ERCoCr-B, a network of Co–M_7C_3 eutectic is developed. Similarly abrasive wear rate of several Triboaloy deposited by plasma transfer arc is presented in Fig. 5.20 [28]. Although, the volume loss of Ni base alloy ERNiCr-C is the minimum, the wear rates of Triboaloys are also quite low. Among various Triboaloy, Triboaloy-800 has minimum wear rate.

The composition of the hardfaced surface also governs the abrasive wear rate. The influence of Mo content of Stellite 6 hardfaced alloy on the abrasive wear rate is shown in Fig. 5.21 due to Shin et al. [55]. The Stellite 6 hardfacing alloys with different Mo contents have been deposited on AISI 1045-carbon steel using plasma-transferred arc (PTA) welding apparatus. With an increase in Mo contents, the $M_{23}C_6$ and M_6C type carbides were formed instead of Cr-rich M_7C_3 and $M_{23}C_6$

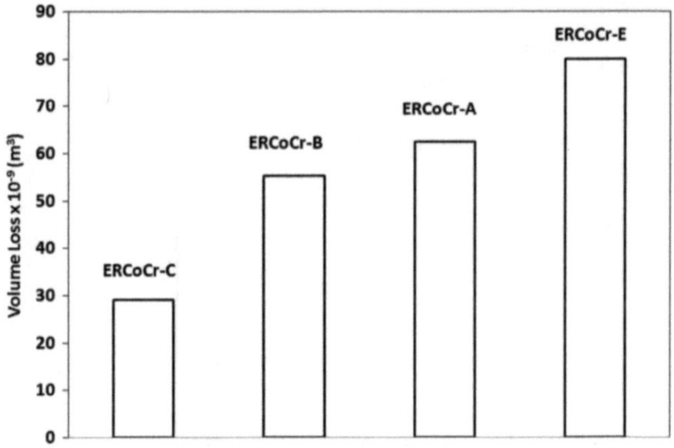

Fig. 5.19 Bar diagram showing the abrasive wear rate of a series of Co base alloys obtained by GTAW process [28]

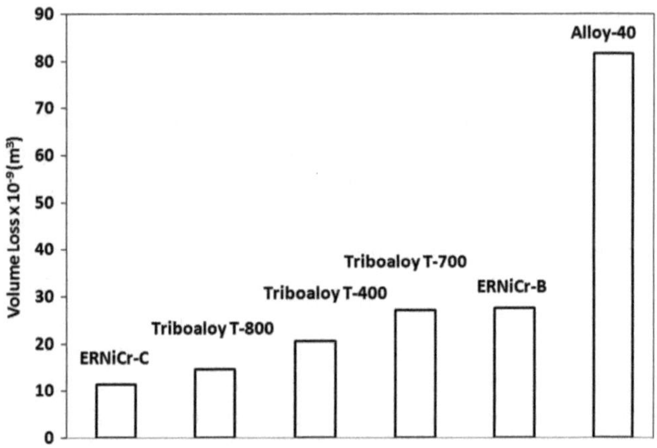

Fig. 5.20 Bar diagram showing the abrasive wear rate of a series of Co base alloys obtained by plasma transfer arc

type carbides observed in the interdendritic region of the Mo-free Stellite 6 hardfacing alloy. The size of Cr-rich carbides in interdendritic region decreased, but that of M_6C type carbide increased with the refinement of Co-rich dendrites. The volume fraction of Cr-rich carbides slightly increased, but that of M_6C type carbide abruptly increased. This microstructural change was responsible for the improvement of the mechanical properties such as hardness and wear resistance of the Mo-modified Stellite 6 hardfacing alloy. This change in mechanical properties, in turn, altered the abrasive wear rate which decreased with increase in Mo content.

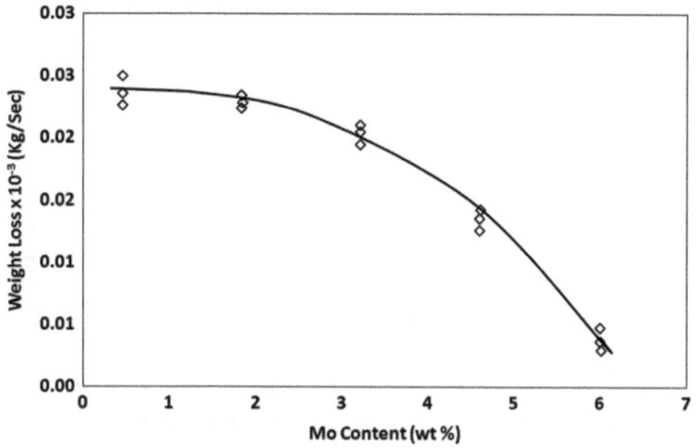

Fig. 5.21 Variation of abrasive wear as function of Mo content [55]

The relative abrasion of a series of nickel base hardfaced surfaces are illustrated in Fig. 5.22. Similar to wear, abrasion resistance of boride-containing alloys is also lower than carbide-containing alloy in general and colmonoy-6 has best abrasion resistance among boride-containing alloys. It is to be stated that unlike wear resistance, colmonoy-98 exhibits very poor abrasion resistance. Further, tungsten containing hardfaced surface also shows excellent abrasion resistance.

5.5.3 Erosive Wear

Another important tribological property for which hardfaced surfaces are used is erosive wear. Solid particle erosion behaviour of different high carbon and high chromium hardfacing electrodes deposited on grey cast iron was studied by Chatterjee and Pal [38] using quartz sand and iron ore as erodent particles. Erosion test was carried out as per ASTM G76 test method. Considerable differences in erosion rates were found among different hardfacing electrodes at normal impact. Both volume fraction of carbides and type of carbides played an important role in the erosion behaviour of the deposits when quartz sand was used as erodent particles. The erosion rate decreased with the total alloying element as shown in Fig. 5.23. On the other hand, only volume fraction of carbides irrespective of carbide type mainly controlled the erosion rate of the same deposits when iron ore was used as erodent particles. This result with iron ore particle is shown in Fig. 5.24. Such difference is attributed to the difference in metal removal mechanisms by the two erodent particles used. Hard quartz sand particles were capable of causing damage to most of the carbides while relatively softer iron ore particles were unable to fracture any carbides present in the microstructures. Furthermore, relatively brittle matrix led to high erosion rate which is significant in case of quartz sand as erodent, but not in case of iron ore particles. Hardness is

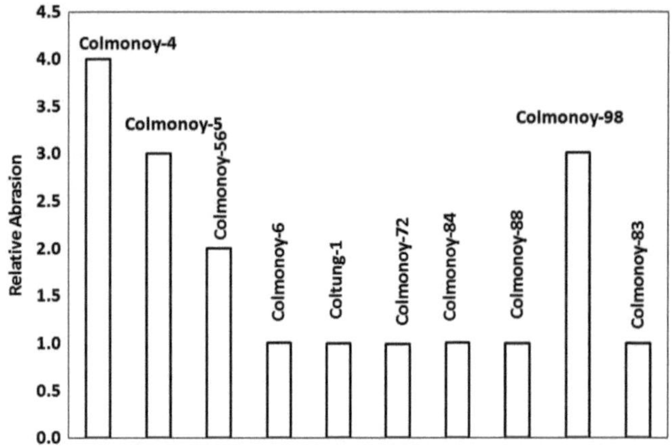

Fig. 5.22 The relative abrasion of a series of nickel base hardfaced surfaces

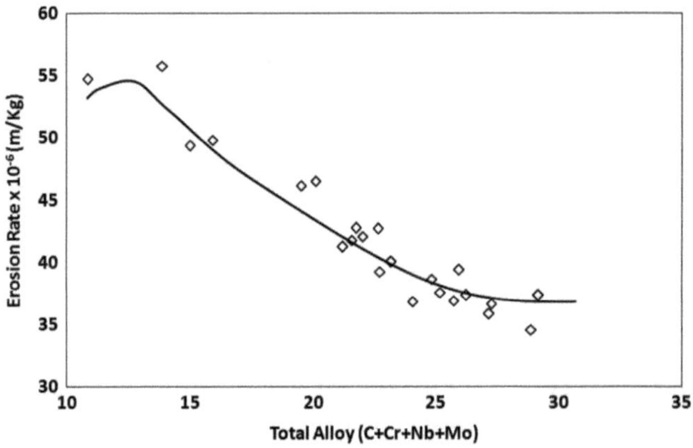

Fig. 5.23 Variation of erosion rate as function of total alloy content [38]

not a true index of erosion resistance of hardfacing deposits. Similarly Stevension and Hutchings [98] also noted that Silica and silicon carbide erodents were able to fracture the carbide particles of hardfaced layer, where as sinter particles were unable to cause any gross fracture of carbides in the hardfaced layer.

Sapate and Rao [99] noted that the dependence of erosion rate on impingement angle was in general weak for weld hardfacing high chromium cast iron alloys. The peak in erosion rate occurred at impingement angle of 60–90° depending upon the alloy and the erodent particle hardness. Although the erosion rate decreased with increase of volume fraction of carbide particle [100], Sapate and Rao noted that the dependence of erosion rate on carbide volume fraction was a strong function of erodent hardness [101]. Large volume fraction of carbides proved beneficial to the erosion resistance with softer erodent particles. With silica sand particles at normal

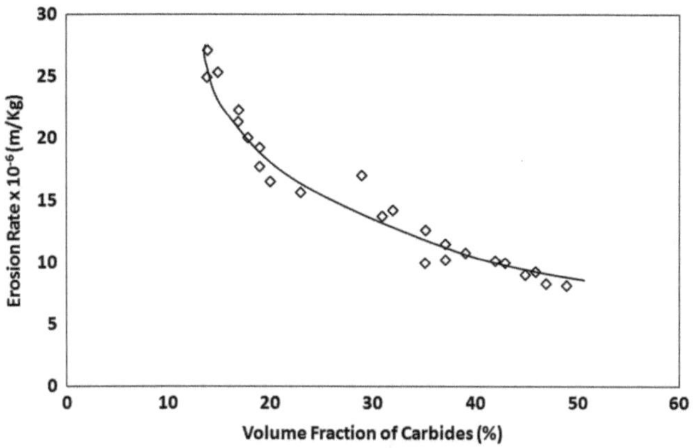

Fig. 5.24 Variation of erosion rate of high carbon high chromium hardfacing layer as function of volume fraction of carbides using iron ore as erodent [38]

impact and with alumina particles, large volume fraction of carbides were detrimental to the erosion resistance. The material removal mechanisms from the carbides were strongly influenced by the erodent particle hardness and the impingement angle. With softer erodent particles, material removal mechanism involved edge chipping whereas with harder erodent particles indentation and gross fracture was predominant erosion mechanism in addition to the edge effect.

Relative erosion of a series of nickel-based alloy shown in Fig. 5.25 indicates that these alloys are not good for erosion. In general, both boride-containing and carbide-containing alloys exhibit poor erosion resistance. However, colmonoy-98 which shows very poor abrasion resistance has excellent erosion resistance. Erosion resistance of tungsten-containing alloy is also very low.

Katsich et al. [102] designed two different types of hardfacing alloys of Fe–Cr–C incorporating Nb, Mo and B to ensure improved erosion performances at elevated temperature. Mild steel was hardfaced with these alloys under optimised gas metal arc welding (GMAW) condition. Erosive wear of these coatings was evaluated for four different temperatures, for two different impact angles and at one impact velocity. Results, as shown in Fig. 5.26 indicate that the erosion rate of these coatings increases with increase of test temperature and impact angles. Among various coatings, Fe–Cr–C coating containing higher amount of Nb, Mo and B exhibits best erosion resistance particularly at elevated temperature. SEM images showing the morphology of the eroded surface and the transverse section of eroded surface of hardfaced Fe–C–Cr alloy are shown in Fig. 5.27. Eroded surface exhibits breaking of carbide and boride hard phases. Limited cracking of the matrix can be seen as the matrix contains high amount of carbide and hence tends to exhibit brittle behaviour. Transverse section exhibits restricted cracking of hard phases and crack-free matrix. The interface between the matrix and the hard phases also appears to be excellent and consequently no pulling out of hard phases is found. Thus material loss is primarily by brittle chipping of both hard phases and hard matrix. Badisch

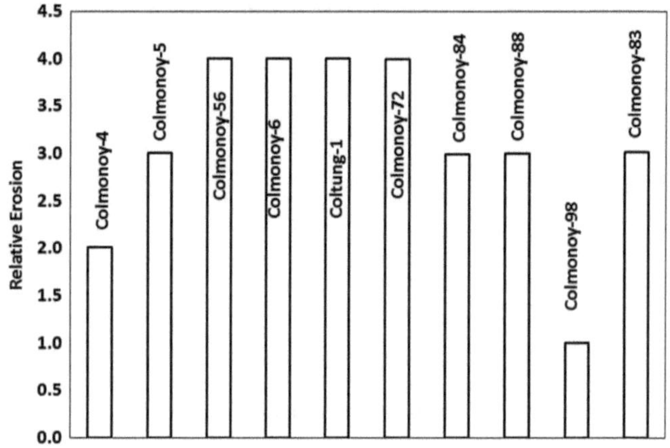

Fig. 5.25 Relative erosion rate for a series of Ni base hardfaced alloy

et al. [103] also noted that during erosion of hardfaced alloys, a coarse microstructure with carbides performs well under glancing impact angle. For Fe–Cr–C hypereutectic alloy, material removal during erosion is by breaking and pulling out of coarse hard phases. Uniform brittle chipping of hard phases and hard matrix causes material loss of Fe–CrC–B complex alloy during erosion.

Yıldızlı et al. [104] surfaced low carbon steel plates with single, double and triple pass bead on-plate welds using the shielded metal arc welding technique with high manganese electrode. The microstructure, hardness and erosive wear resistance of the surfaces were investigated. As the number of the deposit layer increased, the microstructure was found to change from martensite with retained austenite to austenite. Under the condition of erosive wear, the wear resistance of the first layer was the highest of all. Being approximately nine times higher than that of the substrate, the erosion rate of the first layer was suitable for low speed impact erosion tests. However, those of the second and the third layers tend to be arising with impact angle. Being thick and including all layers, triple hardfaced surfaces can be longer lasting in application.

5.6 Application of Hardfacing Surfaces

Hardfacing is carried out on several engineering components. Figure 5.28 illustrates hardfaced surfaces and highly loaded edges of components which are used in agricultural industry for soil treatment [105]. Application of hardfacing surfaces depends on the type of hardfacing materials. Iron-based hardfacing surfaces having pearlitic structure are used as build up of mine car wheels, shafts, rolls, cams and gears, etc. Martensitic iron-based hardfaced surfaces can be used for crane wheels, cable sleeves, undercarriage of tractors and pipe forming rolls. Some martensitic

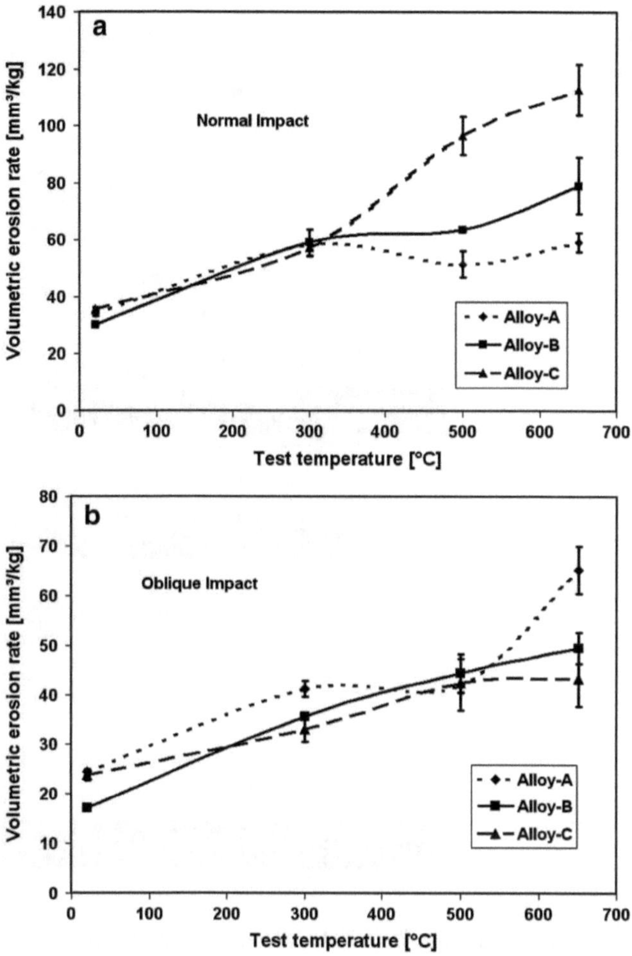

Fig. 5.26 Variation of erosion rate of iron based hardfaced surfaces with temperature under (**a**) normal impact and (**b**) oblique impact [102]

structures contain carbides making it still harder. Such surfaces are used in forging dies, shoes of pincer guide, etc. Tool steel variety of hardfaced surfaces are used for shear blades, trimming dies, cement chutes, earth moving equipment and machine tools components. Hardfaced layer with martensitic stainless steel composition and structures is used for cladding rolls of continuous caster, blast furnace bells. White iron type layers are used for protection in coal pulverising equipment, railway tracks, crushed roll shell, bucket lips, teeth and impact hammer of rock crushing and transporting system.

Cobalt base alloys such as Stellite1 and Stellite 6 are used in seat and gates of valve, shafts and sleeves of pumps, bearings and bearing sleeves. Higher varieties of satellites are used for saw teeth, rotary slitters, cutting edges of knives of carpet,

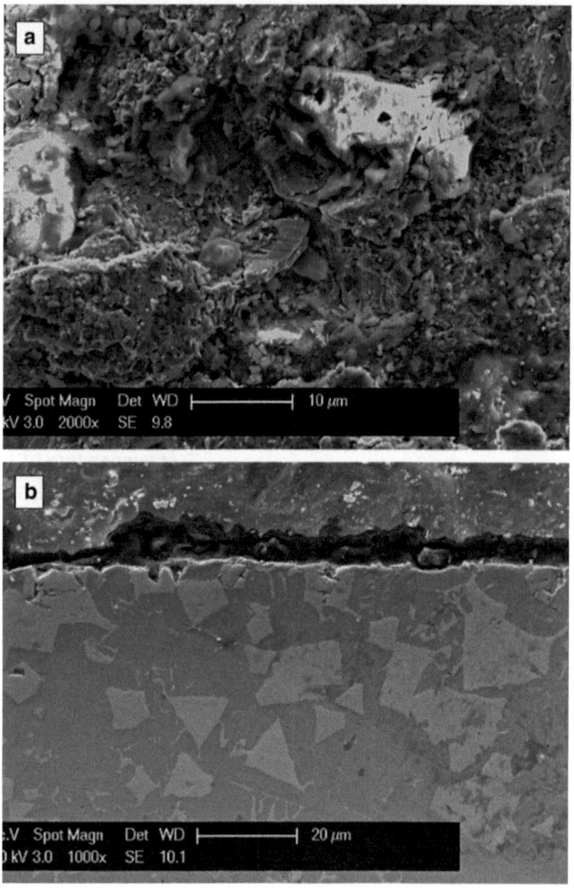

Fig. 5.27 SEM image showing (**a**) the morphology of the eroded surface and (**b**) the transverse section of eroded surface of hardfaced Fe–C–Cr alloy [102]

paper, plastic industries. Stellite 701, 704, 706, etc., are employed for superior wear corrosion applications such as tricone rock bits of oil drilling industries. Triboaloys are also used in application similar to satellite, e.g. mechanical seals, thrust rings, etc.

Among Ni base alloys, Deloro variety alloys are used most extensively. These alloys are used in glass forming plungers, diesel engine valves, wear rings, seal rings, dies, moulds, shafts, bearings, cylinder liner, sleeves, stem guides, etc. Colmonoy varieties of alloys are used for pumps and pumps parts, nuclear industry, plastic extrusion screws, centrifuges and wire drawing capstans. Surfaces by ERNiCr alloys can be used for trims of fluid control valve, mechanical coupling, feed screw, slurry pumps, rebuilding of hot extrusion and forging dies. An image of hardfaced cutting edge used in wood processing is given in Fig. 5.29.

WC–Co is most widely used hardfacing material in recent time. WC–Co hardfacing is used in mining industry. Bucket wheel excavator teeth, shovel teeth, dragline teeth, dragline chain, dragline pins, undercarriage parts, dozer link pads, etc., are hardfaced with these materials in surface mining industry. For deep mining

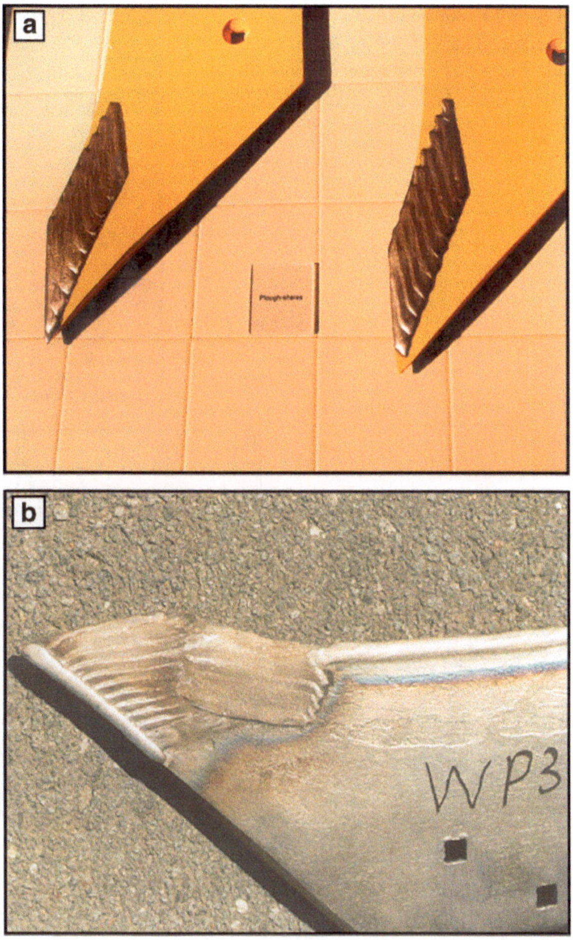

Fig. 5.28 Images of hardfaced surfaces and highly loaded edges in agricultural industry for soil treatment [105]

activity, longwall miners, continuous miners, feeders beakers, cutter drums, crushing hammer, conveyer parts wear parts, etc., are hardfaced with WC–Co. Chutes, hammers, crushers, grinding, pulverising, wear parts and screw conveyers are some implements that are hardfaced with WC–Co. In construction industry, ground engaging tools, cutting edges, dozer blades, grader blades, trencher teeth, tunnelling equipment, wear parts, augers, dozer undercarriage parts, horizontal directional drilling equipment, etc., are hardfaced. Parts of processing industry are mixing and blending parts, screw conveyers, paddles, etc. Sugar mill rolls, sugar mill hammers, disc, injector, point subsoilers, etc., are hardfaced important agricultural equipment.

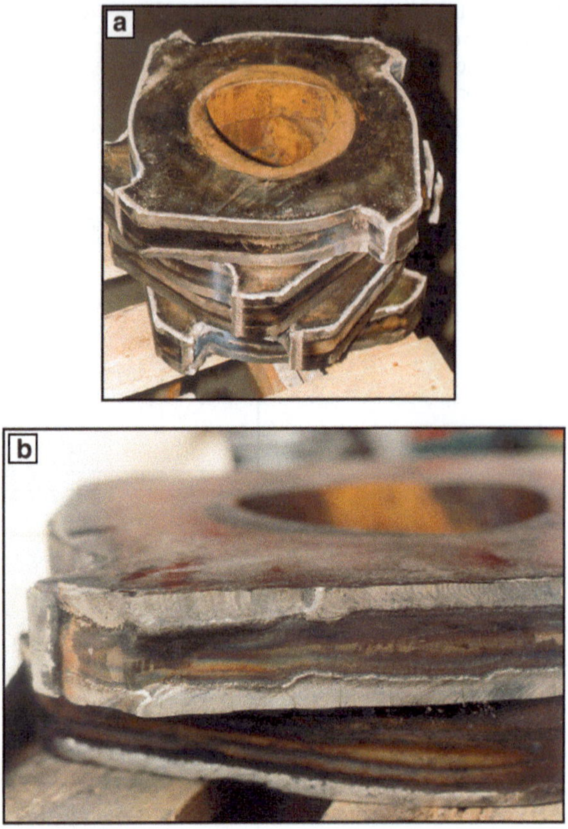

Fig. 5.29 An image of hardfaced part of cutting edge used in wood processing industry [105]

5.7 Direction of Future Research

With hectic activities in research and development on surface engineering all over the world along with the trend of function specific treatment of materials, it is expected that hardfacing will be used extensively in all related field of manufacturing industries. The prerequisite of manufacturing industries with favourable production cost is to use hardfacing as most favoured surface modification technique. Application of hardfacing has been facilitated by availability of flexible equipment. The modern trends are development of single purpose equipment capable of hardening, welding and melting, universal flexibility, adaptability to integration to flexible machine system.

Hardfacing surfaces face direct competition primarily from thermal sprayed coatings. However, hardfaced surfaces produce a diffused or remelted interface with the substrate. In recent times, wide varieties of hardfaced materials have been developed and still further development in materials front having different composition and microstructure will continue. It is imperative to generate extensive database on elevated temperature tribological properties of hardfaced surfaces. It is equally important to carry out extensive study on the nature of interaction between corrosion and tribological degradation of hardfaced layer as the work in this direction has hardly been reported. Thin hardfaced films and application of hardfaced layer have found limited usage, the area of which should be extended. Development of various sensors related to hardfacing operation resulted in better efficiencies and process control. Further development in sensors continues giving improved life of equipment and better surfaces. Research is in progress to develop sensors for monitoring the dilution of hardfaced surfaces. Friction surfacing is another area which exhibits immense potential. Serious attempts are being made to friction surface larger variety of incompatible material and produce more variety of refined structures.

An area which deserves extensive attention is electron beam hardfacing [106]. High energy density and deep penetration capability are the main advantages of electron beam hardfacing and this puts this method ahead of laser surface modification. Thus, this method which is not finding elaborate usage should be explored for variety of materials combination for hardfacing. As this process is a high precision hardfacing process accompanied with low heat input, this process can be employed to many new applications requiring high accuracy, minimum stress and distortion.

In spite of comprehensive research work, there is still wide scope for basic research to establish the potential and limitation of existing methods in order to pave the way for further industrial applications not only in the areas of process technology but also in the areas of materials. Other factors include engineering and economic aspects.

References

1. Brienan EM, Kear BH (1983) In: Bass M (ed) Laser material processing. North Holland Publishing, Amsterdam, p 235
2. Carslaw HS, Jaeger JC (1959) Conduction of heat in solids. Oxford University Press, Oxford
3. Leshchinskiy LK, Sumotugin SS (2001) Weld J 80:25s
4. Gulenc B, Kahraman N (2003) Mater Des 24:537
5. Wilden J, Bergmann JP, Frank H (2006) J Therm Spray Technol 15:779
6. Deuis RL, Yellup JM, Subramanian C (1998) Compos Sci Technol 58:299
7. Nurminen J, Nökki J, Vuoristo P (2009) Int J Refract Met Hard Mater 27:472
8. Man HC, Cui ZC, Yue TM, Chang FT (2003) Mater Sci Eng A 355:167
9. Wang LS, Zhu PD, Cui K (1996) Surf Coat Technol 80:279
10. Chong PH, Man HC, Yue TM (2001) Surf Coat Technol 145:136
11. Powers DE, Schumacher BW (1989) Weld J 68:48
12. Powers DE (1988) In: Metzbower EA, Hauser D (eds) Proc. conf. power beam processing, ASM International, Ohio, pp 25–33
13. Liu Y, Sun C, Chen X (1987) Trans China Weld Inst 8:31
14. Matsui S, Matsummura H, Yasuda K (1987) Weld World 25:16
15. Franchini F (1993) Weld Int 7:206
16. Russel RJ (1980) Weld J 61:21
17. Metzger G, Lison R (1976) Weld J 55:230s

18. Bedford GM (1990) Met Mater 6:702
19. Vitanov VI, Voutchkov II, Bedford GM (2000) J Mater Process Technol 107:236
20. Li JQ, Shinoda T (2000) Surf Eng 16:31
21. Bedford GM, Vitanov VI, Voutchkov II (2001) Surf Coat Technol 141:39
22. Kaskov YM (2003) Weld J 82:42s
23. Kompan YY, Sufonnihov AN, Peprov AN, Svirskii EA (1994) Weld Int 8:986
24. Agarwal A, Dahotre NB (1999) J Mater Eng Perform 8:479
25. Agarwal A, Dahotre NB (1998) Surf Coat Technol 106:242
26. Agarwal A, Dahotre NB (1999) Surf Eng 15:27
27. De A (1993) ASM handbook, vol 6, Welding brazing and soldering. ASM International Materials Park, Ohio, pp 789–807
28. Crook P, Farmer HN (1992) ASM handbook, vol 18, Friction lubrication and wear. ASM International Materials Park, Ohio, pp 758–765
29. Chandel RS (2001) Indian Weld J 26:33
30. Avery HS, Chapin HJ (1952) Weld J 31(10):917
31. Powell GLF (1979) Aust Weld Res 6:16
32. Gulenc B, Kahrarman N (2003) Mater Des 24:537
33. Bayraktar E, Levaillant C, Altintas S (1993) J Physiuqe IV Colloque C7 3:61
34. Pelletier JM, Oucherif F, Sallamand P, Vanes AB (1995) Mater Sci Eng A202:142
35. Ball A (1983) Wear 91:201
36. Menon R (2002) Weld J 81:53
37. Kirchgaßner M, Badisch E, Franek F (2008) Wear 265:772
38. Chatterjee S, Pal TK (2003) Wear behaviour of hardfacing deposits on cast iron. Wear 255:417
39. Powell GLF, Carlson RA, Randle V (1994) J Mater Sci 29:4889
40. Lesko A, Navera E (1996) Mater Char 36:349
41. Atamert S, Bhadeshia HKDH (1990) Mater Sci Eng A 130:101
42. Fischer A (1984) Fortschrittberichte der VDI Zeitschriften. VDI Verlag, Düsseldorf, S 69
43. Buytoz S (2006) Mater Lett 60:605
44. Branagan DJ, Marshall MB, Meacham BE (2005) In: Proceedings of international thermal spray conference, Basel (CD-Rom)
45. Wang XH, Zou ZD, Qu SY, Song SL (2005) J Mater Process Technol 168:89
46. Svensson LE, Gretoft B, Ulander B, Bhadeshia HKDH (1986) J Mater Sci 21:1015abc
47. Lee S, Choo SH, Baek ER, Ahn S, Kim NJ (1996) Metall Mater Trans A27:3881
48. Kesavan D, Kamaraj M (2010) Surf Coat Technol 204:4034
49. Bhaduri AK, Indira R, Albert SK, Rao BPS, Jain SC, Asokkumar S (2004) J Nucl Mater 334:109
50. Persson DHE, Jacobson S, Hogmark S (2003) Wear 255:498
51. Mason SE, Rawlings RD (1994) Mater Sci Technol 10:924
52. Mason SE, Rawlings RD (1989) Mater Sci Technol 5:180
53. Johnson MP, Moorehouse P, Nicholls JR (1990) Diesel engine combustion chamber materials for heavy fuel operation. Institute of Operation, Institute of Marine Engineers, London, p 61
54. Nicholls JR (1994) Mater Sci Technol 10:1002
55. Shin JC, Doh JM, Yoon JK, Lee DY, Kim JS (2003) Surf Coat Technol 166:117
56. Wu W, Wu LT (1996) Metall Mater Trans A27:3639
57. Nicholls JR, Stephenson DJ (1990) Diesel engine combustion chamber materials for heavy fuel operation. Institute of Operation, Institute of Marine Engineers, London, p 47
58. Badisch E, Kirchgaßner M (2008) Wear 202:6016
59. Chermont JL, Osterstock F (1976) J Mater Sci 11:1939
60. Llo S, Just C, Badich E, Wosik J, Danninger H (2010) Mater Sci Eng A527:6378
61. Huang SW, Samandi M, Brandt M (2004) Wear 256:1095
62. Li Q, Lei TC, Chen WZ (2003) Surf Coat Technol 166:117

63. Hoffman JM (2010) Mach Des 14:5
64. Branagan DJ, Marshall MC, Meacham BE (2006) Mater Sci Eng A428:116
65. Zikin A, Hussainova I, Katsich C, Badisch E, Tomastik C (2012) Surf Coat Technol 206:4270
66. Bayhan Y (2006) Tribol Int 39:570
67. Su YL, Chen KY (1997) Wear 209:160
68. Wang XH, Han F, Liu XM, Qu SY, Zou ZD (2008) Mater Sci Eng A 489:193
69. Wang XH, Han F, Qu SY, Zou ZD (2008) Surf Coat Technol 202:1502
70. Foroulis ZA (1984) Wear 96:203
71. Bhansali KJ, Miller AE (1982) Wear 75:241
72. Persson DHE, Jacobson S, Hogmark S (2003) J Laser Appl 15:115
73. Kim CK, Lee S, Jung JY, Ahn S (2003) Mater Sci Eng A 349:1
74. Choo SH, Kim CK, Euh K, Lee S, Jung J-Y, Ahn S (2000) Metall Mater Trans 31A:3041
75. Lee S, Choo S-H, Baek E-R, Ahn S, Kim NJ (1996) Metall Mater Trans 27A:3881
76. Wu W, Wu LT (1996) Metall Mater Trans 27A:3639
77. Lee KY, Kim GG, Kim JH, Lee SH, Kim SJ (2007) Wear 262:845
78. Lee KY, Lee SH, Kim Y, Hong HS, Oh YM, Kim SJ (2003) Wear 255:481
79. Kotecki DJ, Ogborn JS (1995) Weld J 74(8):269s
80. Kim HJ, Yoon BH, Lee CH (2002) Wear 249:846
81. Ghar Z (1971) Met Prog Sept:46
82. Buchely MF, Gutierrez JC, Leon LM, Toro A (2005) Wear 259:52
83. Brookes CA, Green P, Harrison PH, Moxley B (1972) J Phys D 5:1284
84. Moore MA (1980) In: Rigney DA (ed) Fundamental of friction and wear. ASTM, Materials Park, OH, p 73
85. Chang CM, Chen YC, Wu W (2010) Tribol Int 43:929
86. Buchanan VE, Shipway PH, McCartney DG (2007) Wear 263:99
87. Berns H, Fischer A (1985) In: Proceedings of the fifth international conference on wear of materials, Vancouver, BC, pp 625–633
88. Berns H, Fischer A (1983) In: Ludema KC (ed) Proceedings of the fourth international conference on wear of materials, Wear of materials. ASME, New York, pp 298–302
89. Dwivedi DK (2004) Mater Sci Technol 20:1326
90. Coronado JJ, Caicedo HF, Gomez AL (2009) Tribol Int 42:745
91. Berns H, Fischer A (1987) Metallography 20:401
92. Berns H (2003) Wear 254:47
93. Simm W, Freti S (1989) Wear 129:105
94. Katsich C, Badisch E (2011) Surf Coat Technol 206:1062
95. Berns H, Fischer A (1986) Wear 112:163
96. Eroglu M, Zdemir NO (2002) Surf Coat Technol 154:209
97. Huang Z, Hou Q, Wang P (2008) Surf Coat Technol 202:2993
98. Stevension ANJ, Hutchings IM (1995) Wear 186–187:150
99. Sapate SG, Rama Rao AV (2006) Tribol Int 39:206
100. Flores JF, Neville A, Kapur N, Gnanavelu A (2009) Wear 267:213
101. Sapate SG, Rama Rao AV (2004) Wear 256:774
102. Katsich C, Badisch E, Roy M, Heath GR, Franek F (2009) Wear 267:1856
103. Badisch E, Katsich C, Winkelmann H, Franek F, Roy M (2010) Tribol Int 43:1234
104. Yıldızlı K, Eroglu M, Karamış MB (2007) Surf Coat Technol 201:7166
105. Castolino Eutectic company brochures
106. Soon Z, Karppi R (1996) J Mater Process Technol 59:257

Chapter 6
Plating and Tribology

M. Palaniappa and Manish Roy

6.1 Introduction

Surface modification has a long history, dating back to 850 B.C., when it was more an art than science; it has come a long way since then and has diversified into many techniques, thereby acquiring the ability to function under a variety of conditions. The basic idea in surface engineering is to use lean (often cheaper) materials as substrates and rare or strategically important (often expensive) materials as coatings. In such a system approach, one can hope for a synergism of surface and substrate.

Coatings typical of different processes are distinct from each other in a host of ways, such as thickness, hardness, ductility, residual stress state, etc. Whilst all these attributes are important and are indeed the types of parameter upon which coating selection is made, classification on the basis of these properties provides no generic understanding of the coating methods themselves. Understanding of the coating processes results from an examination of the mechanism by which the coating is built up. Figure 6.1 illustrates a scheme under which coating processes may be categorised.

Electrodeposition is a well-established process with proven benefits in controlling corrosion and wear resistance. This process is a convenient way of applying coatings of metals with high melting points such as chromium, nickel, copper, silver, gold, platinum, etc., onto the substrate. The electroplating is defined

M. Palaniappa (✉)
NonFerrous Materials Technology Development Center, Kanchanbagh Post, Hyderabad 500058, Andhra Pradesh, India

Honeywell Technology Solutions Laboratory Pvt. Ltd., 151/1 Doraisanipalya, Bannerghatta Road, Bangalore 560226, Karnataka, India
e-mail: Palaniappa.Murukesan@Honeywell.com

M. Roy
Defence Metallurgical Research Laboratory (DMRL), Kanchanbagh Post, Hyderabad 500058, Andhra Pradesh, India

M. Roy (ed.), *Surface Engineering for Enhanced Performance against Wear*, DOI 10.1007/978-3-7091-0101-8_6,

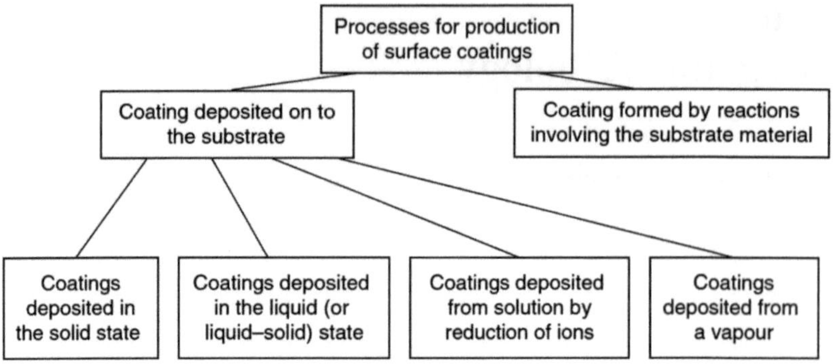

Fig. 6.1 Processes for the production of surface coatings: basic classification of coating types

as deposition of coating on an electrode immersed in an electrolyte by passing electric current through the electrolyte. When the process takes place in aqueous solution it is called electroplating and when it takes place in fused salt at elevated temperature, it is called metalliding. Another variant of electrodeposition is called electrophoretic deposition. In this method metal, polymer or even ceramic can be present in the electrolyte as fine powder or colloid suspension. A charge is applied to the materials. The initiation of potential between anode and cathode causes migration to the surface to be coated. Excellent review of electrodeposition processes can be found in refs. [1–5].

Another important electrodeposition process where no electric current is required is called electroless deposition. As no current is required, it is sometimes termed as autocatalytic process. This coating technique was developed by Brenner and Riddell in 1944. Electroless coatings have unique physicochemical and mechanical properties for which they are being used increasingly. Over the last few decades, a large number of extensive reviews on electroless depositions have been published [6–10].

The utility of electrodeposition can be understood from the range of materials such as metals, alloys, oxide layer, polymers and composite layers that can be deposited. A wide range of engineering components find application of electrodeposited coating. This includes nanostructures for speciality magnetic semiconductors and optical uses, conductive coatings for electronic applications and tribological coatings for mechanical engineering [11–14].

In this chapter, an attempt is made to give a brief introduction of electrolytic and electroless coatings. This is followed by detail description of status of state of art of tribological performances of these coatings.

6.2 Electrodeposited Coatings

Electroplating has several advantages. The coating properties can be modified by changing the plating conditions. Substrate to be coated does not undergo any kind of distortion or metallurgical changes. Parts of the substrate which do not require to be coated can be masked. Coating can be deposited as thick as desired and the deposition rate can be increased by increasing current, temperature, forced circulation of the electrolytes, etc. The process can be automated and the size of the component to be coated is limited by the size of the tank only. It is not very difficult to increase the size of the tank.

Electroplating or electrodeposition processes have certain drawbacks. Parts to be coated by electroplating need special surface preparation. Surfaces are usually prepared by precleaning, intermediate alkaline cleaning, electro cleaning, acid treatment and desmutting. The thickness of electroplated coating is not uniform as it is governed by substrate shape, anode/cathode configuration, composition, conductivity of electrolyte and current density. Electroplating results in evolution of hydrogen which may lead to hydrogen embrittlement. It is not possible to deposit all metals and alloys by electroplating. The bonding between the coating and substrate can be very good but it can never be as good as fusion bonding. Sometimes it is difficult to assess the quality of bonding.

The thickness of electrolytic deposition can be controlled by monitoring the process as described by Faraday's laws. At this point, a metal M can be considered to be deposited from a solution of its ions M^{Z+} as

$$M^{Z+} + ze- \rightarrow M \tag{6.1}$$

Thus under steady state condition rate of deposition can be expressed as

$$\frac{dx}{dt} = \frac{MI\phi}{\rho AzF} \tag{6.2}$$

where I is the current, M is the molar mass, φ is the current efficiency, ρ is the density of plated layer, F is the Faraday constant, A is the deposition area and z is the number of electron transferred per deposited particles.

Microstructures of electrodeposited metals and alloys are characterised with fine grain size, thoroughly intermingled in case of poly phase constituents. It can be fibrous columnar structure or laminar structure or both. In general, coating obtained at low current density is more homogeneous. Coatings obtained at very low or very high current densities are dark and powdery. Hardness of electrodeposited alloys is higher than cast or solution-treated alloys. The morphology and properties of electrodeposited films depend on deposition parameters such as bath composition, plating method, temperature of electrolyte, etc. Chromium, cobalt, copper, cadmium, nickel, gold, silver, zinc, etc. are pure metals that are deposited by electroplating. Among these metals Cr is extensively used for industrial application. Two grades of Cr plating are produced. Hard Cr coatings which have high hardness and abrasion resistance with low corrosion resistance due to high degree of cracking throughout the coatings are one variety. The other variety is bright Cr coatings which have lower hardness and abrasion resistance but improved corrosion resistance. Ni is deposited either as decorative or as functional coating. For decorative coating it can be deposited to a mirror like finish without requiring any further polishing. When used for wear resistance or other functional coatings it is deposited with dull or matt finish. Zinc is typically used as corrosion-resistant coating e.g. galvanised steel. As the standard potential of zinc is lower than that of iron, zinc acts as sacrificial anode to protect iron and steel. Zinc is plated in rack and barrel electrochemical reactor from zinc cyanide alkaline or acidic electrolyte solutions. Cd plating is also done primarily for corrosion protection. While Cd plating is used for protection in marine environment, Zn plating is applied against corrosion in industrial environment. Cadmium is toxic and should be avoided where parts will be in contact with food. Cd coating is susceptible to hydrogen embrittlement and hence proper care should be taken during deposition. Precious metals such as gold, silver and platinum can be plated either by aqueous solution electroplating [15] or fused salt electrodeposition [16]. Silver and gold plating are primarily used for decorative purpose. In recent times, silver has replaced gold in many electronic devices and it is used in metallic leadframes that support silicon chips. However, in many applications where long-term service is required, gold is still used. Aqueous solution electroplating or fused salt electrodeposition are used for depositing platinum which is used as diffusion barrier layer in Ni base super alloys.

6.2.1 *Continuous Electrodeposition*

Development of continuous electrogalvanising has resulted in new applications for zinc, zinc–nickel and zinc–iron alloys which are ideal for deep drawing, automotive body panel, etc. Tin plating is another continuous electrodeposition process and tin plating is used for long-term storage of food. Typical tin-plated layer consists of five layers such as inner most steel layer, tin iron intermetallic layer, an exclusive tin layer, a thin passivation layer of chromium oxide and a lubricated outer layer.

6.2.2 *Fused Salt Electroplating*

Fused salt electroplating or metalliding is carried out when the plating material cannot be deposited by conventional process but can be deposited by this process and also when deposition rate is controlled to match the diffusion rate of the deposition species in the substrate to develop a diffusion coating. The molten salt medium consists of a soluble form of metal to be plated dissolved in molten salt solvent such as alkali metal halide. In this process the element to be plated is made as anode and the substrate is made as cathode. As the process is diffusion controlled, diffusion gradient on the surface can be seen. Although this process found limited application for refractory metals and ceramic coatings, it is also used for platinum group metals. Small scale, small size specialised and strategically important components can be coated because of the fact that this is a one-step diffusion process.

Table 6.1 List of the components used in the electroless coating bath and their functions

Ingredients	Function
Oxidising agent	Provides metal ions
Reducing agent	Provides electrons
Complexants	Stabilises the electrolyte
Accelerators	Activates reducing agents
Stabilisers	Prevents breakdown of the electrolyte
Buffer	Monitors pH of the electrolyte
Stabiliser	
Wetting agent	Increases wettability of electrodes

6.3 Electroless Coatings

Electroless plating is an autocatalytic process where the substrate develops a potential when it is dipped in electroless solution that contains a source of oxidising agent, reducing agent, complexing agent, stabiliser and other components. Compared to electrodeposition, layers obtained via the electroless plating process have a very homogeneous distribution regardless of the substrate geometry. During the deposition of electroless films, the growth of the film starts at isolated locations on the substrate. The whole substrate is then covered by lateral growth. The films are characterised by the presence of crystalline and microcrystalline structure. Some of them are amorphous. The appearance of a number of phases may be due to the compositional inhomogeneities prevailing in the film in the as-deposited state. The presence of inhomogeneities in electroless films may energetically as well as kinetically favour different local equilibriums in different regions of the film leading to the formation of different phases depending on the actual composition. In addition to a source of coating materials (nickel chloride, Ni sulphate, Cu sulphate, etc.) and a reducing agent (sodium hypophosphite/sodium borohydride), many other components are required for the occurrence of deposition in the electroless coating bath. A list of the components used in the electroless coating bath and their functions are summarised in Table 6.1.

6.3.1 Reducing Agents

Several reducing agents have been used in electroless coating of alloys. Four types of reducing agent have been used for electroless nickel bath including sodium hypophosphite, amineboranes, sodium borohydride and hydrazine. Hypophosphite baths are the most common types of commercially used electroless nickel baths due to higher deposition rates, increased stability and greater simplicity of bath control. The mechanism of the electroless Ni–P deposition reactions taking place is still not well understood. The borohydride ion is the most powerful reducing agent available for electroless plating especially for Ni plating. The reduction efficiency of sodium borohydride is much higher than that of dimethylamine borane and sodium hypophosphite [17]. Besides the high reduction efficiency, borohydride reduced baths are preferred to dimethylamine borane based baths in terms of cost-effectiveness of operation. However, control of pH is important to avoid the spontaneous decomposition of the bath solution and to decrease the cost of operation. It is generally used as alkaline baths. Use of amineboranes in electroless plating solutions has been limited to two compounds: *N*-dimethylamine borane (DMAB)-$(CH_3)_2NHBH_3$, and Hdiethylamine borane (DEAB)-$(C_2H_5)_2NHBH_3$ [18, 19]. DEAB is used primarily in European establishments, whereas DMAB is used generally in the USA. DMAB is readily soluble in aqueous solutions while DEAB should be mixed with a short-chain aliphatic alcohol such as ethanol before mixing into plating bath. The amineboranes are effective reducing agents over a wide range of pH but due to evolution of hydrogen, there exists a lower limit of pH up to which the plating process can be carried out [18]. Nickel, in the deposit increases as pH of the bath increases. Usually, the amineborane baths have been used in the pH range of 6–9. Operating temperatures for these baths range from 323 to 353 K. However, they can be used at temperature as low as 303 K. Accordingly, amineborane baths are very useful for plating non-catalytic surfaces such as plastics and nonmetals, which are their primary applications. The rate of deposition varies with pH and temperature but is usually 7–12 μm/h. Hydrazine has also been used to produce electroless deposits [20]. These baths operate in the temperature range of 363–368 K and pH range of 10–11. Their rate of deposition is approximately 12 μm/h. Because of the instability of hydrazine at high temperatures, these baths tend to be very unstable and difficult to control. The deposit is brittle and highly stressed for practical applications.

6.3.2 Complexing Agents

One of the difficulties of reduction reactions or chemical plating is the maintenance of the bath composition. As the plating proceeds, continuous lowering of the rate of reduction of coating element occurs. The solutions cannot be replenished due to the formation of some unwanted compounds. Because of the reduction in rate of deposition, accelerators like salts of carbonic acids, soluble fluorides and inhibitors like thiourea, etc. can also be added to avoid the total decomposition of the bath. These additives known as complexing agents exert a buffering action that prevents the pH of the solution from falling too fast. They reduce the concentration of free ions of plating materials by forming meta-stable complexes.

6.3.3 Stabilisers

Presence of dust or other nucleating agents may result in breaking down of the bath reagents and spontaneous decomposition of bath may take place. Heavy metals such as Bi, Pb, Cd and Te are added to the bath as stabilisers [21]. These stabilisers are added in concentrations of only a few parts per million. It also reduces the deposition on the sides and bottom of the bath container. Only traces of stabiliser are required. Sometimes addition of heavy metal may result in poisoning of the bath. There are several varieties of additives available to overcome this problem of poisoning [22]. Adequate stabiliser replenishment is one of the most difficult problems in electroless plating.

6.3.4 Surfactants

Surfactants are wetting agents that lower the surface tension of a liquid, allowing easier spreading and lower the interfacial tension between two liquids or a liquid and solid surface. In an electroless plating bath, presence of surfactant promotes the coating deposition reaction between the bath solution and the immersed substrate surface. Surface finish of the coated layer is also found to improve [23]. Further, hardness of deposits increased with the addition of the surfactants. Surfactants can be anionic (sodium dodecyl benzene sulphonate, sodium lauryl sulphate) and cationic (triethanolamine).

6.3.5 Buffers

Electroless plating requires buffering agents such as sodium or potassium salts or carboxylic acids to maintain a steady pH in the bath [24]. A high buffer capacity is desirable to support the high concentration of hydrogen ions [25]. Automatic monitoring and adjusting devices are generally employed for this purpose.

6.3.6 Accelerators

Although stabilisers and complexants are useful in proper plating, they usually slow down the plating process. In order to overcome this drawback, accelerators are added. Normally fluoride ions are used as accelerator and these fluoride ions get incorporated within the metal lattice [26]. These ions also help in improving the hardness and stress resistance of the deposit. Saturated unsubstituted dicarboxylic acids, aliphatic short-chain saturated amino acids and short-chain saturated aliphatic acids are normally used as accelerators [27].

Electroless nickel deposit is one of the most widely investigated coatings. It is often mistaken for two other commercial nickel-plating methods, immersion and vapour decomposition of nickel carbonyl. Immersion deposition (at 70 °C) provides a poorly adherent and non-protective coating while vapour decomposition (at 180 °C) is expensive and hazardous. Consequently, electroless nickel coating has gained widespread acceptance in many industries.

As-deposited electroless nickel is a metastable, supersaturated metal–metalloid alloy. The equilibrium phase diagrams for Ni–P and Ni–B, exhibit no solubility of phosphorus or boron in nickel at ambient temperatures. Under equilibrium conditions, therefore the alloys consist of essentially pure nickel and the intermetallic compound Ni_3P or Ni_3B. The conditions existing during plating, however, do not permit the formation of the intermetallic compounds. Even the growth of tiny crystals would involve the movement of large numbers of atoms by surface diffusion in order to achieve correct stoichiometry of three nickel atoms to one phosphorus or boron atom. This movement cannot occur before the next layer of atoms has been laid down; thus the phosphorus or boron atoms are trapped between nickel atoms, resulting in supersaturation.

The crystal structure of nickel is face-centred cubic (f.c.c.), in which each atom has 12 neighbours. The entrapment of phosphorus or boron makes it impossible for this atom arrangement to extend over large surfaces. The volume of material within which the f.c.c. atom arrangement can be maintained is called a grain. Grain sizes are thus very small in electroless nickel. If the f.c.c. structure cannot be maintained at all, the structure is equivalent to that of a liquid and is considered to be amorphous.

The properties of electroless nickel coatings are directly attributed to their microstructural characteristics. The phosphorus and boron content of electroless nickel deposits control their microstructure and properties [28]. The detailed structure of electroless nickel deposits is not well understood but as-plated electroless nickel coatings have been reported to be either crystalline, amorphous or a combination of the two. In spite of some contradictory results in the literature, the general trend in the understanding of electroless nickel deposits is that as-plated electroless nickel coatings containing 1–5 wt.% phosphorus (low phosphorus) are crystalline; those containing 6–9 wt.% phosphorus (medium phosphorus) consist of mixed crystalline and amorphous microstructures; whereas those containing 10–13 wt.% phosphorus (high phosphorus) are amorphous. The coating crystallises on heat treatment to nickel and various forms of nickel phosphides or borides. The degree of crystallinity affects resultant properties and is a complex function of a number of factors [29], namely, phosphorus content, heating rate, heat treatment temperature, time at heat treatment temperature and previous thermal history. Zhang and Yao [30] conducted a comprehensive study on the microstructure of electroless nickel coatings. They also studied the effect of heat treatment on the transformation of amorphous to crystalline microstructure of electroless nickel coatings. Deposits under different heat treated conditions show continuous well-defined rings which are characteristic of a microcrystalline structure after annealing at 200 °C for 1 h. The grain size was about 10 nm. Annealing at 400 °C for 1 h gave rise to the precipitation of b.c.t. Ni_3P and Ni_3B. Heating at temperatures higher than 400 °C can lead to precipitation of Ni_2B and Ni_7B_3 also [31]. Increasing annealing temperature to 600 °C resulted in coarsening of crystals.

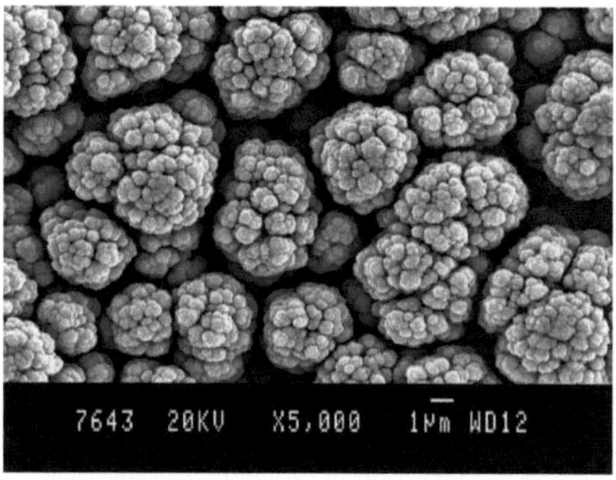

Fig. 6.2 SEM image showing the surface morphology of electroless Ni–P plating

SEM image showing the surface morphology of electroless Ni–P plating is given in Fig. 6.2. Nodular feature resembling that of a cauliflower is the main characteristics of the microstructure of electroless coating. Typical TEM image of electroless Ni–W–P coating along with its diffraction pattern is shown in Fig. 6.3. The TEM image confirms crystalline structure and very fine grain size. The grain size of the coating is around 5–15 nm. The diffraction pattern is also in conformity with the image and indicates crystalline structure. X-ray diffraction pattern obtained from electroless nickel coatings is illustrated in Fig. 6.4. X-ray diffraction profiles in the as-plated condition from the region of (111) reflection of nickel showed an amorphous profile with a wide angular range of approximately 40–60° in the case of Ni–P and Ni–W–P. The diffraction from Ni–W–P deposit containing higher amount of W indicates a comparatively sharp (111) reflection along with a very faint (200) reflection indicating a partially crystalline deposit. Theoretically, a disorder in arrangement of atoms manifests itself as a broad peak in X-ray diffractograms. The observed diffractogram of the electroless Ni–P deposit can be explained on the basis of the mechanism of formation of these deposits. During electroless deposition, phosphorus atoms are randomly captured on the nickel atoms and the rate of segregation of these atoms determines the crystallinity of the deposit. Among these two atoms, the rate of diffusion of phosphorus is relatively small compared to that of nickel. Hence if the deposit contains higher amount of phosphorus, then a larger number of phosphorus atoms must be moved from a given area per unit time during deposition to achieve segregation of nickel and phosphorus.

One of the outstanding characteristics of electroless nickel coatings is their high hardness, especially after heat treatment. There are three major parameters affecting the hardness of these coatings, namely, phosphorus content, time and the temperature of the post-plating heat treatment. The hardness of an electroless nickel deposit is especially important when superior wear resistance is required. The hardness of electroless nickel deposits, as with many other properties, is directly affected by the phosphorus content. Increasing the phosphorus content of the

Fig. 6.3 TEM image of electroless Ni–W–P coating along with its diffraction pattern

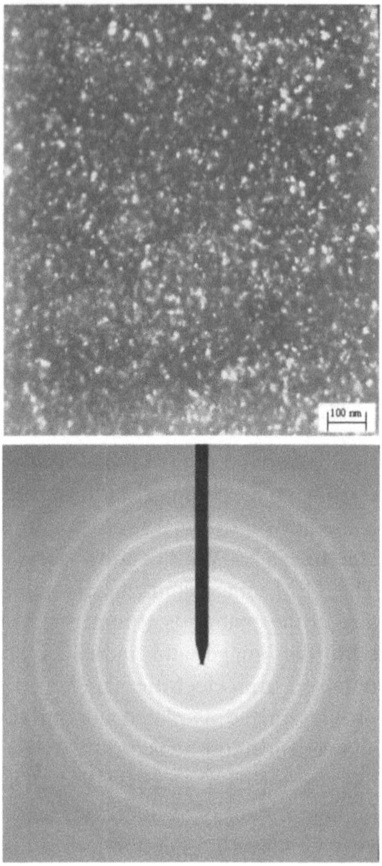

deposit lowers the hardness of the coating in as-plated condition. Hardness value is maximum at about 4 wt.% and it decreases with further increasing phosphorus and reaches a minimum at 11 wt.% phosphorus. Post-heat treatment has a significant impact on the hardness of the electroless nickel deposit. As a matter of fact, one of the outstanding characteristics of electroless nickel coatings is the possibility of obtaining very high hardness values (980–1,050 VHN) through an appropriate heat treatment. This provides excellent wear and erosion resistance. The parameters of the post-heat treatment (time and temperature) are defined by the phosphorus content. This transformation occurs at 400 °C for low phosphorus and 330–360 °C for high phosphorus content alloys [32]. However, in the region between 4.5 and 11 wt.% phosphorus, there is another transformation that occurs at 250–290 °C and causes the precipitation of fine phosphide particles throughout the coating. Many other sources have also reported that the maximum hardness is obtained after a heat treatment at 400 °C for 1 h [33]. Ni–B alloys are very hard and heat treatment can make them harder than hard chromium. They possess outstanding resistance to wear and abrasion.

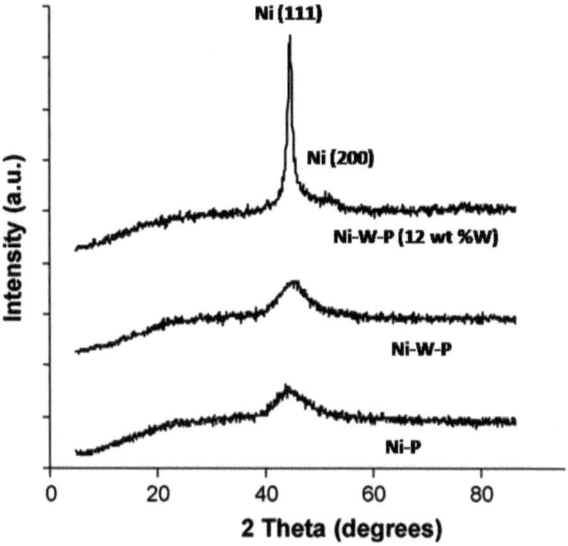

Fig. 6.4 XRD pattern obtained from electroless nickel phosphorous coatings

6.4 Tribological Properties

6.4.1 Electroplated Coating

Electroplated Cr coating is extensively used for protection against wear, erosion in industrial applications [34–36]. The wear mechanisms of electrodeposited Cr coating are primarily abrasive and adhesive in nature. If the Cr coating is subjected to thermal cycling, the hardness and hence the wear resistance of the coating decreases due to generation of thermal stresses [37]. It was reported [38–41] that the wear and corrosion resistances of electroplated Cr coatings are associated with their hardness and cracks. Heydarzadech Sohi and Kashi [34] have conducted the comparative tribological behaviour of hard and crack-free electroplated Cr coatings and found that with regards to the steel counterpart, hard electroplated Cr coatings with high hardness exhibited better wear resistance than crack-free electroplated Cr coatings. Arieta and Gawne [42] have conducted the wear resistance of Cr coatings under lubricated condition and found that the wear resistance of electroplated Cr coatings correlated to the oil wettability of the Cr coatings, which was greatly dominated by the coating's surface structure and cracks. Additionally, many interests were also focused on the corrosion resistance of electroplated Cr coatings. Snavely and Faust [43] found that the corrosion resistance of electroplated Cr coatings would be greatly affected by the crack density on the surface of electroplated Cr coatings. However, Huang et al. [44] has pointed out that the corrosion resistance of Cr coatings is affected slightly by the crack density.

Susceptibility to thermal stresses can be reduced by depositing the coating by pulse plating technique. Electrodeposited nanocrystalline Ni exhibited 125–175 times increase in wear resistance and 40–50 % reduction in friction coefficient [45]. Alloying Co with Ni can significantly improve hardness and wear resistance and can reduce friction coefficient [46]. Amorphous layers of Ni–W and Ni–Mo alloys show high hardness and high wear resistance [47]. Zinc alloys containing cobalt and iron result in diminished friction coefficient [48] whereas addition of Ni increases friction coefficient [49]. Bozzini et al. [50] noted that electrodeposited Au–Cu–Cd failed by microploughing and adhesive failure. Weston et al. [51] examined the tribological behaviour of nanostructured cobalt and cobalt–tungsten alloy electrodeposits sliding against a hard steel counterbody and compared the behaviour of these couples with that associated with sliding of electrodeposited chromium. The wear rates of the cobalt–tungsten alloy electrodeposits were at least an order of magnitude lower than those of the unalloyed cobalt electrodeposits. The wear rate of the cobalt–tungsten was also lower than that of the chromium electrodeposit. Haseeb et al. [52] noted that electrodeposited nanocrystalline Ni–W alloys have somewhat lower friction coefficient against steel counterbody as compared to that of the nickel–steel pair. Addition of tungsten to nickel is also seen to result in an improvement in wear resistance. Sriraman et al. [53] evaluated the hardness and sliding wear resistance of electrodeposited nanocrystalline Ni–W alloys. Crystallite size reduced with an increase in current density due to an increase in the W content. Ni–W coatings with 6–8 at.% W exhibited superior wear resistance. Electrodeposited composite coatings are found to be very useful for friction reduction [54].

6.4.1.1 Influence of Test Condition on Wear Rate

Applied load and sliding velocity are two important test parameters that govern wear behaviour of electrodeposited coating. As discussed earlier, chapter wear rate in plasticity dominated regime can be given as

$$W = k\frac{L}{H} \tag{6.3}$$

where k is known as wear coefficient, L is applied load and H is hardness of wearing material. Thus wear rate increases with increases of applied load as described in the work of Grigorescu et al. [55]. They found wear rate of Ni, Ni–Co and hard Cr coating increases with increase of applied load. Similar observation is also made by Panagopoulos et al. [56] as shown in Fig. 6.5. They investigated the sliding wear behaviour of zinc–nickel electrodeposited coatings on mild steel substrates using a spherical ball of hardened steel and alumina. For the studied range of normal load, the weight loss of various coatings seems to increase with increasing normal load. It is also observed that the weight loss of the zinc–nickel coatings against hard steel is lower than that against alumina. The main wear mechanism of the zinc–nickel coating sliding against stainless steel was noted to be severe shearing of the surface layers of the coating due to the ploughing action of the steel pins. For

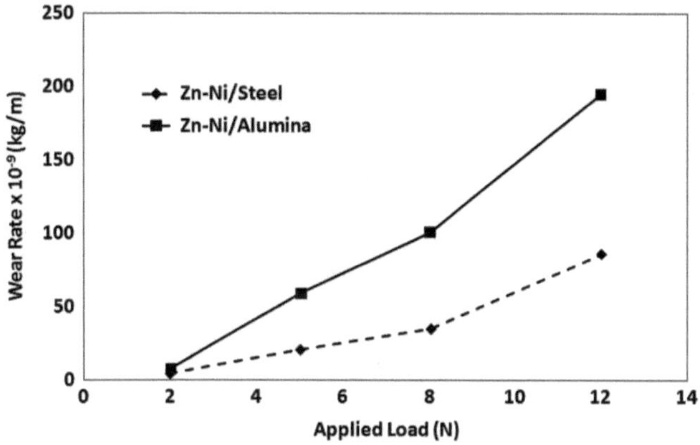

Fig. 6.5 The influence of applied load on the wear rate of electrodeposited zinc–nickel coating [56]

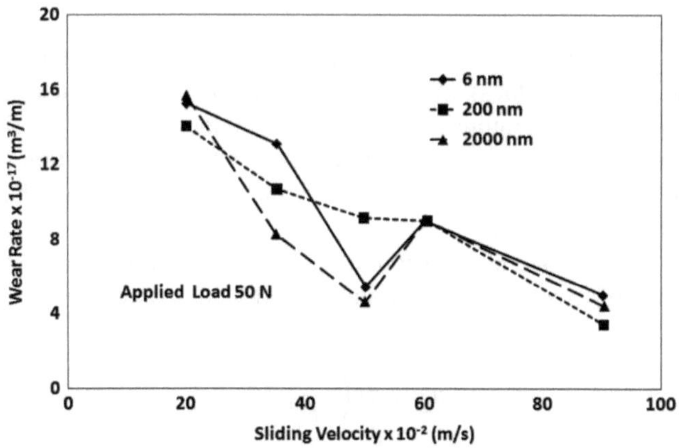

Fig. 6.6 Variation of wear rate of Ni–P–W multilayer coating with sliding velocity [57]

the wear experiments of zinc–nickel coatings against alumina spheres, a surface delamination mechanism is proposed to be the predominant wear mechanism of the coatings.

The influence of sliding velocity on wear rate is obtained by Panagopoulos et al. [57] for electrodeposited multilayer coating as described in Fig. 6.6. It is observed that decrease in sliding speed, while keeping the other parameters constant, leads to an increase in wear volume. Such behaviour is noted when wear takes place in mild oxidation wear regime. The wear rate in this regime can be given as [58, 59]

$$W = \frac{LC^2 A_o}{vZ_c H_o} \exp\left(-\frac{Q}{RT_f}\right) \tag{6.4}$$

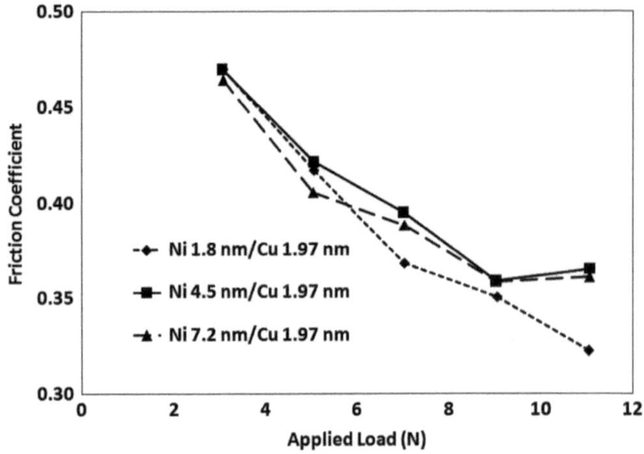

Fig. 6.7 Variation of friction coefficient of Ni/Cu multilayer coatings with applied load [60]

where A_o is Arrhenius constant, Q is the activation energy for oxidation, L is applied load, H_o is hardness of the oxide scale, Z_c is the critical thickness of spalling, T_f is the temperature at the place of contact of the asperities, v is the sliding velocities, C is a constant which depends on the composition of the oxide scale and R is the molar gas constant. It is clear from Eq. (6.4) that wear rate increases with decrease in v.

The applied load governs the friction coefficient. As discussed earlier chapters the dependence of friction coefficient on applied load is obtained as

$$\mu_a = \pi\tau\left(\frac{3R}{4E}\right)^{\frac{2}{3}} L^{-\frac{1}{3}} \tag{6.5}$$

where μ_a is the coefficient of friction, τ is the average shear strength during sliding and L is the applied load. E is the effective elastic modulus and R is the effective radius of curvature. Thus, the coefficient of friction is inversely related to the cube root of the applied load. The observation of Ghosh et al. [60] follows the above relation. The electrodeposition of Ni/Cu multilayer coatings with varied Ni sublayer thickness (t_{Ni}) ranging 1.8–11 nm, keeping Cu sublayer thickness (t_{Cu}) fixed at 1.97 nm, has been carried out from sulphate-based electrolyte. Friction coefficient and sliding wear test of the deposited multilayer coatings have been performed against standard WC ball in ball-on-plate geometry under dry condition. The friction coefficient was found to be dependent on load and t_{Ni}. A typical sinusoidal variation of friction coefficient with sliding distance was noticed indicating stick–slip behaviour under adhesive–abrasive wear process. The variation of friction coefficient with applied load for various coatings is shown in Fig. 6.7. It is clear from this graph that all multilayer coatings exhibit reduction in friction coefficient with increase in load. This behaviour is in conformity with

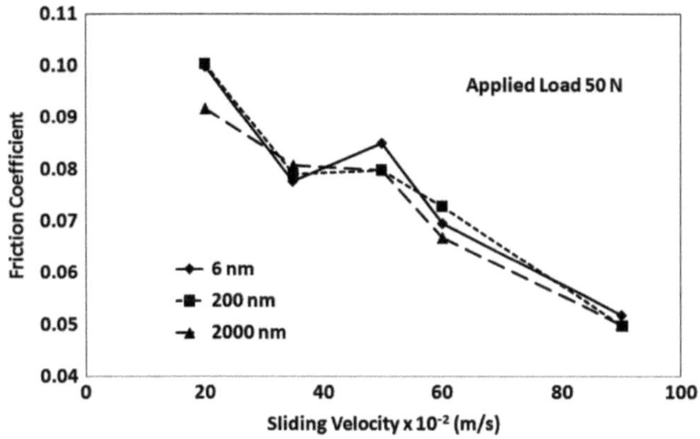

Fig. 6.8 Variation of friction coefficient of Ni–P–W multilayer coating with sliding velocity at an applied load of 50 N [57]

Eq. (6.5). When compared with pure Ni and Cu coatings [61], coating with comparable layer thickness exhibits similar to pure Ni behaviour at low load and pure Cu behaviour at higher load. In this case, due to very small sublayer thickness of both the components, the multilayer effect is not prominent because of the proximity of alternate interfaces. Thus, friction coefficient follows pure component behaviour at respective loads. The friction coefficients of multilayer coatings although are not significantly different, have shown an interesting trend, i.e. with increase in t_{Ni}/t_{Cu} ratio, average friction coefficient decreases.

Finally the effect of sliding velocity on the friction coefficient is shown in Fig. 6.8 due to Panagopoulos et al. [57] for Ni–P–W multilayered alloy coatings sliding against hardened steel discs. It is seen that a decrease in sliding speed leads to a gradual increase in the friction coefficient. This result suggests a strong dependence of the friction coefficient on the sliding velocity. Further, the thickness of the multilayered coatings has no effect on the results. The above behaviour can also be explained by equation given below [58].

$$\mu = K_1 - K_2 \mathrm{Ln}(v) \tag{6.6}$$

where μ is coefficient of friction and v is sliding velocity. K_1 and K_2 are constants. Thus friction coefficient is expected to decrease with increase in sliding velocity in accordance to Eq. (6.6) and the behaviour depicted in Fig. 6.8 is in conformity with the Eq. (6.6).

6.4.1.2 Influence of Coating Condition on Tribological Properties

It is well known that pulse electroplating is more effective than the traditional DC electroplating to make composite and alloy due to its independent, controllable parameters and the high instant current [62]. When there are particles among the

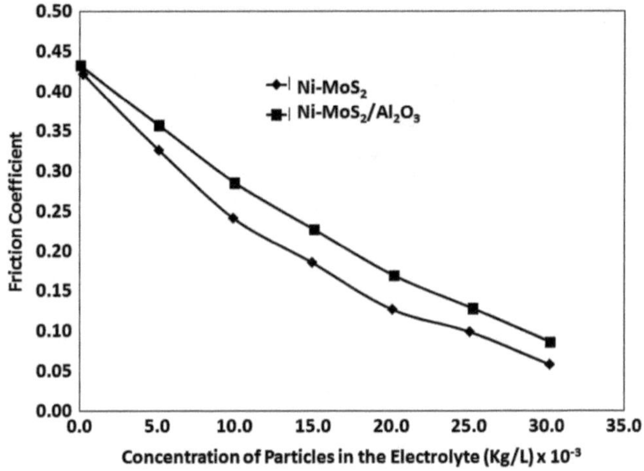

Fig. 6.9 The influence of concentration of particles in the electrolyte on the friction coefficient of electrodeposited Ni–MoS_2 and Ni–MoS_2/Al_2O_3 coating [78]

plating bath, the particles will be absorbed onto the plate. Such kind of coating plate is called composite plating. Many researchers have confirmed that the multi coating possess of many specific properties due to the different particles in the coating [63–67]. It has been reported that low-friction composite materials such as Ni–polyethylene [68], graphite–brass [69], Ni–PTFE [70, 71], graphite–bronze [72], Ni–P–carbon-nanotube [73] and good wear resistance composite materials such as Ni–Al_2O_3 [74], Ni–P–SiC [75], Ag–ZrO_2 [76], Ni–ZrO_2 [77] can be obtained via the electroplating technique. Work due to Huang and Xiong [78] presented in Fig. 6.9 and Fig. 6.10 indicates the tribological behaviour of the electroplated Ni–MoS_2 and Ni–MoS_2/Al_2O_3 coating. An increase in the amount of both MoS_2 and MoS_2/Al_2O_3 particles in the solution led to a decrease of the coefficient of friction (see Fig. 6.9) and increase weight loss in wear test (see Fig. 6.10). The friction coefficient of Ni–MoS_2 coating is lower than that of Ni–MoS_2/Al_2O_3 coating. As MoS_2 particles concentration in solution reaches to 20 g/l, weight loss increases sharply with the increase of concentration. This can be attributed to the soft nature of the MoS_2 particles. By contrast, weight loss increases slowly when the MoS_2/Al_2O_3 particles increases, due to the hard nature of the Al_2O_3 particles which were coated on the MoS_2 particles. The wear morphology demonstrates that the worn scars of the Ni–6 vol.% MoS_2/Al_2O_3 coating are much shallower and the surface is smoother than the Ni–6 vol.% MoS_2 coating.

The effect of surfactant content in the electrolyte on tribological performances of electrodeposited coating is studied by Chen et al. [79] who electrodeposited Ni–Al_2O_3 composite coatings from a Watts-type bath containing particles in suspension, which were dispersed with the aid of a cationic surfactant hexadecyl-pyridinium bromide (HPB). They investigated the effect of the surfactant on the tribological properties of composite coatings. The influence of concentration of

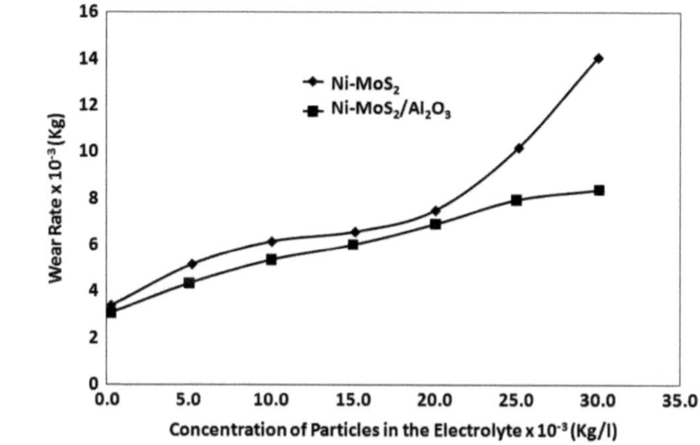

Fig. 6.10 Variation of wear rate with concentration of reinforcing particles in the electrolyte for electrodeposited Ni–MoS_2 and Ni–MoS_2/Al_2O_3 coating [78]

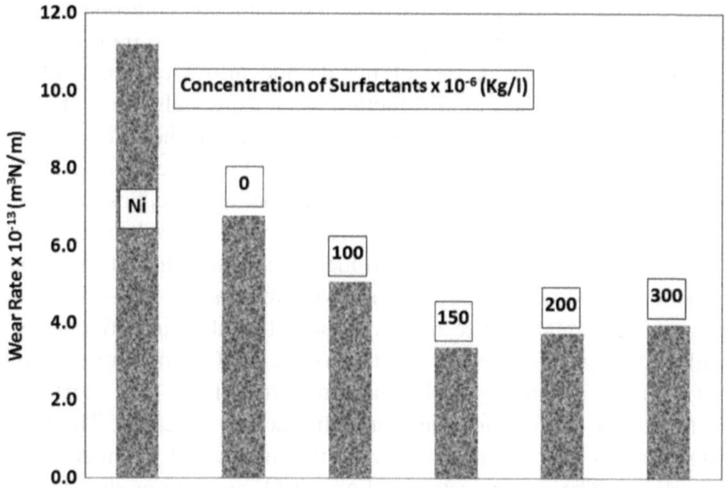

Fig. 6.11 Bar diagram showing the effect of concentration of surfactants on the wear rate of electrodeposited Ni–Al_2O_3 coating [79]

surfactant HPB on the wear resistance of coatings under dry sliding condition is illustrated in Fig. 6.11. It is observed that the Ni–Al_2O_3 composite coatings exhibited lower wear rates when compared with pure Ni coating. The wear resistance of composite coatings would reach maximum at 150 mg/l of HPB and decreased slightly with further increase in the concentration of surfactant continuously. Similarly initial decrease of wear rate with increase of surfactant and subsequent increase of wear rate with surfactant concentration is reported by Gull et al. [80]. However, they found the friction coefficient increases with surfactant concentration.

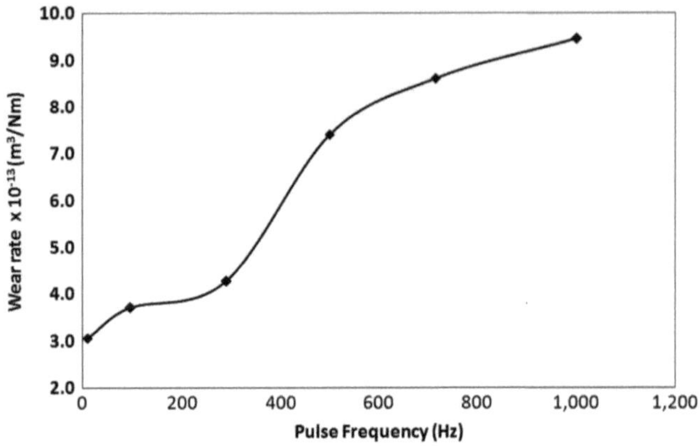

Fig. 6.12 Influence of pulse frequency on the wear rate of electrodeposited Ni–Al_2O_3 coating [79]

It is well known that pulse electrodeposition is one of the most effective methods in fabrication of metals and alloys due to its independently controllable parameters and higher instantaneous current densities when compared to traditional DC electrodeposition [78]. The properties of metals and alloys can be controlled and improved by modifying their microstructure when using pulse current. As a result, the effects of pulse electrodeposition parameters on the microstructure and properties of coatings have been reported in many literatures. The effect of pulse frequency on the wear rate of electrodeposited coating is reported by Chen et al. [79]. Nickel matrix composite coatings reinforced with sub-microsized Al_2O_3 particles were produced by pulse electrodeposition and the effect of pulse frequency on the microstructure, hardness and wear resistance of Ni–Al_2O_3 composite coatings were investigated. The results showed that the pulse frequency significantly influenced the preferred orientation of Ni–Al_2O_3 composite coatings; the texture of the coatings progressively changed from a strong (111) preferred orientation to a random orientation when pulse frequency increased. The hardness of composite coatings decreased slightly with the increase of volumetric content of alumina particles. The wear resistance of Ni–Al_2O_3 coatings as shown in Fig. 6.12 decreased with the increase of incorporated alumina particles under dry sliding wear condition and this was largely influenced by the microstructure of Ni matrix due to the presence of adhesive wear.

Gul et al. [80] prepared Ni/Al_2O_3 metal matrix composite (MMC) coatings from a modified Watt's type electrolyte containing nano-Al_2O_3 particles by direct current (DC) plating method to increase the surface hardness and wear resistance of the electrodeposited Ni. They reported the effect of current density on wear rate and friction coefficient. They found that both wear rate and friction coefficient decreases with increase in current density. Their result on friction coefficient is provided in Fig. 6.13. The increase in current density leads to decrease in grain size, increase in reinforcing particle concentration and lattice distortion. All these phenomena put together is responsible for observed effect.

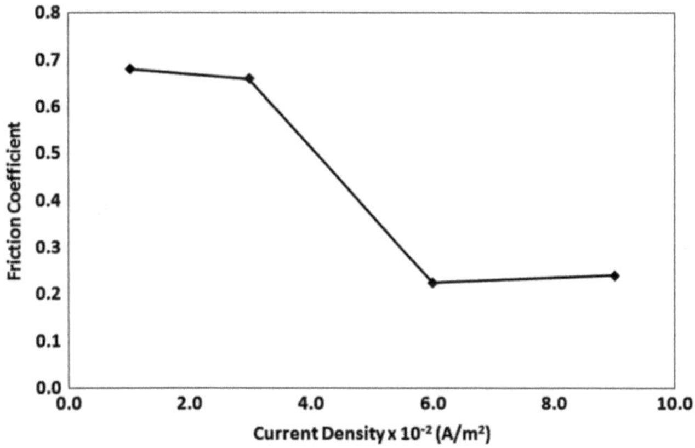

Fig. 6.13 The influence of current density on the friction coefficient of electrodeposited Ni–Al_2O_3 coating [80]

6.4.1.3 Influence of Coating Characteristics on Tribological Properties

Influence of alloying of the electroplated coating on tribological properties is studied by Panagopoulos et al. [55, 81]. They investigated the wear behaviour of electrodeposited zinc coatings alloyed with iron and nickel on copper substrates against stainless steel. They found that the friction coefficient of Zn coating decreased with increasing load and with addition of alloying elements as shown in Fig. 6.14. Further, the friction coefficient was least when 16 % iron was added. They also found that addition of nickel did not alter the friction coefficient as compared to iron. The influence of Fe and Ni addition on friction coefficient of Zinc is shown in Fig. 6.14. It is known that several components contribute to the value of friction coefficient. The most important of those components are (1) the adhesion between the contact materials, (2) the ploughing of the softer material by the asperities of the harder one and (3) the surface roughness of the two counterfaces. The main wear mechanism of these coatings was noted to be plastic deformation and shearing of the surface layers of the coatings due to the ploughing action of the harder steel spheres. As the hardness of the coating increases with alloying addition, the ploughing action by the asperities of harder stainless steel decreases giving rise to lower friction coefficient. The higher friction coefficient of Zn–Ni coating may be attributed to higher adhesion between Zn–Ni film with stainless steel counterface.

The friction coefficient (μ) of rough contacting surfaces are given by

$$\mu = \mu_{\text{ad}} + \mu_{\text{grov}} \tag{6.7}$$

where μ_{ad} is friction coefficient due to adhesion and μ_{grov} is friction coefficient due to grooving. Friction coefficient due to adhesion can be obtained following the equation suggested by Roychoudhury and Pollock [82] as

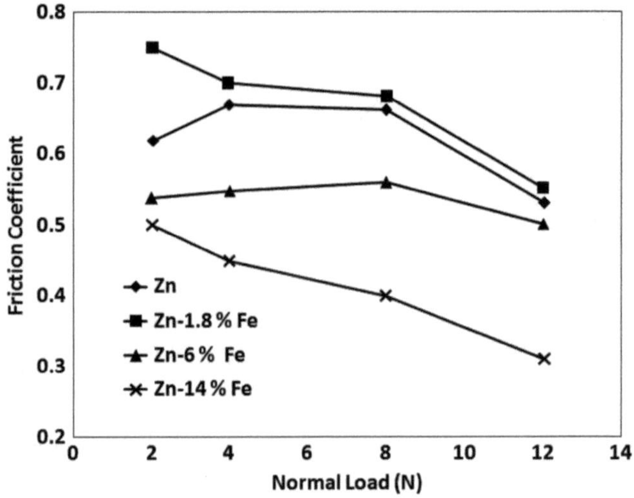

Fig. 6.14 Dependence of friction coefficient on normal load and amount of alloying element of various electrodeposited Zn alloy coating [81]

$$\mu_{\mathrm{ad}} = \frac{\tau_m}{H}\left[\left\{1 - \frac{\gamma_{\mathrm{ad}}}{H \times S}\right\}^{-2} - 1\right] \tag{6.8}$$

where τ_m is the average shear strength of the adhesion junction or the coating, H is the hardness, γ_{ad} is the work of adhesion and S stands for standard deviation of asperity height. Further the friction coefficient due to grooving is given by

$$\mu_{\mathrm{grov}} = \frac{\tau_{\mathrm{c}} \times A}{L} \tag{6.9}$$

where L is the applied load, A is the area sheared plastically by an abrasive particle and τ_{c} is the critical shear strength of the coating.

At low applied load as is the case in Fig. 6.14, friction coefficient due to adhesion and friction coefficient due to grooving are comparable. At a given applied load, grooving components of all coatings are equal according to Eq. (6.9). However, addition of iron to the coating increases the hardness and thereby decreases the friction coefficient due to adhesion as per Eq. (6.8). This behaviour is reflected in Fig. 6.14. Further, for a given coating friction coefficient due to adhesion is same. However, friction coefficient due to grooving decreases with applied load according to Eq. (6.9). Thus the decrease of friction coefficient with applied load as shown in Fig. 6.14 is also consistent with theoretical estimation. Similar trend for applied load is noted in Fig. 6.15 also. It is to be noted that unlike addition of iron, addition of nickel does not decrease the friction coefficient. This is related to the fact that the hardness does not increase significantly with addition of nickel and also the work due to adhesion is more with the coating containing nickel than with the coating containing iron.

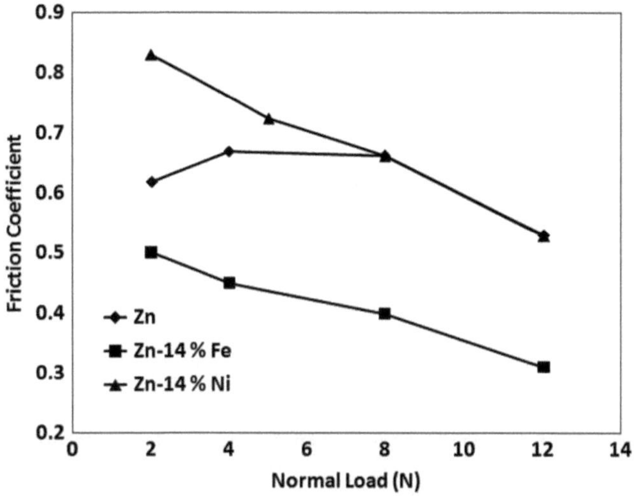

Fig. 6.15 Effect of normal load on friction coefficient of various electrodeposited Zn alloy coating [81]

Capel et al. [83] examined the wear behaviour of two alloy systems, namely Co–W and Co–W–Fe by electrodepositing onto steel substrates. It was found that under some conditions, the Co–W alloys exhibited a wear resistance and a friction coefficient similar to that of hard chrome. Moreover, the corrosion resistance of the Co–W alloys was higher than that of the hard-chrome plate. The Co–W–Fe alloys exhibited a high as-deposited hardness. All the Co–W–Fe coatings were more corrosion resistant than the hard chrome. However, even the most wear resistant Co–W–Fe coating exhibited a high friction coefficient and more wear than the hard chrome.

The influence of reinforcing particles on the wear behaviour can be estimated using the relation proposed by Zum Ghar et al. [84] as given below.

$$\frac{1}{W} \infty \frac{d^{3/2}V}{\lambda} \tag{6.10}$$

where W is the wear resistance, d is mean diameter of reinforcing particles, V is the volume fraction of reinforcing particles and λ is the mean free path of the particles in the matrix. Thus, according to Eq. (8.12) wear resistance will increase or in other word wear rate will decrease with increase in volume fraction of reinforcing particles. Such behaviour has been noted in the work of Vaezi et al. [85] for electrodeposited Ni coating reinforced with SiC nanoparticles. Their result is illustrated in Fig. 6.16. Ni–SiC nanocomposite coatings with different contents of SiC nano-particulates were prepared by means of the conventional electrodeposition in a nickel-plating bath containing SiC nano-particulates to be co-deposited with them. The microhardness, wear and corrosion resistance of the nanocomposite

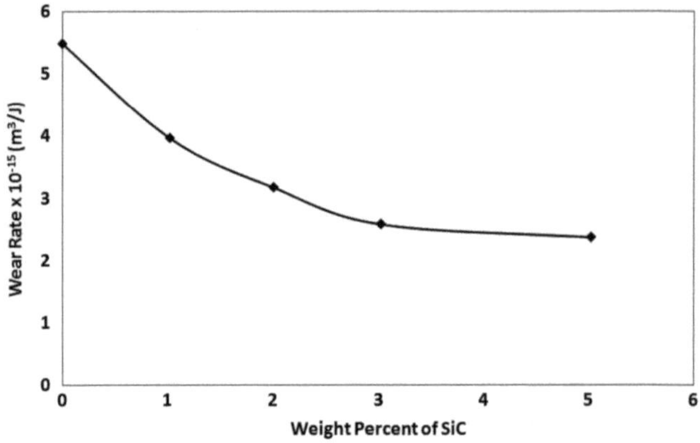

Fig. 6.16 Variation of wear rate with amount of reinforcing particles for electrodeposited Ni–SiC nanocomposite coating [85]

coatings also increased with increasing content of the SiC nano-particulate in the coating. Similar observation is also reported by Xue et al. [86] for Ni–CeO_2 coating, Hou et al. [87] and Garcia et al. [88] for Ni–SiC composite coating. Xue et al. developed Ni–CeO_2 composite coatings by electrodeposition method from a nickel sulfamate solution containing CeO_2 particles. They reported that the composite coating with a lower CeO_2 content of 2.3 wt.% shows somewhat increased wear resistance than the pure Ni coating, while the composite coating with a higher CeO_2 content has much better wear resistance than the pure Ni coating. Study by Hou et al. [87] attempted to investigate the wear resistance of Ni–SiC composites manufactured by electro-codeposition process. The volume fraction of SiC is found to be closely related to the concentration of submicron SiC and the surfactant in the bath. It is found that the surfactant not only disperses the SiC particles well in the electrolyte therefore leading to better homogeneously distributed SiC particles in the matrix but also enhances the SiC vol.% embedded in co-deposition layer. It is confirmed that the wear resistance of co-deposited layer increases with increasing submicron-size SiC volume fraction. Garcia et al. [88] co-deposited SiC particles of three different sizes, namely 5, 0.7 and 0.3 mm, with nickel from Watts' solutions. It is found that for a given number density of particles in the plating solution, the number density of particles in the coating increases with decreasing particle size. The friction and wear behaviour of these composite coatings was evaluated in uni- and bi-directional sliding tests against corundum balls. The best sliding wear resistance was obtained with Ni–SiC composite coatings containing 4–5 vol.% submicron SiC particles.

Another reason for decrease in wear rate of composite coating with increasing volume fraction of reinforcing particles is decrease in friction coefficient of composite coating. The friction coefficient decreases with increase in amount of reinforcement. The influence of quantity of reinforcing particles on the friction coefficient of electroplated Ag–ZrO_2 and Ni–WC coating is shown in Fig. 6.17 [50].

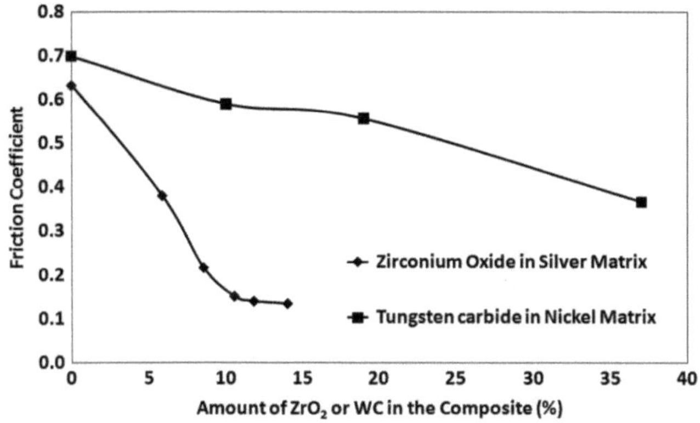

Fig. 6.17 Variation of friction coefficient with amount of reinforcing particles in electrodeposited Ag–ZrO_2 and Ni–WC coatings [53]

Clearly friction coefficient decreases with increase of particle concentration and this in turn, reduces wear rate. Benea et al. [89] also reported reduction of friction coefficient particles of Ni coating reinforced with SiC particles. Low wear rate and low friction coefficient of nanocomposite WC/CoNi and WC/CoNiP coatings also noted by Wang et al. [90]. The wear studies by Srivastava et al. [91] indicated that the wear volume loss was less for micron size SiC reinforced Ni–Co composites compared to nano size SiC reinforced composites. Interestingly, Wang et al. [92] reported that the wear resistance of pure nanocrystalline electrodeposited Ni plating was superior to the wear resistance of nanodiamond-reinforced nanocrystalline Ni coating.

Nanocrystalline electrolytic coating not only exhibits improved wear resistance but also exhibits increase in wear resistance with decreasing grain size. Decrease in grain size results in increase in strength and ductility according to Hall–Petch equation as given below:

$$\sigma_o = \sigma_i + k^/ D^{-\frac{1}{2}} \tag{6.11}$$

where σ_o is yield strength, σ_i is friction stress opposing movement of dislocation, $k^/$ is unpinning constant and D is diameter of the grain. Wear data are expressed by the Taber Wear Index (TWI), which is the weight loss in mg per 1,000 cycles:

$$\mathrm{TWI} = \frac{W_\mathrm{f} - W_\mathrm{i}}{N} \times 1{,}000 \tag{6.12}$$

where W_f and W_i are the initial and final weight of the specimen and N is the total number of cycles. The work due to Jeong et al. [45] clearly demonstrates the validity of above statement. The result of their work is provided in Fig. 6.18 which shows increase of hardness and decrease of Taber Wear Index with decreasing grain size.

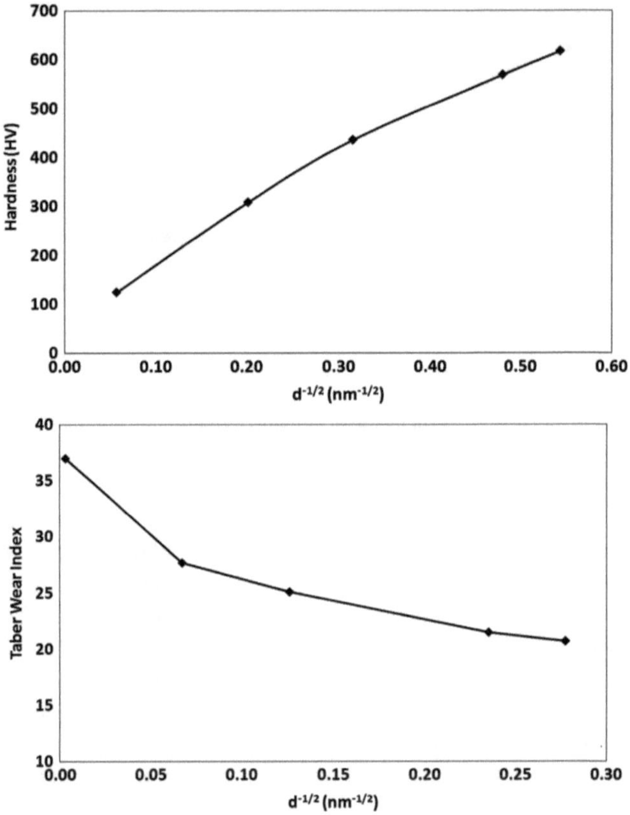

Fig. 6.18 The influence of grain diameter on the hardness and Taber Wear Index for nanocrystalline Ni coating [45]

The SEM image showing the morphology of the worn surface of electrodeposited Ni–Al_2O_3 composite coating is shown in Fig. 6.19 [79]. The worn surface is characterised by severe plastic deformation indicating adhesive wear as the material removal mechanism. Some continuous grooves along the wear track can also be seen. Presence of such grooves confirms ploughing action during sliding. No delamination crack can be seen on the worn surface.

6.4.2 *Electroless Plating*

Electroless nickel coatings are very smooth and lubricious [9] due to their unique nodular microstructure resembling that of a cauliflower [93] as shown in Fig. 6.2. A beneficial characteristic of electroless nickel deposition is the superior wear resistance of the coatings. However, addition of various particles and elements and different surface treatments lead to a change of the conventional microstructure of

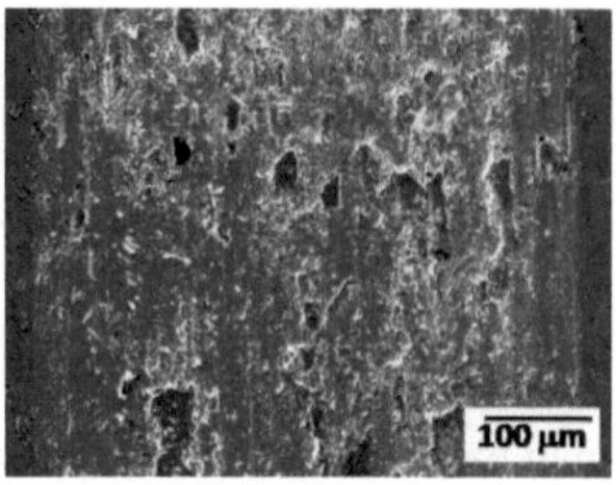

Fig. 6.19 The SEM image showing the worn surface of electrodeposited Ni–Al_2O_3 composite coating [79]

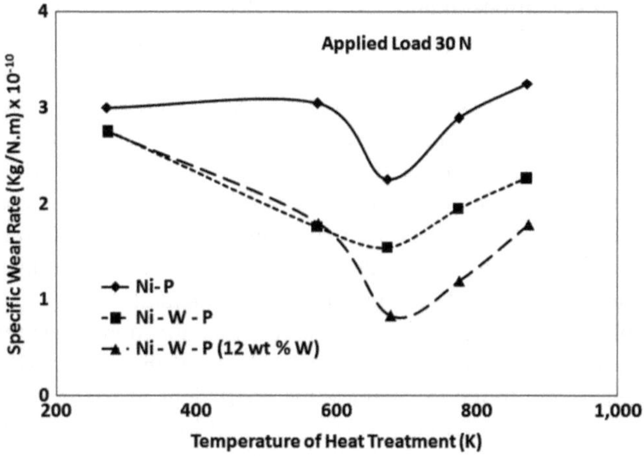

Fig. 6.20 Variation of specific wear rate of Ni–P coatings with heat treatment temperature [97]

the coating resulting in modified friction behaviour. Heat treatment in general results in a reduction of the friction coefficient of various electroless coatings compared to the as deposited coatings. Theoretically, there is a correlation between wear resistance and hardness of a surface. However, the wear properties of a surface are affected by numerous parameters such as the nature of the applied stress, surface morphology, etc. The wear resistance of electroless nickel deposits depends on both phosphorus content and the type of post-heat treatment applied. The higher the hardness, the lower is the weight loss due to wear. According to some researchers, higher the hardness, less the deformation and consequently there is less intimate contact and less welding and friction [94–96]. Hard surfaces are also less likely to have shear fracture occur at the original weld, resulting in lesser wear. However, Gawne and Ma [94] showed that electroless Ni–P coatings heat treated at 600 °C exhibited lower wear rate than coatings heat treated at 400 °C even though the hardness of latter deposit was high. The influence of heat treatment on the specific wear rate of electroless Ni–P coating containing various amount of W is shown in Fig. 6.20 [97]. It is clear from Fig. 6.20 that Ni–W–P coatings show higher wear resistance. This is due to the tungsten which induces higher hardness by solid solution strengthening. Specific wear rate is seen to be the lowest in the case of coating subjected to heating at 673 K. This is due to the maximum hardness of the coatings achieved by heating at this temperature. When heated beyond this temperature, the specific wear rate is found to increase due to the softening of coatings by grain coarsening. Wear of electroless Ni–P coatings also depends on the type of counterface material used. The lowest wear rates are obtained when both contacting surfaces have heat treated electroless Ni–P coatings. Principal advantages of electroless Ni–B deposits are their superior hardness and wear resistance in as-deposited condition [31, 93]. Electroless Ni–B coatings are more wear resistant than tool steel and hard chromium coatings and can replace gold in the electronics industry [31, 98]. The columnar structure of Ni–B coatings is useful in retaining lubricants under conditions of adhesive wear [99].

The heat-treated Ni–B deposits are found to present a virtually incompatible surface for the hard counterface material and reduce friction coefficient [100]. Several studies have been performed on Ni–P coatings [101, 102]. The electroless Ni–P based composite coatings possess better wear resistance than Ni–P alloy coatings [103]. Amplitude of the friction coefficient oscillation of Ni–P coating after tempering is found to be lower than that of as-received Ni–P coating [104]. It is experienced that adhesion is the main cause of wear of electroless coatings against steel but no direct correlation between wear rate and hardness has been observed during the tests [105]. It has been experimentally observed that regardless of phosphorus content, the composite coatings are more wear resistant than base Ni–P alloy as a consequence of high hardness of the co-deposited particles. Heat treatment is also found to decrease the coefficient of friction of Ni–P–ZrO_2 coatings [106]. The laser-irradiated surface showed lower friction coefficient than that of the furnace-annealed surface with similar hardness [107]. The addition of hard particles such as B_4C and SiC is found to be responsible to increase the friction coefficient of electroless nickel coatings. The coefficient of friction of electroless Ni–P coating is found to increase after addition of B_4C particles [108, 109] due to the fact that particles of B_4C are separated during testing and acted as abrasives after being trapped between the specimen surface and the test pin [110]. The friction coefficient of electroless Ni–P–B_4C composite coating with 25 vol.% of B_4C is approximately doubled with increasing the loads from 15 to 60 N [109]. In a recently developed Ni–P–Al_2O_3–ZrO_2–Al_3Zr composite coating, it is observed that the wear rates of aluminium and steel substrates are significantly higher than that of the corresponding electroless Ni–P–Al_2O_3–ZrO_2–Al_3Zr-coated pin samples. The co-deposition of SiC particles in Ni–P coating also increases the friction coefficient of the coating. Wu et al. [110] found that co-deposition causes an increase of about 10 % in friction coefficient when compared to pure Ni–P coating. On the other hand, friction coefficient showed a drastic reduction when soft particles viz. PTFE, graphite [111] and MoS_2 were introduced in the coating. The incorporation of PTFE [112] and graphite [113] in Ni–P matrix decreased the friction coefficient markedly and facilitated the stable state of the whole wear course. Wu et al. [112] noted that the friction coefficient of Ni–P coating to decrease as high as 70 % after the introduction of PTFE in its matrix. The PTFE-rich mechanically mixed layer (PRMML) formed on the worn surface is responsible for the good anti-friction properties. The SiC particles when mixed with

PRMML, plays a load-bearing role in protecting PRMML from shearing easily. A continuous supply of PTFE to the tribo-surface is an important precondition for the formation of PRMML and the exertion of its anti-friction properties. The low friction coefficient of Ni–P matrix with PTFE particles causes a little change of the temperature and further ensures the stable state of the whole sliding. The relatively high carbon content in the form of graphite was responsible for the lower friction coefficient Ni–P–Gr compared to Ni–P–Gr–SiC coating. In case of wear of graphite-reinforced Ni–P coatings, a graphite-rich film (GRF) formed on the worn surface of the Ni–P matrix with graphite and this seemed to be responsible for good tribological behaviour. A tribolayer is found to cover the worn surface. Since it consists of materials from the composite and the counterface as well as environment, it was denoted as a 'mechanically mixed layer' (MML) or graphite-rich mechanically mixed layer (GRMML). Because of the softness and lamellar structure of graphite, the micro-hardness of Ni–P–Gr coating was deteriorated and the subsurface cavities containing graphite particles tended to deform with the subsurface matrix, squeezing out graphite to the wearing surface during the sliding process under high pressure. The graphite smeared on the wearing surface layer by layer and transferred to the mating surface, changing the test system [113]. The presence of the GRMML decreased the direct contact area between the testing sample and the counterface, therefore improving the wear resistance of the hybrid composite. Although PTFE promotes lubricity in the coating, Ramalho and Miranda [113] found PTFE not to be effective in the friction reduction. They explained this anomaly by the fact that the phosphorus content by itself leads to low friction and the PTFE lubrication effect is not enough to balance the effect of increasing the surface roughness induced by the PTFE particles. Hence, the friction coefficient increased. Friction coefficient of the Ni–P–nano-MoS_2 composite coating decreased greatly compared to those of Ni–P electroless coatings [114]. The wear rate and the friction coefficient of the coatings showed a decreasing trend with the increase of the volume fractions of IF–MoS_2 [115] as shown in Figs. 6.21 and 6.22. Moreover, Ni–P–(IF–MoS_2) composite coating displayed excellent friction properties in vacuum implying the good stability of IF–MoS_2 in different environments. The favourable effects of IF–MoS_2 nanoparticles on the friction properties are attributed to their unique fullerene-like structure. Further decrease of wear rate with increase of volume fraction of IF–MoS_2 is consistent with Eq. (6.10). Hosseinabadi et al. [116] studied the wear of electroless Ni-P and Ni–P–B_4C composite coatings. Auto catalytic reduction of Ni in nickel sulphate and sodium hypophosphate bath including suspended B_4C particles with different concentration was used to create composite coatings with 12, 18, 25 and 33 vol.% of B_4C particles. Coatings 35-μm thick were heat treated at 673 K for 1 h in an argon atmosphere and the wear resistance and friction coefficient of heat-treated samples were determined by block-on-ring tests. All wear tests were carried out at 297 K, 35 % moisture, 0.164 m/s sliding speed and about 1,000 m sliding distance. Results show that an electroless Ni–P–B_4C composite coating with 25 vol.% of B_4C has the best wear resistance against a CK45 steel counterface.

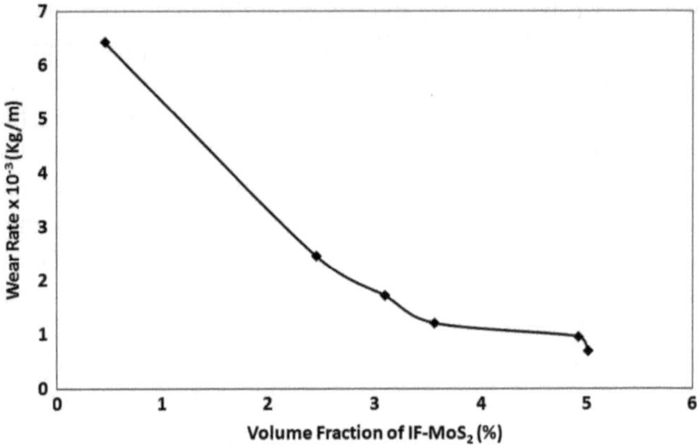

Fig. 6.21 The effect of the volume fraction of IF–MoS_2 on the wear rate of electroless Ni–P–(IF–MoS_2) composite coatings [115]

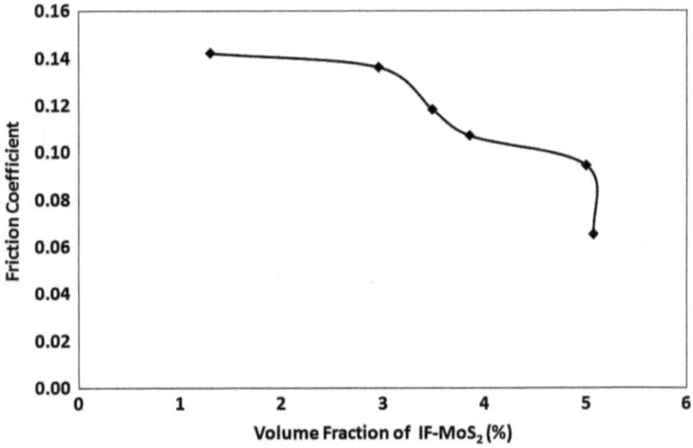

Fig. 6.22 The effect of the volume fraction of IF–MoS_2 on the friction coefficient of electroless Ni–P–(IF–MoS_2) composite coatings [115]

Chen et al. [117] studied the tribological behaviour of carbon nano tubes (CNT) incorporated electroless nickel coatings and compared them with Ni–P–graphite and Ni–P–SiC. They found that Ni–P–CNT showed the best lubricating property which can be related to the self-lubricating property of the CNTs. CNTs comprise concentric cylindrical layers or shells of graphite-like sp^2-bonded cylindrical layers or shells, where the intershell interaction is predominately van der Waals, can easily slide or rotate each other, leading to a low friction coefficient. Also friction coefficients decrease with increase of load from 10 to 30 N. Incorporation of SiO_2 nano particles contributed to the improvement of the friction-reducing ability of electroless Ni–P matrix [118]. Lee [119] studied the friction behaviour of Ni–P coatings in lubrication with brine solution and found that the strong passivity of the coating acted as a lubricator and reduced the friction coefficient. Parametric studies

Fig. 6.23 SEM images of the worn surface of electroless Ni–P coating [97]

involving the variation of bath composition to reduce the friction coefficient of Ni–P coatings have also been conducted [120] and these studies revealed that annealing temperature and bath temperature have significant influence in controlling the friction characteristics of electroless Ni–P coating.

Ternary alloy coatings are used to provide enhanced performance in specific properties over conventional electroless coating systems. By incorporating a third element in significant or trace levels, the basic structure and physical properties of the coating can be altered. Deposition of a ternary alloy means that all components should be plated simultaneously. The distribution of the third element through coating thickness is interesting not only from the standpoint of electroless alloy formation but also because it may affect coating properties, particularly adhesion if there are some specific features of its profile at the interface [121]. There are many limitations to the production of ternary alloys. It is clear that the main metal (nickel) must be capable of being catalytically deposited. The reduction of the second metal which participates in the alloy is determined by its standard electrochemical potential as well as its catalytic properties in relation to the reduction process. A few metals which do not possess the relevant catalytic properties in relation to the reduction process and are not catalytic poisons can also be co-deposited with nickel. Incorporation of certain elements like tungsten increases the hardness [122] and wear resistance [97] of conventional Ni–P deposits due to solid solution strengthening. SEM images of the worn surface of electroless Ni–P coating is presented in Fig. 6.23 [97]. Worn surface indicates mild adhesive wear. Material removal in patches is noticed. Presence of shear dimples along worn surface indicates ductile fracture of the worn surface. Some delamination crack can also be seen on the worn surface. Thus delamination wear of the coating has also taken place.

6.5 Applications of Electrolytic and Electroless Coatings in Tribology

Most widely used electroplated coating for wear resistance application is hard Cr plating. Cr-plated piston ring exhibits five times higher life than a ring without plated. The shock absorber rods and struts of automobiles are Cr plated. Hydraulic

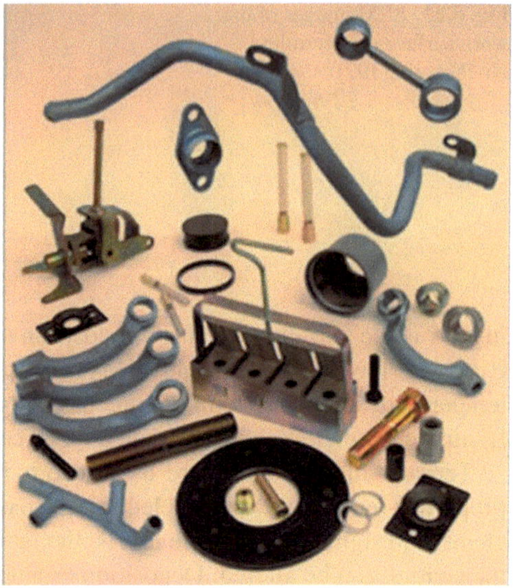

Fig. 6.24 Several electrodeposited components (Trenton Electroplating, UK)

shafts of all kinds of equipments are 20–30 μm thick Cr plated. A variety of forming tools are coated with hard chromium plating in order to reduce wear and friction. Steel and beryllium copper dies for moulding of plastics are usually coated with hard chromium plating. Deep drawing tools are often plated with hard chromium. Various types of gauges are also plated with hard chromium in order to improve their service life. Service life of many cutting tools such as taps, reamers, flash plates, etc. are extended by chromium plating. Hard chromium is also used in printing plates and stereotypes. Texture roller of alluminium automobile engine parts are iron or Ni plated. For reliable electrical contact indium plating is used in electronic industry. Multiple layers of Cu/Ag or Cu/Co plating is used for rocket thurst alignment. Several electrodeposited components coated in Trenton Electroplating, UK are shown in Fig. 6.24.

Electroless Ni–P coatings are applied to various travellers. The ring traveller as shown in Fig. 6.25 with Ni–P coatings results in increase in spindle speed which in turns increases the yarn tenacity and strength, increases packing coefficient and promotes higher compactness and interlocking of the fibres. Application of Electroless Ni–P ternary alloy coating provides better wear resistance, smooth surface finish, better elongation properties, increase in speed and increased the life-time of the traveller. If no lubricating film exists, the contact area of the traveller/ring might reach 800 °C giving rise to micro-welding, premature wear and tear of the ring, erratic traveller operation, differences in the yarn tension and reduction in the yarn quality. Several other varieties of travellers are also coated with electroless low friction coatings.

Electroless coatings have found widespread applications in many industries, such as printed circuit boards, magnetic storage media, microelectronics, radio-electronics, computer engineering, aviation, aerospace, petroleum, chemistry, machinery, textiles, automotives and metallization of plastic [123], etc. From the

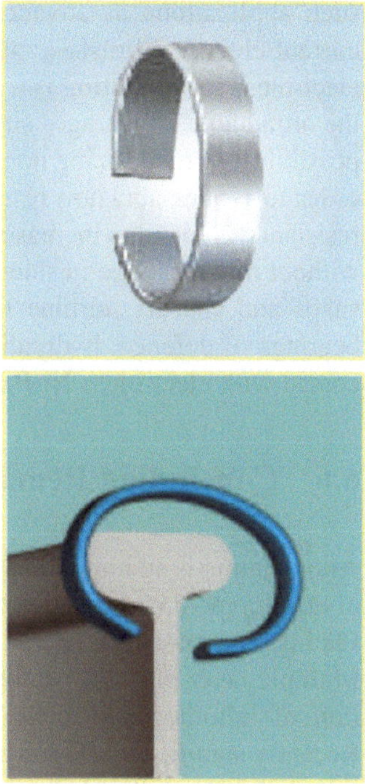

Fig. 6.25 The ring traveller coated with electroless Ni–P film

tribological view point, electroless coatings are mainly applied due to their hardness and wear resistance, lubricity (low friction coefficient) and corrosion-resistant properties. Electroless nickel phosphorus coatings have been successfully applied on GFRP substrate for making blades of offshore wind turbine installations [115]. Depositing Ni–P coatings on GFRP substrate increases its wear–corrosion resistance besides imparting conductivity to resist damage by lightning strikes.

Electroless nickel coating has also been used to improve the wear and wear–corrosion properties of CFRP materials which are frequently utilised in the aerospace, aviation and automobile industries due to their light weight [119]. The potential of Ni–P coatings for marine applications has been investigated by various researchers [124, 125]. Composite coatings of electroless Ni–P–PTFE exhibited improved performances of aluminium alloys in the bearing regions of lightweight external gear pumps [126]. Das et al. [127] reported the application of 4 μm thick electroless Ni–P–Si_3N_4 on the surface of ferrous-based bearings that can be successfully used for water-lubricated applications in pH 10 water for 9 years under the actual application environment. The excellent wear resistance plus low friction coefficient of the Ni–P–Cgraphite–SiC coating is found suitable for applications in moulds, automobile parts and other fields [128]. Klingenberg et al. [129] suggested that all of the electroless Ni–P, Ni–Co–P and Co–P processes with occluded diamond particles have the potential to impart the required adhesion, hardness and tribological properties, while reducing the environmental impact of chromium plating processes [127]. The high electrical and thermal conductivity properties, combined with the in situ lubricating ability of graphite-containing composites, which mix further with ceramic SiC particles, makes them potential candidates for such applications as advanced high speed–high load bearing, high speed–high current electrical brushes, etc. [113]. Among the various metallization processes, electroless metal plating is a preferred way to produce metal-coated fabrics due to the attractable advantages such as uniformity of coverage, excellent conductivity, possibility of metallizing non-conductors and flexibility. Ni–P plating is applied in order to render abrasion resistance to EMI shielding fabrics [130]. The abrasion resistance is found to increase with the incorporation of SiO_2 particles in the coating without much change in shielding effectiveness of the fabric. Bearing journal, seal snaps and spacers, turbine front bearing cases of aircraft industriy, tank tarred bearings of defence, hydraulic shaft and cylinders of material handling plants are coated with electroless Ni-P coatings.

6.6 Concluding Remark

Alloy plating is an important process development [131, 132]. However, as the pace of development is very slow, most of the work becomes matter of academic interest. Plating of brass, bronze and stainless steel has been tried several decades ago. Multiple layer plating is another emerging technology [133]. This involves deposition of inhomogeneous alloy consisting of lamellae of different composition. Recently significant effort is put to develop multilayered coatings by less expensive and well-industrialised electrodeposition technique using either single bath or duel bath [134–136].

Environmental hazard is one of the most important drawbacks of electroplating industry particularly with effluent treatment. The most widely used method for extracting metallic elements from effluent is to precipitate metals in sludges as hydroxides or sulphides. The solid waste is then disposed to landfills. In electroplating shop, metals and their compounds are inadvertently ingested. Solution containing metal ions are sometimes allowed on the skin for prolong period. Sometimes metal-containing compounds are taken into the body by breathing. All these cause harmful health hazards such as cancer, skin rash, etc. Research should be carried out to optimise the process condition so as to minimise such effects.

It has emerged that electroless coatings can serve as viable replacements to the conventional electroplating in suitable situations. Properties of these coatings such as hardness, low friction, wear resistance and corrosion resistance have given possibilities of wide range of modification and their usage in tribological applications. In addition, the provision of uniform deposition and the ability to coat any materials have served as an added advantage to their application in various areas. The advantages of modifying the properties of electroless coatings by suitable surface treatments (heat treatment, laser treatment, etc.) and the incorporation of various elements (copper, tungsten, etc.) and particles (SiC, TiO_2, Si_3N_4, etc.) have been utilised by various researchers to evaluate the suitability of these coatings for various applications. The electroless coatings are mainly applied for wear resistance and corrosion resistance applications. Further modification possibilities of electroless nickel coatings lead to more advanced tribological application of the coatings.

References

1. Dibari GA, Chatham NJ (2002) Met Finish 100:34
2. Dietz A (2002) Materialwissenschaft Werkstofftechnik 30:581
3. Landolt D (1994) Electrochim Acta 39:1075
4. Winand R (1994) Electrochim Acta 39:1091
5. Van der Biest OOV, Vandeperee LJ (1999) Annu Rev Mater Sci 29:327
6. Okinawa Y, Hoshino M (1998) Gold Bull 31:3
7. Bhatgadde LG (1997) Trans Met Finish Assoc India 6(3):229
8. Bauddrand D (2004) Plat Surf Finish 91(8):19
9. Balaraju JN, Narayanan TSNS, Seshadri SK (2003) J Appl Electrochem 33:807
10. Sahoo P, Das SK (2011) Mater Des 32:1760
11. Rasmussen FE, Ravnkilde JT, Tang PT, Hansen O, Bouwstra S (2001) Sensor Actuator A 92:242
12. Fujita T, Nakamichi S, Ioku S, Maenaka K, Takayama Y (2007) Sensor Actuator A 135:50
13. Miura S, Honma H (2003) Surf Coat Technol 169–170:91
14. Mohamed Ali MS, Takahata K (2010) Sensor Actuator A 163:363
15. Palaniappa M, Jayalakshmi M, Balasubramanian K (2011) J Mater Eng Perform 20:1028
16. Blair A (1994) Surface engineering, vol 5, ASM handbook. ASM, Materials Park, OH, p 247
17. Krishnaveni K, Sankarnarayan TSN, Seshadri SK (2005) Surf Coat Technol 190:115
18. Mallroy GO (1979) Product finishing magazine. Allied Kellite Products Division
19. Stallman K, Speakhardt H (1981) Metalloberfl Angew Electrochem 35:979
20. Dini JW, Coronado DR (1967) Plating 38:385
21. El Mallah AE, Brenner A (1978) Met Finish 76(11):62
22. Duncan RN (1986) In: Proceedings of the 3rd AESF electroless plating symposium, American Electroplating and Surface Finishing Society, Orlando, FL, p 1
23. Elansezhian R, Ramamoorthy B, Kesavan Nair P (2008) Surf Coat Technol 203:709
24. Durney L (1984) Electroplating engineering handbook, 4th edn. Van Nostrand Reinhold, New York
25. VanSlyke DD (1922) J Biol Chem 55:525
26. Gutzeit G (1960) Plating 1:63
27. Fieldstein N, Lancsek TS (1971) Trans Inst Met Finish 49:156
28. Palaniappa M, Seshadri SK (2007) Mater Sci Eng A 460–461:638
29. Ashassi-Sorkhabi H, Rafizadeh SH (2004) Surf Coat Technol 176:318
30. Zhang YZ, Yao M (1999) Trans Inst Met Finish 77:78
31. Duncan RN, Arney TL (1989) Plat Surf Finish 76:60
32. Duncan RN (1996) Plat Surf Finish 83:65
33. Riedel W (1991) Electroless plating. ASM, Materials Park, OH
34. Heydarzadech Sohi M, Kashi AA (2003) J Mater Process Technol 138:219
35. Wang L, Gao Y, Xue Q, Liu H, Xu T (2006) Surf Coat Technol 200:3719
36. Lausmann GA (1996) Surf Coat Technol 86–87:814
37. Hadavi SMM, Abdollah-Zadeh A, Jamshidi MS (2004) J Mater Process Technol 147(3):385
38. Yang D, Jiang C (1998) Plat Surf Finish 85(1):110
39. Ning SK, Ning HX, Hai ZJ, Ren WJ (1996) Wear 196:295
40. Onate JI (1989) Met Finish 87(3):25
41. Darbeïda A, von Stebut J, Barthole M, Belliard P, Lelait L, Zacharie G (1994) Surf Coat Technol 68–69:582
42. Arieta FG, Gawne DT (1995) Surf Coat Technol 73:105
43. Snavely CA, Faust CL (1950) J Electrochem Soc 97:3
44. Huang C-A, Lin W, Liao MJ (2006) Corros Sci 48(2):460
45. Jeong DH, Ganzalez F, Palumbo F, Aust KT, Erb U (2001) Scr Mater 44:495
46. Wang L, Gao Y, Xue Q, Liu H, Xu T (2005) Appl Surf Sci 242:326
47. Stepanova LI, Purovskaya OG (1998) Met Finish 96:50

48. Panagopoulos CN, Geogarakis KG, Petroutzakou S (2005) J Met Process Technol 160:234
49. Ramanauskas R, Quintana P, Maldonado L, Pomes R, Pech-Canul MA (1997) Surf Coat Technol 92:16
50. Bozzini B, Fanigliulo A, Lanzoni E, Martini C (2003) Wear 255:903
51. Weston DP, Shipway PH, Harris SJ, Cheng MK (2009) Wear 267:934
52. Haseeb ASMA, Albers U, Bade K (2008) Wear 264:106
53. Sriraman KR, Ganesh Sundara Raman S, Seshadri SK (2006) Mater Sci Eng A 418:303
54. Surender M, Basu B, Balasubramanian R (2004) Tribol Int 37:743
55. Grigorescu IC, Gonzalez Y, Rodrigues O, Vita YD (1995) Surf Coat Technol 76–77:604
56. Panagopoulos CN, Georgarakis KG, Agathocleous PE (2003) Tribol Int 36:619
57. Panagopoulosa CN, Papachristosa VD, Christoffersen LW (2000) Thin Solid Film 366:155
58. Lim SC, Ashby MF (1987) Acta Metall 35:11
59. Roy M (2009) Trans Indian Inst Met 62:197
60. Ghosh SK, Limaye PK, Bhattacharya S, Soni NL, Grover AK (2007) Surf Coat Technol 201:7441
61. Ghosh SK, Limaye PK, Swain BP, Soni NL, Agrawal RG, Dusane RO, Grover AK (2007) Surf Coat Technol 201:4609
62. Choo RJC, Toguri JM, El Sherik AM, Erb U (1995) J Appl Electrochem 25:384
63. Li J, Wu YY, Wang DL, Hu XG (2000) J Mater Sci 35:1751
64. Musiani M (2000) Electrochim Acta 45:3397
65. Landolt D, Marlot A (2003) Surf Coat Technol 169–170:8
66. Brooks I, Erb U (2001) Scr Mater 44:853
67. Yang C, Yang Z, An M, Zhang J, Tu Z, Li C (2001) Plat Surf Finish 88(5):116
68. Abdel Hamid Z, Ghayad IM (2002) Mater Lett 53:238
69. Ghorbani M, Mazaheri M, Khangholi K, Kharazi Y (2001) Plat Surf Finish 148:71
70. Ebdon PR (1988) Plat Surf Finish 75(9):65
71. Pena-Menoz E, Bercot P, Grosjean A, Razrazi M, Pagetti J (1998) Surf Coat Technol 92:16
72. Ghorbani M, Mazaheri M, Khangholi K, Kharazi Y (2004) Surf Coat Technol 187:293
73. Arai S, Endo M (2003) Electrochem Commun 5:797
74. Szczygiel B, Kolodziej M (2005) Electrochim Acta 50:4188
75. Hou KH, Hwu WH, Ke ST, Ger MD (2006) Mater Chem Phys 100:54
76. Gay PA, Bercot P, Pagetti J (2001) Surf Coat Technol 140:147
77. Wang W, Hou FY, Wang H, Guo HT (2005) Scr Mater 53:613
78. Huang ZJ, Xiong DS (2008) Surf Coat Technol 202:3208
79. Chen L, Wang L, Zeng Z, Zhang J (2006) Mater Sci Eng A434:319
80. Gul H, Kılıc F, Aslan S, Alp A, Akbulut H (2009) Wear 267:976
81. Panagopoulos CN, Agathocleous PE, Papachristos VD, Michaelides A (2000) Surf Coat Technol 123:62
82. Roy Chowdhury SK, Pollock HM (1981) Wear 66:307
83. Capel H, Shipway PH, Harris SJ (2003) Wear 255:917
84. Zum Ghar KH, Eldis GT (1980) Wear 64:175
85. Vaezi MR, Sadrnezhaad SK, Nikzad L (2008) Colloids Surf A 315:176
86. Xue YJ, Jia XZ, Zhou YW, Ma W, Li JS (2006) Surf Coat Technol 200:5677
87. Hou KH, Ger MD, Wang LM, Ke ST (2002) Wear 253:994
88. Garcia I, Fransaer J, Celis JP (2001) Surf Coat Technol 148:171
89. Benea L, Bonora PL, Borello A, Martelli S (2002) Wear 249:995
90. Wang CB, Wang DL, Chen WX, Wang YY (2002) Wear 253:563
91. Srivastava M, William Grips VK, Jain A, Rajam KS (2007) Surf Coat Technol 202:310
92. Wang L, Gao Y, Xue Q, Liu H, Xu T (2005) Mater Sci Eng A 390:313
93. Delaunois F, Lienard P (2002) Surf Coat Technol 160:239
94. Gawne DT, Ma U (1989) Wear 129(1):123
95. Kanani N (1991) Trans Inst Met Finish 70:14
96. Yu L-G, Zhang X-S (1993) Thin Solid Films 229:76

97. Palaniappa M, Seshadri SK (2008) Wear 265:735
98. Baudrand DW (1981) Plat Surf Finish 68:57
99. Gawrilov GG (1979) Chemical (electroless) nickel plating. Portcullis, Surrey
100. Krishnaveni K, Narayanan TSNS, Seshadri SK (2005) Surf Coat Technol 190:40
101. Apachitel I, Duszczyk J, Katgerman L, Overkemp PJB (1898) Scr Mater 38:1347
102. Henry J (1984) Met Finish 82:17
103. Balaraju JN, Seshadri SK (1999) Met Finish 97:8
104. Li Z, Chen Z, Liu S, Zheng F, Dai A (2008) Trans Nonferrous Met Soc China 18:819
105. Gawne DT, Ma U (1987) Wear 120:125
106. Gay PA, Limat JM, Steinmann PA, Pagetti J (2007) Surf Coat Technol 202:1167
107. Tsujikawa M, Azuma D, Hino M, Kimura H, Inoue A (2005) J Metastab Mater 24–25:375
108. Araghi A, Paydar MH (2010) Mater Des 31:3095
109. Hosseinabadi ME, Dorcheh KA, Vaghefi SMM (2006) Wear 260:123
110. Wu Y, Liu H, Shen B, Liu L, Hu W (2006) Tribol Int 39:553
111. Palaniappa M, Veera Babu G, Balasubramanian K (2007) Mater Sci Eng A47(1):165
112. Wu Y, Shen B, Liu L, Hu W (2006) Wear 261:201
113. Ramahlo A, Miranda JC (2005) Wear 259:828
114. Hu X, Xiang P, Wan J, Xu Y, Sun X (2009) J Coat Technol Res 6:275
115. Zou TZ, Tu JP, Zhang SC, Chen LM, Wang Q, Zhang LL (2006) Mater Sci Eng A426:162
116. Hosseinabadi ME, Dorcheh KA, Moonir Vaghefi SM (2006) Wear 260:123
117. Chen XH, Chen CS, Xiao HN, Liu HB, Zhou LP, Li SL (2006) Tribol Int 39:22
118. Dong D, Chen XH, Xiao WT, Yang GB, Zhang PY (2009) Appl Surf Sci 255:7051
119. Lee CK (2009) Mater Chem Phys 114:125
120. Sahoo P (2009) Mater Des 30:1341
121. Armyanov S, Steenhaut O, Krasteva N, Georgieva J, Delplancke J-L, Winand R, Vereecken J (1996) J Electrochem Soc 143(11):3692
122. Palaniappa M, Seshadri SK (2007) J Mater Sci 46:6600
123. Lee CK (2008) Surf Coat Technol 202:4868
124. Gao R, Du M, Sun X, Pu Y (2007) J Ocean Univ China 6:349
125. Wang J, Yan F, Xue Q (2009) Tribol Int 35:85
126. Veronesi P, Sola R, Poli G (2008) Int J Surf Sci Eng 2:190
127. Das CM, Limaya PK, Grover AK, Suri AK (2007) J Alloys Compd 436:328
128. Wu YT, Lei L, Shen B, Wu WB (2006) Surf Coat Technol 201:441
129. Klingenberg ML, Brooman EW, Naguy TA (2005) Plat Surf Finish 92:42
130. Hu Z, Rongli L (2008) Sen'i Gakkaishi 64:372
131. Law M (1984) Finishing 8:30
132. Krugger J, Nepper JP (1986) Metalloberflache 40:107
133. Ruff AW, Lashmore DS (1991) Wear 151:245
134. Yahalom J, Zadoc O (1987) J Mater Sci 22:499
135. Lashmore DS, Dariel MP (1988) J Electrochem Soc 135:1218
136. Bonhote C, Landolt D (1997) Electrochim Acta 42:2407

Chapter 7
Laser Surface Modification for Protection Against Wear

S.V. Joshi and Manish Roy

7.1 Introduction

With the advent of laser technology during the past few decades, surface modification using lasers has been used extensively. It was way back in the 1960s when Maiman [1] first invented a working ruby laser. This was followed by the diode laser proposed by Basov [2]. In 1962, diode laser was demonstrated independently by Hall et al. [3] at General Electric, Nathan et al. [4] at IBM and by Quist et al. [5] at MITs Lincoln Laboratory. The first CO_2 laser was invented by C.K.N. Patel soon thereafter in 1964 while working in Bell Labs [6]. In the same year Nd–YAG laser was also invented in Bell Labs. Ewing and Brau [7] came up with the excimer laser in 1974 in Avco Everett Laboratory. A free electron laser was first demonstrated by Madley's group at Stanford University. As a result of above developments, today optical energy is available in an easily controllable form and has therefore been gainfully initialized for various materials processing operations. The mastery gained over the use of laser initially as a tool for routine operations such as cutting drilling and welding has now enabled realisation of various surface modification approaches based on high power lasers.

Although widely perceived as being expensive, the laser based techniques offer several distinct advantages over other conventional surface modification processes. All variants of laser surface modification are characterised by very fast heating and cooling rates resulting in a rapidly solidified layer, in which both the microstructure and the distribution of the alloying elements could be tailored as required by

S.V. Joshi (✉)
International Advanced Research Center for Powder Metallurgy and New Materials, Balapur, Hyderabad 500005, Andhra Pradesh, India
e-mail: SVjoshi.uoh@gmail.com

M. Roy
Defence Metallurgical Research Laboratory, Kanchanbagh, Hyderabad 500058, Andhra Pradesh, India

M. Roy (ed.), *Surface Engineering for Enhanced Performance against Wear*,
DOI 10.1007/978-3-7091-0101-8_7,

suitably controlling the operating parameters. The metastable and non-equilibrium phases that can form as a consequence of rapid quenching, offer the possibility to develop layers with novel microstructures and properties superior to those of traditional processes. Laser energy is highly directional and it can be controlled precisely and with the use of appropriate optics, the beam can also be shaped into a spot, line or area beam as required. Due to directionality of laser light, it is possible to utilise a laser beam to treat locations that are difficult to approach. Laser light can also be of very short duration which allows deposition of large amount of energy into a very narrow region. Moreover, laser-based processes are chemically clean, extremely environment friendly and generally do not require any post-processing treatment. The fact that the process is very and can be automated easily is a significant-added benefit. Laser surface-modified components also have minimum risk of distortion as the heat input is highly localised and the part being treated serves as a substantially large heat sink to prevent overheating of the bulk component. With the advent of high power diode laser, the concerns regarding maintenance have also been largely alleviated to make laser considerably user friendly.

The above unique advantages offered by a laser make it an unparallel tool for surface modification. Surface engineering is a subject of growing industrial interest as it allows fabrication of improved components with an ideal combination of surface and bulk properties to enhance performance, extend life, improve aesthetics, etc. Realising its relevance in the current competitive era, engineering industries have become increasingly receptive to improved, sophisticated and advanced surface modification techniques for development of high performance coatings. Growing awareness of laser surface modification and its immense potential have made laser surface processing an integral part of modern technology and industry. In view of the above, this chapter is intended to provide an overview of laser surface modification specifically for tribological applications.

7.2 Laser Surface Modification Processes

7.2.1 Process Fundamentals

During laser surface modification, several process parameters significantly influence the microstructure and mechanical properties of the modified layer. The prominent among them are traverse speed, power density, extent of overlap, etc. As the power density is influenced by beam spot area, the beam diameter/shape is important. The nature of surface treated in respect of absorbance of the incident laser energy is also a determining factor. Apart from the above generic factors, the powder feed rate (or amount of powder preplaced) are important too, in case of laser alloying. Additionally changes in these parameters result in change in depth, width and microstructure of the modified region as well as its functional properties. An increase in beam power is found to increase the width and depth of the modified layer [8, 9]. Higher scan speed and lower laser power lead to faster cooling rates giving rise to a finer microstructure. For example, a higher scan speed has been reported to result in changes from dendritic microstructure to cellular microstructure during laser carburizing of Ti–6Al–4V [10]. The laser power density (q) for a continuous wave laser and a pulsed laser is given by following equation:

$$q = \frac{P}{A} \tag{7.1}$$

and

$$q = \frac{P}{A\tau} \tag{7.2}$$

where P is the power of the laser, A is the area of the beam and τ is the duration of the pulse of the laser. The radius (r_s) of the beam spot is related to the wavelength (λ), beam quality factor, K, the focus distance of the lens (f) and the beam diameter (D) by the following expression:

$$r_s = \frac{2\lambda}{\pi} \frac{f}{D} \frac{1}{K} \tag{7.3}$$

The type of laser employed plays a crucial role in governing the nature of the modified layer. It is widely acknowledged that if the surface is to be modified by co-deposition of an alloying element or by cladding, a continuous laser is better suited than a pulsed laser [11]. On the contrary, it has been reported that, for laser gas nitriding of Ti–6Al–4V, a pulsed laser provides a better surface finish than that achieved using a continuous laser [12]. During co-deposition by laser cladding, the clad layer thickness increases with laser power for a given traverse speed and clad thickness [13], whereas the clad layer thickness decreases with traverse speed when other process parameters are kept constant [14]. There is clear evidence that the clad layer thickness increases monotonically with powder feed rate [13]. The extent of overlapping and rescanning are two key parameters which govern the nature of modified layer. Rescanning with a defocused beam during laser carburizing improves the surface roughness and homogeneity of the modified layer [12].

The heat input is essentially governed by the ratio of laser power to traverse speed. A key parameter known as the side bead angle is particularly important in multipass cladding. A side bead angle greater than 120° generally results in good integrity in the overlap zone between consecutive passes. Angles lower than 120° lead to increase impurity trapping and less fusion in the overlapped region. The side bead angle increases rapidly with powder depth in case of preplaced powder and powder feed rate in case of dynamic powder feeding. This parameter is not strongly affected by either laser power or travel speed. Thermophysical properties such as relative melting point, vapour pressure of the elements are known to play a crucial role. In case of high vapour pressure alloying species high evaporation loss is noted [15, 16]. Alloying low melting point element with a high melting point substrate is fraught with difficulties relating to bobble formation and entrapment in the resolidified near surface region [17]. Liquid state miscibility is another governing factor, for example, it is much easier to alloy in systems like Pd–Ni but difficult to alloy in system like Ag–Ni.

Dilution, another important parameter in laser surface modification, is defined as the percentage of the total volume of the surface-modified layer contributed by melting of the substrate. It is usually estimated by area measurement of the cross section and is given as ratio of the area of the melted substrate to the sum of the area of the deposit and the area of the melted substrate. It can also be calculated by measuring the composition of the as processed surface layer as

$$\text{dilution} = \frac{\rho_{\mathrm{A}}(\%N_{\mathrm{A+S}} - \%N_{\mathrm{A}})}{\rho_{\mathrm{S}}(\%N_{\mathrm{S}} - \%N_{\mathrm{A+S}}) + \rho_{\mathrm{A}}(\%N_{\mathrm{A+S}} - \%N_{\mathrm{A}})} \tag{7.4}$$

where ρ_{A} is the density of the alloying material, ρ_{S} is the density of the substrate, N_{A} is the weight percent of the element N in alloying material, N_{S} is the weight percent of the element N in the substrate and $N_{\mathrm{A\,+\,S}}$ is the weight percent of the element N in total surface layer. In order to assess the influence of process parameters on dilution, it is imperative to define the parameter specific energy. Specific energy is defined as the ratio of the laser power to the product of the beam width and travel speed. Dilution has been widely reported to increase with specific energy. As expected, the increase in dilution with specific energy is much more rapid in cases where surface treatment is carried out with powder preplaced to a shallow depth as compared to deeper powder layer depth. For a given powder layer depth and specific energy, the dilution increases with increasing power density.

7.2.2 Process Classifications

Laser surface modification encompasses a large variety of processing techniques. If laser surface modification is carried out without addition of any consumables, the process can be designated as laser transformation hardening and/or laser diffusion homogenization. If the modification results in superficial melting of the substrate followed by rapid solidification, it may be called laser glazing [18]. When extraneous materials are added to the substrate surface during laser processing, it can be classified as laser surface alloying or laser cladding [19–22]. In terms of processing, alloying and cladding are very similar and the differentiation between the two is largely based on the extent of dilution—somewhat arbitrarily the process is termed as cladding if the dilution exceeds 10 %. Laser surface modification can also be classified under different heads on the basis of number of steps involved, namely single-step laser deposition process and two-step laser deposition process. The one-step laser deposition technique consists of dynamically injecting the powder into the laser beam–substrate interaction zone where the powder is heated (melted) and subsequently deposited on the substrate. The most recent industrial application of one-step laser deposition is rapid prototyping, which enables direct production of components having complicated three-dimensional shapes. Two-step laser deposition consists of laser treatment of a pre-deposited coating. The pre-deposition can be carried out by any of the numerous thin and thick coating techniques available or, in its simplest form, by merely brushing on the alloying/cladding material mixed with a suitable binder. Popular among the pre-deposition processes is thermal spraying, which is often simplest to apply a reasonably thick uniform layer of the alloying/cladding material.

7.2.2.1 Laser Heat Treatment

Laser heat treatment is a process which uses the rapid heating and cooling rates produced on the metal surface exposed to scanning laser beams and offers several advantages over conventional surface hardening processes. Laser heat treatment processes can broadly be classified into laser transformation hardening and laser remelting. Transformation hardening is more commonly applied laser heat treatment and is induced by heating a material above a critical temperature and then rapidly quenching it rather than allowing equilibrium phases to form. Steels and cast irons are good candidate materials for laser transformation hardening. On the other hand, laser remelting operations involve melting of the surface and are carried out at lower scan speeds or higher power levels. Although a relatively low power density is utilised for both forms of heat treatment, the cross-sectional area of the beam used to scan the part being heat treated is normally much larger than the focused beam used for welding and cutting to enable large area processing at practically meaningful speeds. Therefore higher beam power is required for laser heat treatment. A schematic representation of laser transformation hardening process is provided in Fig. 7.1. During transformation hardening, the laser beam is defocused or oscillated to cover the desired area with a power density of 10^3–10^4 W/mm^2 and a relative motion between the work piece and the beam of 5–50 mm/s. An ideal power distribution is one which gives a uniform temperature over the area to be treated. It is relevant to note that a laser beam is extremely flexible and it offers unparallel versatility to enable time share and power share as well as single focus, dual focus and defocused configurations. The flexibility is illustrated by schematics in Fig. 7.2 and further reinforced by ability to shape beams through use of appropriate optics. Typical microstructures of several varieties of steels with varying carbon content after laser transformation hardening are provided in Fig. 7.3. Very fine microstructure due to transformation hardening is evident. The advantages and disadvantages of laser transformation hardening compared with other hardening processes such as induction hardening and flame hardening is summarised in Table 7.1.

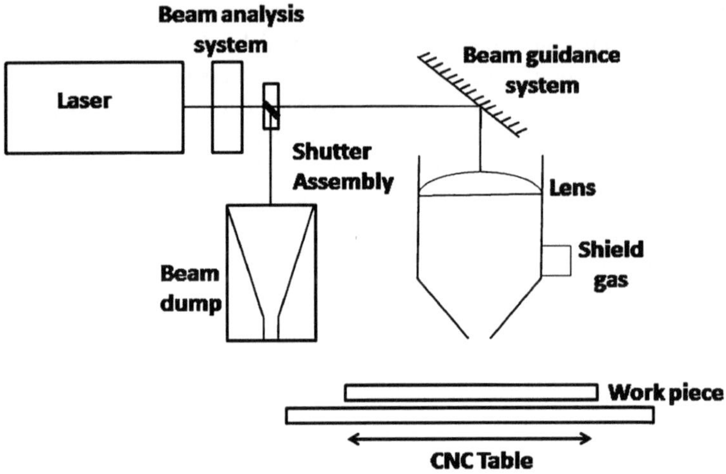

Fig. 7.1 Schematic representation of laser transformation hardening system

7.2.2.2 Laser Glazing

Laser glazing involves surface melting followed by rapid self-quenching resulting in a fine or amorphous microstructure. Experimental arrangement for laser glazing is similar to laser transformation hardening. The surface to be melted is shrouded by an inert gas. In this process, the most important variables are reflectivity of glazed surface, shape of the beam, shrouding of the melt pool, apart from the laser power and scan speed. It is difficult to control the reflectivity of a molten surface. As the material becomes hot, reflectivity decreases due to increased concentration of phonons. Figure 7.4 shows typical microstructures of as-plasma-sprayed and laser-glazed WC–Co coatings at two different scan speeds. Laser glazing can potentially reduce the porosities of the coated layer and also refines the microstructures to achieve improved mechanical properties. The extent of refinement and property enhancement is however dependent on the laser power and scan speed.

7.2.2.3 Laser Surface Alloying

Laser surface alloying is a process in which a component of the alloy is injected into the molten pool on the substrate during processing. Although alloying element is generally added in powder or wire form, introduction in the gaseous form is also possible. Alloying requires a greater laser power density, as substantial energy is expended to create a melt pool on the substrate and often also melt the injected powder on the substrate surface. The alloying process enables formation of metallic and ceramic alloys, such as nitrides or borides. The process starts with melting of a substrate by laser irradiation. On the surface of the melt pool, there is a temperature

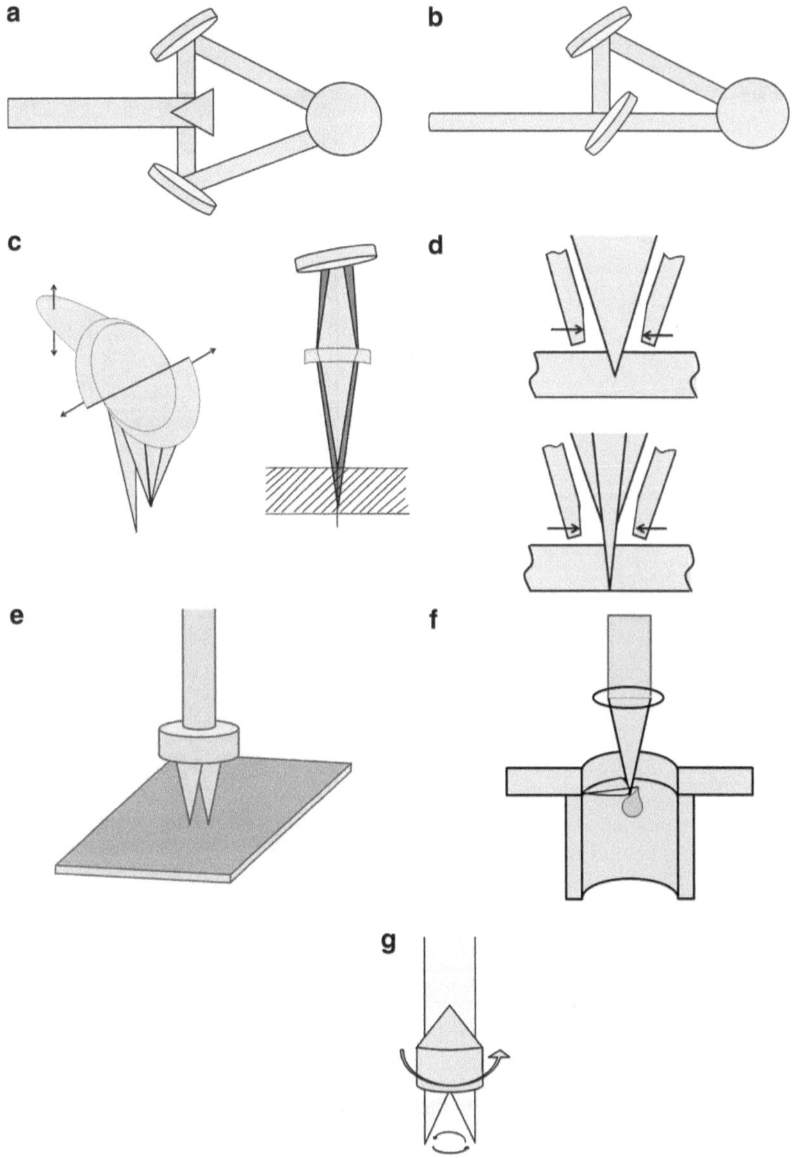

Fig. 7.2 Various optics options illustrating the flexibility of a laser beam for materials processing

distribution, which results in the differential surface tension distribution. The shear stress, which is equal to the gradient in surface tension, pulls the material from the centre and causes convection currents within the melt pool. In case of the injection of solid particles into the melt, the convection permits good mixing with the substrate material within the alloyed layer. The particles are sometimes also melted

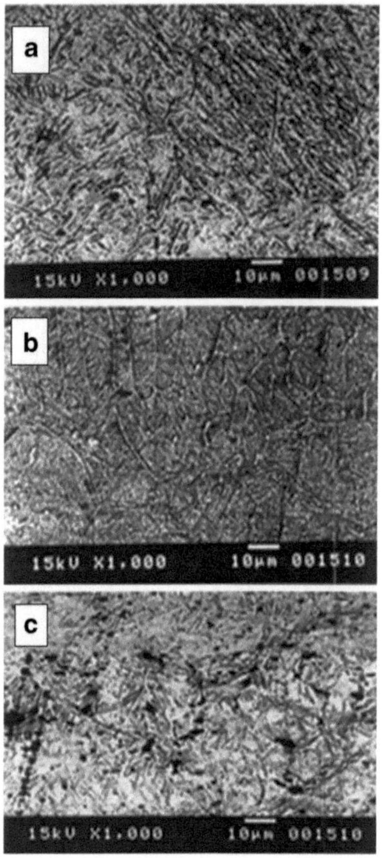

Fig. 7.3 SEM images of several varieties of steels after laser transformation hardening

Table 7.1 The advantages and disadvantages of laser transformation hardening with other hardening processes

Characteristics	Laser hardening	Induction hardening	Flame hardening
Distortion	Very low	High	Very high
Quenching requirement	No	Yes	Yes
Process rate	High	High	Low
Case controllability	Easily controllable	Controllable	Non controllable
Post processing	Not required	Required	Required
Thickness limitation	No	Depends on job	Yes
Equipment cost	Very high	High	Very low
Process flexibility	High	Low	Very low
Special requirement	Absorbent coating	Depends on job	Not required
Automation	Possible	Not possible	Not possible
Selective treatment	Possible	Not possible	Not possible

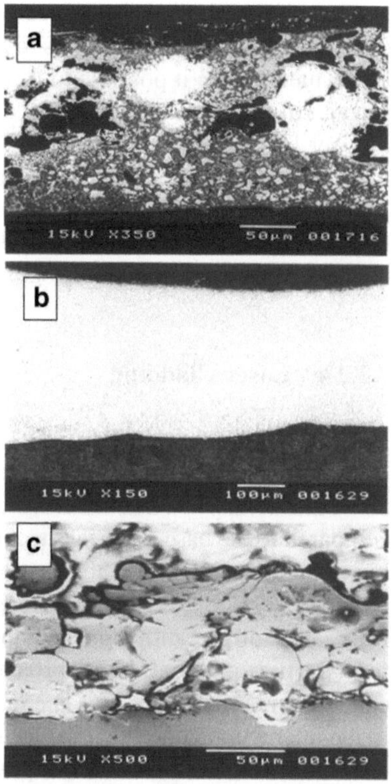

Fig. 7.4 SEM images of as-plasma-sprayed and laser-glazed WC–Co coatings

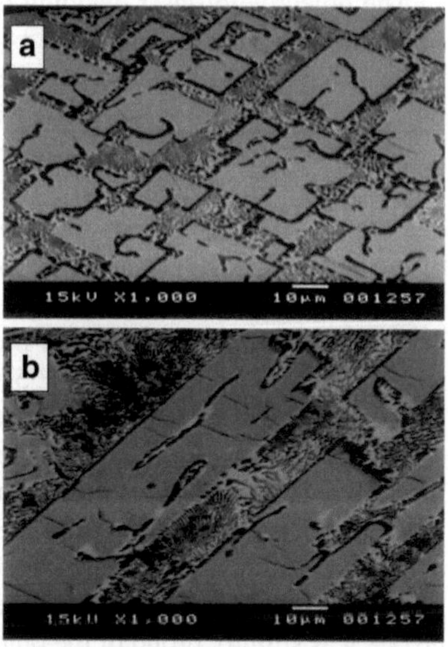

Fig. 7.5 SEM images showing the microstructure obtained by laser boronising of steel

and reaction with the substrate can take place. The reaction slows down and stops soon after the laser beam moves to the next position. The subsequent rapid cooling of the melt makes it possible to form metastable or high-temperature phases as the product of the reaction. However, the cooling rate can also be slowed by lowering of laser beam speed over the substrate. The microstructure of the layer resulting from laser alloying of the steel substrate with iron boride is shown in Fig. 7.5. Through such alloying, the surface properties of the substrate can virtually be tailored (for wear resistance, corrosion resistance, etc.) to meet the requirements dictated by any given application.

7.2.2.4 Laser Cladding

Laser cladding can be defined as addition of extraneous material with limited substrate melting resulting in development of a metallurgically bonded coating on the substrate to yield a surface with completely different composition and structure compared to the base material. In contrast, laser surface alloying involves relatively lower amount of consumable material addition giving rise to a compositionally modified surface layer. As previously mentioned, laser cladding is differentiated from laser surface alloying depending on the extent of dilution of the surface layer from the base material. If the surface layer is diluted to an extent less than 10 %, it is called laser surface alloying. The manner in which the extraneous material is added in the interaction zone often varies in different processes. The coating materials are commonly added in the form of powder or wire in case of both laser alloying and laser cladding.

The objective of laser cladding is to deposit a completely different layer on the substrate with a sound interfacial bond without any dilution. Laser cladding can be done either with preplaced powder (two-step laser processing) or with blown powder as shown in Fig. 7.6 or by wire feeding. The physics of the cladding process using coaxial injection of powder has been analysed in reported literature [23–25]. It was found that there is a certain minimum power necessary to produce the coating. This power corresponds to the beginning of substrate melting. The velocity of the injected powder particles was determined using laser Doppler velocimetry (LDV), and values of 1–2.5 m/s has been reported. The laser power density necessary for in-flight melting of the particles has been estimated and theoretically found to be in the range of q = 5–7 kW/cm^2 in the case of Stellite 6 alloy. The cladding of the same material was analysed theoretically [24] for a powder injection system and it has been found that the powder efficiency can be as high as 69 % for a linearly polarised CO_2 laser beam at high angles of incidence (the angle between the normal to a molten pool and the laser beam axis). Lin [26] and Hayhurst et al. [27] found that the powder stream can be mildly focussed with a coaxial nozzle. Lin used a circumferential powder feeder and obtained best focus when the central nozzle was slightly proud of the outer nozzle to encourage some couette flow of the powder. Hayhurst et al. [27] employed four independent powder injection streams 90° apart, which on colliding merge into a single central stream. One variation of

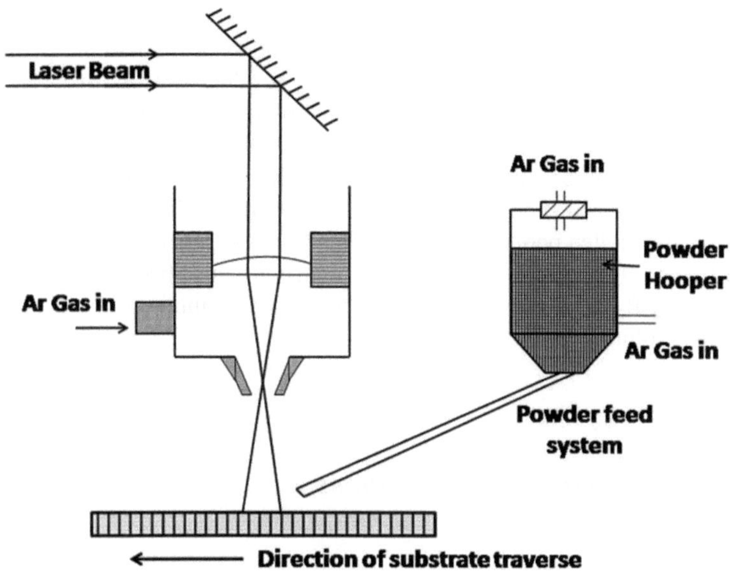

Fig. 7.6 Schematic representation of laser cladding by blown powder

laser cladding is known as bi-thermal laser cladding [28] in which powder is fed into the focal spot of the laser beam and melts instantaneously. The molten powders are then allowed to impact the substrate which is situated at lower power density. Employing this technique it was possible to coat chromium on aluminium. Such coating is rather difficult by conventional process.

7.2.2.5 Laser Composite Processing

Laser hard phase dispersion is a coating process that consists of injecting hard second-phase particles into a melted substrate. These particles remain solid during laser processing. After solidification, the hard particles are dispersed in a matrix of the substrate material throughout the modified layer. This process, which produces metal matrix composites (MMCs), was initiated in the 1970s [29] and is in many ways similar to alloying. As there is convection in a molten pool, the reinforcement powder injection angle and spot need to be carefully optimised. Kloosterman and De Hosson [30] found that the powder injection spot should be centred inside a laser beam. Any other position produces a less homogenous dispersion of particles in the modified layer. The depth of particle penetration depends on the injection velocity, which is determined, in turn, by the carrier gas flow. An increase in powder feed rate can transform a hard phase dispersion process into a cladding process if correspondingly higher power to ensure complete melting is employed. To achieve a condition wherein the substrate melts, but the hard particles remain solid, the temperature of the melt must be well below the melting point of the hard phase. If this is not possible, one can increase the traverse speed to limit the time of particle dissolution in the melt. However, rapid solidification of the melting pool, which includes particles of hard phase, can generate residual stresses that can lead to cracking of the coating. Cracks can be avoided by preheating the substrate prior to coating, although such heating essentially makes the process a two-step one. In order to achieve a good surface-modified layer, the particles must be wetted by the substrate material and should also be strongly bound to it. The particles should not dissolve in the molten pool during processing. Carbides are most frequently used as the hard phase. Injection of B_4C in a steel matrix has also been shown to produce Fe_3B and $Fe_{23}B_6$ [31], while injection of SiC in an aluminium matrix resulted in $AlSiC_4$ as a solution product [32].

7.2.3 *Process Governing Equations*

7.2.3.1 Heat Flow Equations

The governing differential equation for conduction of heat in stationary medium when there is no heat loss due to convection or radiation is

$$\Delta^2 T = \frac{1}{\alpha}\frac{\partial T}{\partial t} \tag{7.5}$$

During transformation hardening using continuous laser, the temperature at any point of hardening can be derived using the moving point source by Ashby and Easterling [33]. In this model, it is assumed that the work piece is semi-infinite, homogeneous, isotropic having constant thermal properties. This model does not account for heat loss due to transformation, radiation and convection. According to this model, temperature at any point beneath the heat source is given by

$$T - T_o = \frac{Aq/v}{2\pi k[t(t+t_o)]^{1/2}} \exp\left\{-\frac{1}{4\alpha}\left[\frac{(z+z_o)^2}{t} + \frac{y^2}{t+t_o}\right]\right\} \tag{7.6}$$

where T_o is the initial temperature, A is the absorptivity at the sample surface, q is the input energy, k is the thermal conductivity (W/m/K), v is the scan speed (m/s), z is depth below the surface (m), z_o is a length constant, y is distance in Y-direction, t is time (s), t_o is the time at the start of power (s) and α is thermal diffusivity (m^2/s). This model can be used conveniently for explaining transformation laser surface hardening.

The moving Gaussian surface heat source model was first solved by Cline and Anthony [34] and Devis et al. [35]. Employing this model it is possible to derive an expression for the depth of hardness assuming constant thermal properties, no latent heat effect and no surface heat loss by convection or radiation. Accordingly the depth of hardness can be obtained as

$$d = 0.76D \times \left[\frac{1}{(\alpha' C/q) + \pi^{1/4}(\alpha' C/q)^{1/2}} - \frac{1}{(\alpha' C/q_{\min}) + \pi^{1/4}(\alpha' C/q_{\min})^{1/2}} \left(\frac{q_{\min}}{q} \right)^3 \right] \tag{7.7}$$

where $\alpha' = \frac{\rho u D C}{2k}$, $q = \left[\frac{2P(1-r_f)}{D\pi^{3/2}k(T_c - T_o)} \right]$, $q_{\min} = (0.40528 + 0.21586C)^{1/2}$

with $q_{\min}$ = the minimum absorbed power for which any hardening can occur and

$$C = C_\infty - 0.4646(C_\infty - C_0)(\alpha C_0)^{1/2} \tag{7.8}$$

The constants vary with materials and values determined employing this equation fits quite well with experiments [36].

7.2.3.2 Solidification Equations

Considering the mass balance on the solidification front, the gradient of the solute in the liquid at the solidification interface is obtained as

$$\left[\frac{dC_L}{dx} \right]_{x=0} = -\frac{R}{D_L} C_L^*(i - k) \tag{7.9}$$

Constitutional super cooling will be absent if the temperature gradient in the liquid at the interface (G) satisfies the condition

$$G \geq \left(\frac{dT_L}{dx} \right)_{x=0} \tag{7.10}$$

The value of this gradient can be obtained as

$$\left[\frac{dT_L}{dx} \right]_{x=0} = m_L \left[\frac{dC_L}{dx} \right]_{x=0} \tag{7.11}$$

where C_L is liquidus composition, x is the distance from the interface, T_L is the liquidus temperature in °C, R is the rate of solidification, D_L is diffusivity in m^2/s, C_L^* is the liquidus composition in equilibrium with solidus composition C_S^*, k is partition coefficient, m_L is the slope of the liquidus and G is thermal gradient in °C/m.

Combining Eqs. (7.9) and (7.11) and assuming $C_S^* = C_L^*$ at equilibrium, criterion for general constitutional supercooling is obtained as

$$\frac{G}{R} \geq -\frac{m_L C_S^* (1-k)}{k D_L} \tag{7.12}$$

When the ratio of *G*/*R* is very large, planar front solidification takes place. When this ratio is very small, regimes for 'absolute stability' comes into play.

Scales of solidification structures can be obtained by employing Fick's second law as

$$D_L \frac{\delta^2 C_L}{\delta y^2} = \frac{\delta C_L}{\delta t} \tag{7.13}$$

$$\frac{dC_L}{dt} = \left(\frac{dC_L}{dT}\right)\left(\frac{dT}{dx}\right)\left(\frac{dx}{dt}\right) = -\frac{GR}{m_L} \tag{7.14}$$

Substituting (dC_L/dt) from Eq. (7.14) and integrating across the cell width λ following expression is obtained:

$$\left[\frac{\delta C_L}{\delta y}\right]_{y=0} = -\frac{GR\lambda}{m_L D_L} \tag{7.15}$$

and

$$\Delta C_{L_{max}} = -\frac{GR\lambda^2}{2 m_L D_L} \tag{7.16}$$

Thus cell spacing is inversely related to the square root to the product of *G* and *R* or the cooling rate. Thus, higher the product of *G* and *R*, finer is the microstructure.

7.2.3.3 Fluid Flow Velocity in the Melt Pool

There are many forces those act in the melt pool. The largest among them is variation of surface tension due to steep thermal gradient. Anthony and Cline [34] have derived a simplified expression for fluid flow velocity driven by surface tension gradient during laser processing as

$$U_x(y,t) = \frac{1}{\eta}\frac{d\sigma}{dT}\Delta T\left[(D_u t)^{1/2} - y\right] \tag{7.17}$$

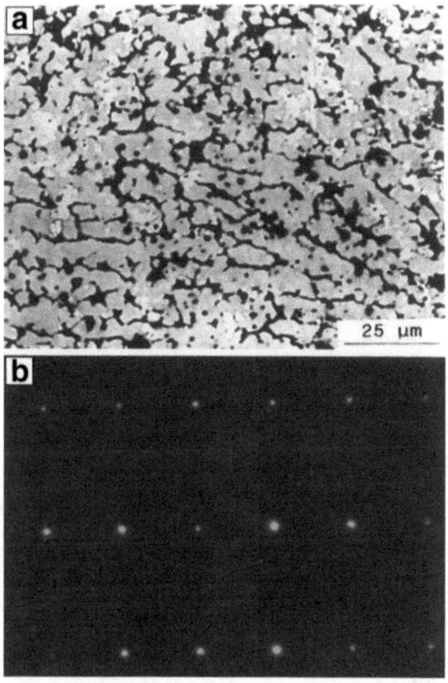

Fig. 7.7 Back scatter electron image of laser-Ti alloyed LM-9 with corresponding electron diffraction pattern [37]

Where $U_x(y,t)$ is the liquid velocity in the x direction at any given depth y and at any time t, η is the viscocity, $d\sigma/dT$ is temperature coefficient of the surface tension of the liquid and D_u is the diffusivity of the velocity profile.

7.3 Characterisation of Laser Surface-Modified Layer

7.3.1 The Microstructure

The microstructure of laser-treated surfaces is often of paramount interest and has, consequently, been very widely studied for all types of laser surface-modified layers. For example, Das et al. [37] characterised the microstructure of laser-glazed and laser surface Ti and Ni-alloyed aluminium alloy LM-9. The alloyed layer can be divided in two sub regions. The alloying element was mainly confined in the upper sub region while the lower sub region was laser melted but contained very little alloying element. Alloying resulted in segregation of alloying elements in the alloyed region. A typical SEM image of the Ti-alloyed regions is illustrated in Fig. 7.7 along with the corresponding electron diffraction pattern. Ni was found to be more severely segregated than Ti. As homogenization of alloying elements during laser surface alloying takes place by surface tension driven convection, such segregation was attributed to higher density of nickel compared to titanium.

Certain constituents have been reported to refine the microstructure in laser-alloyed layer. For instances, the addition of CeO_2 refine the microstructure in laser-remelted alloy spray coatings, decrease the dendrite arm spacing, strengthen and

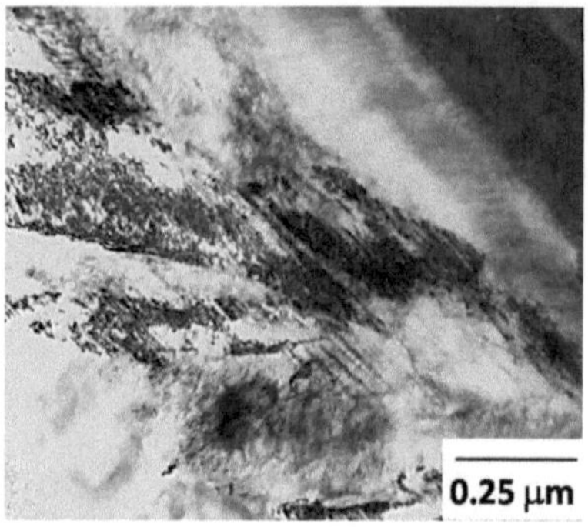

Fig. 7.8 TEM micrograph of the laser-remelted alloy spray coatings with CeO_2 having a martensitic structure [38]

purify the grain boundary, promote martensitic transformation and improve the morphology and distribution of eutectics and compounds. Martensitic as well as austenitic structure with lower density of dislocations can be found in laser-remelted alloy spray coatings with CeO_2. The martensitic structure has a high density of dislocation and fine twins. The morphology of martensitic structure includes lath-like, leaf-like, blocky and chrysanthemum-like shapes. Typical TEM micrographs of the laser-remelted alloy spray coatings with CeO_2 having martensitic structure is shown in Fig. 7.8 [38]. Figure 7.9 exhibits the eutectic martensite along with $M_{23}(CB)_6$ and the corresponding electron diffraction pattern.

Laser gas alloying is an important pathway for surface modification. Ti base alloys can be gas nitrided or gas carburised using a laser. Many fine TiN dendrites can be noted in the gas-nitrided laser surface-alloyed coatings. These dendrites are very hard and serve to enhance the wear resistance of the Ti–6Al–4V alloy. During laser surface melting under a nitrogen atmosphere, the nitrogen atoms dissolve in the high temperature Ti–6Al–4V melt, resulting in alloying of the melt-pool with nitrogen. After solidification of the alloyed melt bath, the wear-resistant 'in situ' composite reinforced by the as-solidified TiN dendrites is produced in the laser-melted surface layer. The laser beam scan rate has a significant influence on the microstructure of the laser surface-alloyed composite coatings. A lower beam scan rate results in a laser surface-alloyed composite coating having a higher volume fraction of TiN reinforcement due to the longer gaseous nitrogen and liquid interaction time. The average volume fraction of TiN dendrites decreases from approximately 62 to 40 % as the beam scan rate increases from 6 to 10 mm/s [39]. Similar observation is also reported by several other authors. With increasing scan speed, the cooling rate increases and the microstructure becomes finer as per Eq. (7.16). The influence of laser processing parameters on the microstructure is significant as illustrated in the work of Roy et al. [9]. Figure 7.10a, b shows the effect of scan power on the dendritic microstructure near the surface of the alloyed region obtained by laser nitriding commercially pure Ti. The dendritic structure is coarser for the melt pool obtained at higher power (2.1 KW) than the melt pool obtained at lower power (1.5 KW). It is also seen that the dendrites grow from the surface of the pool to the centre prior to the solidification of the centre. This indicates that the surface has solidified prior to the centre of the pool. Otherwise,

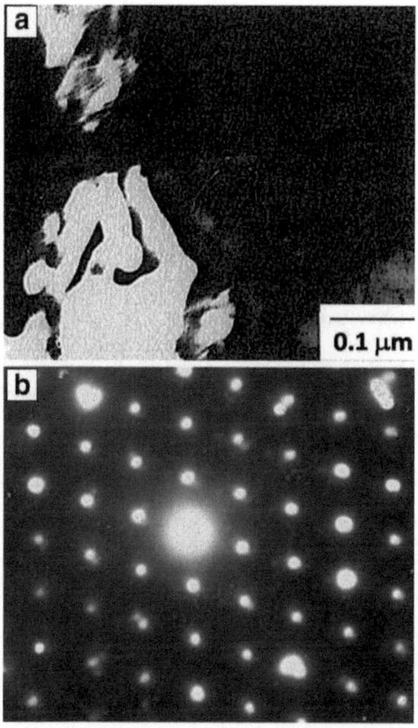

Fig. 7.9 Eutectic martensite along with $M_{23}(CB)_6$ and the corresponding electron diffraction pattern of the laser-remelted alloy spray coatings with CeO_2 [38]

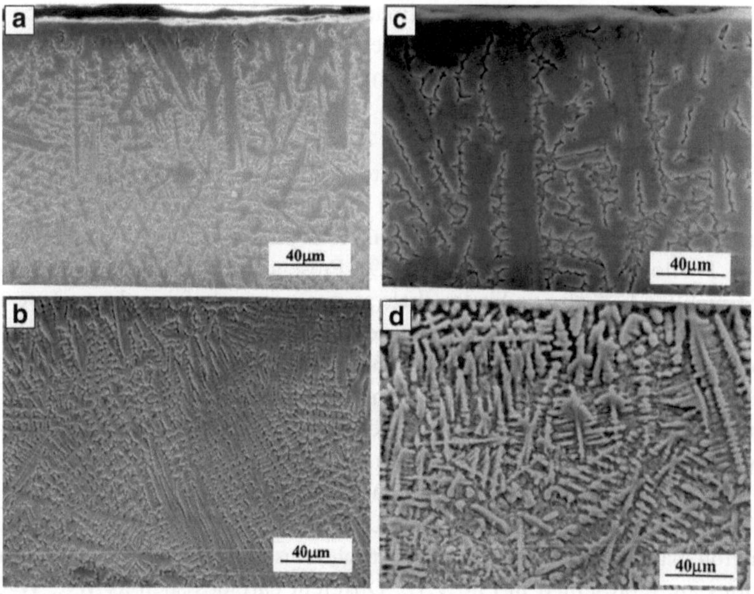

Fig. 7.10 Influence of laser processing conditions on the dendritic microstructure of Ti alloy [39]

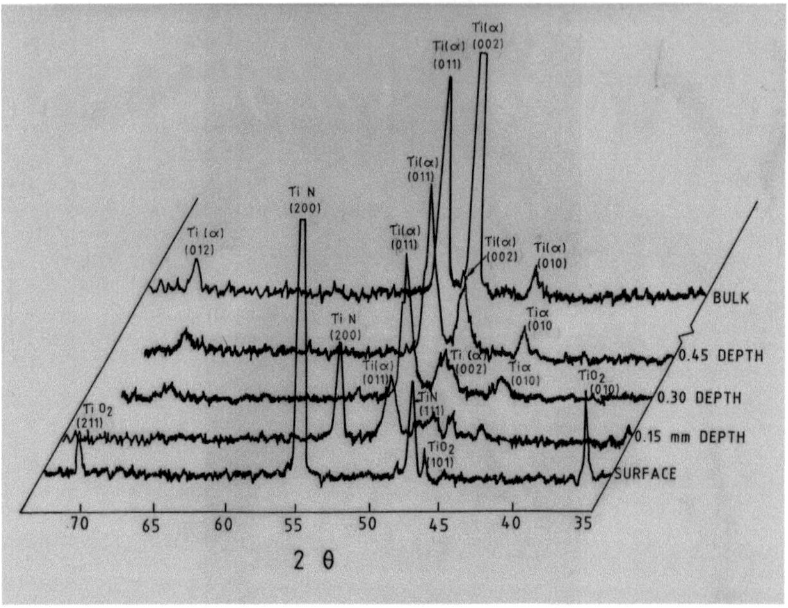

Fig. 7.11 Typical XRD patterns obtained at different depths in case of laser gas-nitrided Ti alloy [39]

in the alloyed melt pool, solidification starts at the surface and at the interface between the melted layer and unmelted substrate nearly at the same time. Such behaviour can be attributed to the formation of TiN layer near the surface of the melt pool. The temperature of the surface reaches below the melting point of high melting TiN phase almost at the same time as that of the region near the interface having low melting Ti (α). It can also be seen that the dendritic structure becomes finer with increasing laser scan speed as evident in Fig. 7.10c, d. This example is pertinent to note while investigating microstructure in laser-alloyed layer.

The X-ray diffraction technique is an important method to identify the phases present in laser-modified layers. A typical XRD pattern obtained from a laser gas-nitrided Ti alloy is shown in Fig. 7.11 [39]. Presence of hexagonal Ti and bcc TiN can be identified. It also indicates that, as one proceeds from the surface to the interior, the amount of TiN phase decreases and that of Ti (α) phase increases. Such composition gradient is often noted in laser-alloyed layer. The XRD pattern also reflects the changes in phase constitution as well as in the changes of quantity of various phases in the laser-modified layer on changing of processing conditions. The influence of laser beam scan speed on the XRD pattern is demonstrated by Staia et al. [40], who modified the surface of an A-356 alloy with a powder of 96 % WC, 2 % Ti and 2 % Mg in a two-step process with the help of a laser. The X-ray diffraction patterns obtained for the samples laser treated at different traverse velocities are represented in Fig. 7.12. Different carbides such as WC, W_2C and WC_x could be identified. The presence of small amount of WC_x can probably be

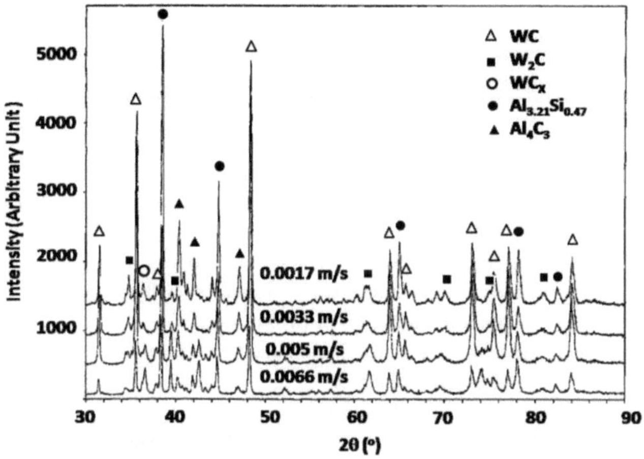

Fig. 7.12 X-ray diffraction patterns obtained for samples laser treated at different traverse velocities [40]

produced by partial dissolution of WC in the molten pool during laser processing and its subsequent solidification. Decarburisation of WC to W_2C is also a distinct possibility and widely reported. Decarburisation is accompanied by liberation of carbon which diffuses into the matrix and allows the formation of Al_4C_3. No other carbides are detected from the X-ray analysis. Formation of $Al_{3.21}Si_{0.47}$ is also noted. The oxidation reaction which takes place in the melt pool promotes the oxidation of aluminium and silicon in the form of Al_2O_3 and SiO_2, respectively, although these are not detected during the X-ray analysis. As illustrated in the above example, laser alloying and laser cladding can lead to formation of multiple phases completely governed by the process conditions.

7.3.2 Mechanical Properties

With regard to the mechanical properties, hardness is the most widely reported parameter. As the modified layer generally has a gradient of composition, especially in laser-alloyed layer, a gradient in hardness is inevitably obtained. The gradient in the alloyed layer is also responsible for scanty reported literature on fracture toughness (indentation toughness) of the modified layer. Laser transformation hardened steels also exhibit a graded hardness in the heat affected zone (HAZ) underneath the hardened layer. However, such gradient is normally not noticed in laser-cladded layer. Hardness of a wide variety of laser-modified layers has been reported in literature. Although different systems investigated and the varying process conditions preclude a comparison, a compilation of hardness obtained for various modified layers is made in Table 7.2 [38, 39, 41–55]. Depending on the resulting microstructure and phase constitution, the hardness of the modified

Table 7.2 Compilation of hardness of various laser modified layers

Name of the investigators	Laser processing system	Hardness (HV)
Jiang et al. [39]	Nitriding of Ti–6Al–4V	1,100 ± 25
Man et al. [41]	Ti and Sic alloying of Al6061 alloy	540 ± 20
Kwok et al. [42]	NiCrSiB alloying of UNSS31603	380 ± 20
Tassin et al. [43]	Laser surface processing of 316L SS with mixtures of Ti and Sic and Cr_2C_3 and Cr	450 ± 20
Fu and Batchelor [44]	Laser alloying with Ni and Cr of AA6061	320 ± 20
Wang et al. [38]	Laser melted spray coating of amorphous Fe based alloy	850 ± 50
Khedkar et al. [45]	Chromium laser alloyed Cr–Mo steel	591
Shamanian et al. [46]	Austenite steel cladding on ductile cast iron	1,200
Grenier et al. [47]	Ti laser gas carburised	1,800 (KHN)
Guo et al. [48]	Ti laser alloyed with Al	600 ± 50
Staia et al. [49]	Cast Al alloy laser alloyed with WC–Al–Mg powder	600
Datta Majumder and Manna [50]	Pure Cu laser alloyed with Cr	220
Datta Majumder and Manna [51]	304 SS laser alloyed with Mo	700
Datta Majumder and Manna [52]	Ti laser alloyed with Si and Al	700–750
Manna et al. [53]	Plain carbon steel laser cladded with Fe–BC–Si–Al–C	1,150
Persson et al. [54]	316l SS laser cladded with Stellite	700

surfaces can widely vary but the above table gives an idea of the large volume of data that exists. However, there is genuine shortage of data related to toughness of the modified layers. The variation of microhardness as function of depth beneath the laser-melted surface is shown in Fig. 7.13 [39]. Figure 7.13a shows the influence of laser power on the hardness profiles, whereas Fig. 7.13b details the role of scan speed on hardness. Clearly the depth of molten pool increases with increase of laser power and decrease of scan speed.

7.4 Tribology of Laser-Modified Surfaces

Study of tribology of laser-treated surfaces concerns with sliding wear, abrasive wear and erosive wear. Reported investigation on fretting wear is very limited and hence it is not discussed in this section. Most of the study with abrasive wear is on three body abrasion. The applied load and the abrasive give rise to high stress abrasion and low stress abrasion. Solid particle erosion and cavitation erosion are the main subject matters of erosive wear. Other forms of erosion such as rain drop erosion, liquid metal erosion, etc. of laser treated surfaces are not touched upon.

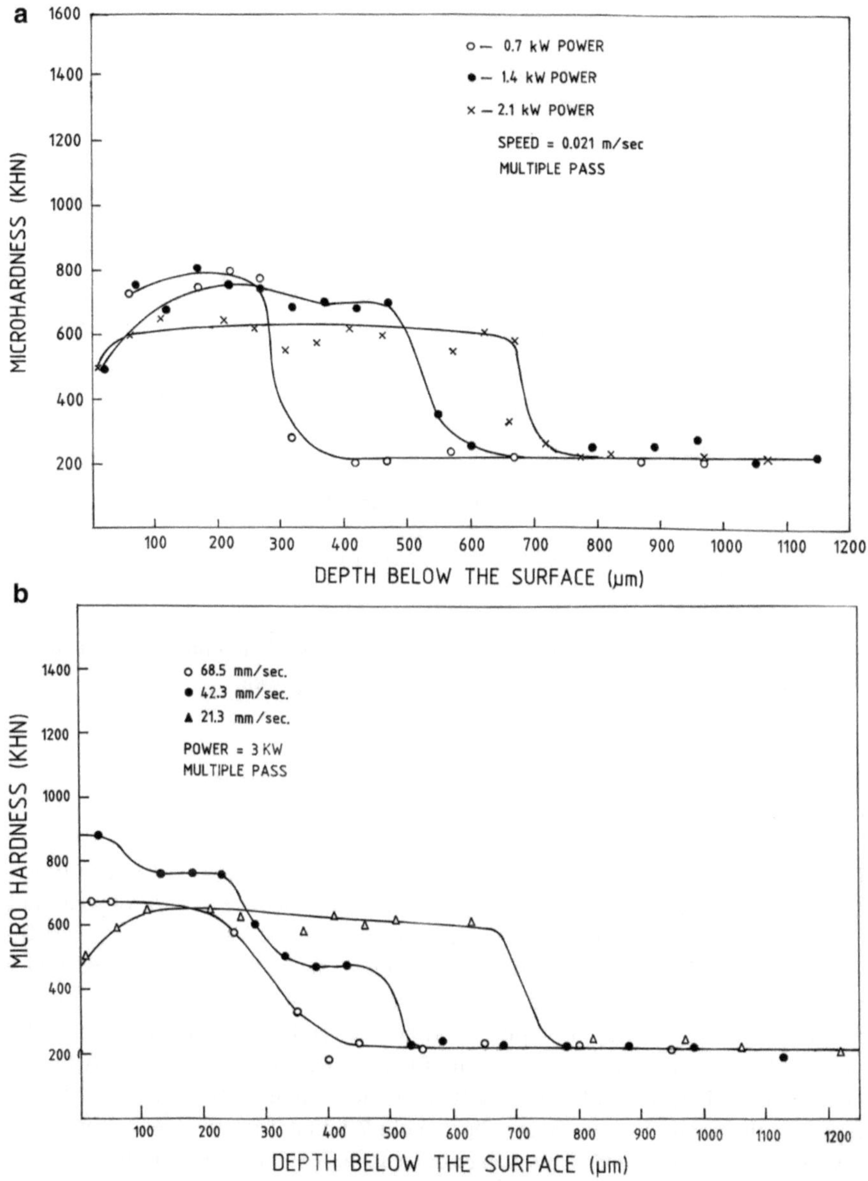

Fig. 7.13 Variation of microhardness as function of depth beneath the laser-melted surface [39]

7.4.1 Sliding Wear

Although tribology of laser modified surfaces has been studied quite comprehensively, sliding wear of laser-treated surfaces has received the maximum attention.

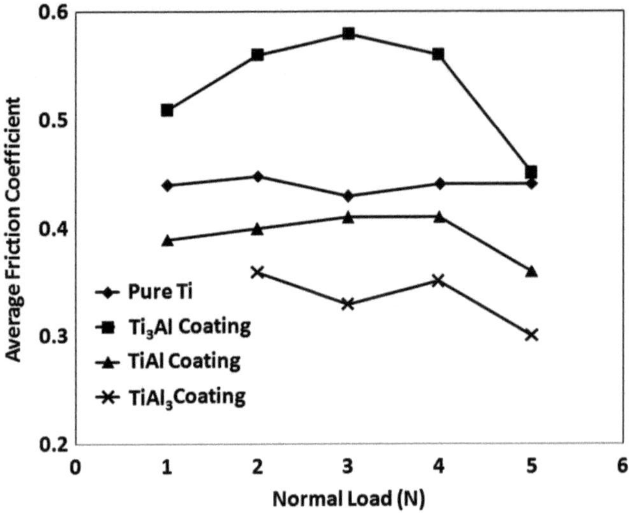

Fig. 7.14 Variation of friction coefficient as function of normal load for various titanium aluminide coatings [48]

The sliding wear characterization is also confined primarily to unidirectional sliding wear and much less study corresponds to reciprocating wear.

7.4.1.1 Friction Response

The impressive amount of data available on the friction response of laser surface-modified layers permits an attempt to generalise the influence of various wear and/or laser processing conditions on friction behaviour of laser-modified surfaces. The influence of applied load on the friction coefficient can be found in the work of Geo et al. [48]. Titanium aluminide coatings were in situ synthesised on a pure Ti substrate by laser alloying a preplaced Al powder and the average friction coefficient of the resulting layer as a function of load is presented in Fig. 7.14. Although $TiAl_3$ exhibited the lowest friction coefficient, the friction coefficient did not follow any specific trend. Rather, it can be concluded that friction coefficient is not governed by applied load. This behaviour can be explained by the fact that the friction coefficient (μ) in case of rough contacting surfaces is given by

$$\mu = \mu_{\text{ad}} + \mu_{\text{grov}} \tag{7.18}$$

where μ_{ad} is the friction coefficient due to adhesion and μ_{grov} is friction coefficient due to grooving. The adhesion component can be expressed by the following relation [56]:

$$\mu_{ad} = \frac{\tau_m}{H} \tag{7.19}$$

where H is the hardness of the softer material and τ_m the shear strength of the contacting asperities and the shear strength is known to depend on the work of adhesion of the mating materials [57]. Further, the friction coefficient due to grooving is given by

$$\mu_{grov} = \frac{\tau_c \times A}{L} \tag{7.20}$$

where L is the applied load, A is the area sheared plastically by an abrasive particle and τ_c is the critical shear strength of the modified layer. From the above equations, it is clear that at low applied load, the grooving component dominates while when load is increased the adhesion component dominates. Thus, the friction coefficient remains more or less unaltered. Similarly, Dutta Majumder et al. [53] have also noted the load independent friction coefficient of Ti laser surface alloyed (LSA) with Si, Al and Al + Si. In this study, the tribological characteristics of the laser-alloyed samples were investigated to ascertain the variation in wear depth as a function of load and time using a computer-controlled reciprocating ball-on-disc wear testing machine fitted with an oscillating hardened steel ball. A detailed post-wear microstructural analysis was also conducted to determine the mechanism of wear and the role of alloying elements in improving the resistance to wear. The authors have reported that LSA with Si is more effective in improving the wear resistance of Ti than that by Si + Al or Al alone. The enhanced performance in Si surface-alloyed samples has been attributed to the presence of uniformly distributed Ti_5Si_3 in the alloyed zone. Work due to Tassin et al. [43] shown in Fig. 7.15 indicates increase of friction coefficient with increase of hardness. This behaviour is in contrast to Eq. (7.19) which suggests that the friction coefficient should decrease with increase of hardness signifying that the adhesion component plays a nominal role.

Similarly, the influence of sliding speed on friction coefficient is shown in Fig. 7.16. The friction coefficients of laser-processed titanium aluminide coatings were found to decrease drastically with increasing sliding speed. Under identical dry sliding wear conditions, the wear resistance of the titanium aluminide coatings was in the following order: Ti_3Al coating > TiAl coating > $TiAl_3$ coating. The decrease of the friction coefficient with increase of sliding velocity follows the trend described by the equation given below [58].

$$\mu = K_1 - K_2 \mathrm{Ln}(v) \tag{7.21}$$

where μ is the coefficient of friction and v is the sliding velocity. K_1 and K_2 are material-specific constants.

Roughness of the modified surface also plays a crucial role in determining the friction coefficient. Higher roughness typically corresponds to more asperities slope

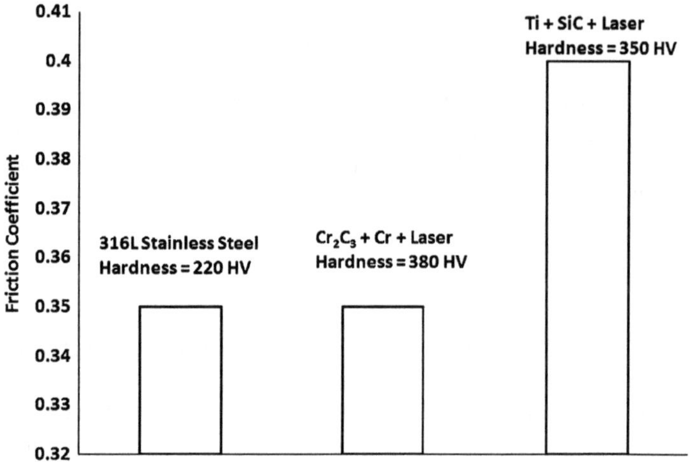

Fig. 7.15 Friction coefficient of various laser surface-alloyed stainless steel showing increase of friction coefficient with increase of hardness [43]

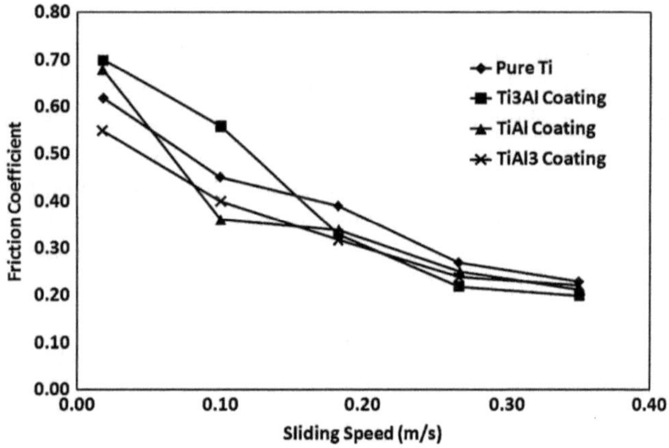

Fig. 7.16 Influence of sliding speed on the friction coefficient of titanium aluminide coatings [48]

angle, which, in turn, increases the friction coefficient. There are two well-established mechanisms responsible for higher friction coefficient being noted in films having higher roughness. The first one is the ratcheting mechanism, where relative motion between two surfaces is achieved by asperities riding over each other. The second is the energy loss mechanism where asperities push each other. In the initial stage of sliding, there can be other operative mechanisms resulting in different trend. Work due to Staia et al. [49] indicated a decrease in friction coefficient with increase in roughness. Their work is presented in Fig. 7.17. This observation, however, is in direct contrast with that reported by many others [53]. The contradicting results of Staia et al. [49] are related to debonding of reinforcing

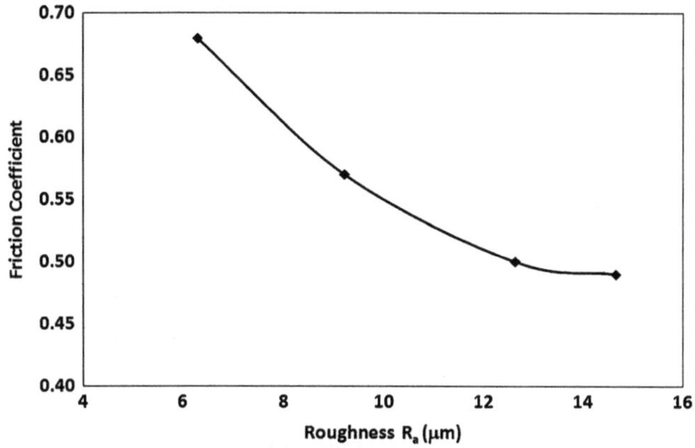

Fig. 7.17 Influence of surface roughness on friction coefficient [49]

WC particles which subsequently act as abrasive particles and thereby increasing friction coefficient.

7.4.1.2 Wear Rate

The wear rates of laser-processed surfaces are determined by wearing conditions. As is the case with other modified surfaces, the wear rate of laser-modified surface increases with applied load as long as the wear is plasticity dominated as per Eq. (7.22) given below:

$$W = k\frac{F}{H} \tag{7.22}$$

where k is the wear coefficient, F is the applied load and H is the hardness of the wearing material. The friction and wear properties of the titanium aluminide coatings at different normal loads and sliding speeds were investigated. It was found that the hardness of the titanium aluminides coatings was in the following order: Ti_3Al coating > TiAl coating > $TiAl_3$ coating. Friction and wear tests revealed that, at a given sliding speed of 0.10 m/s, the wear volume of pure Ti and all titanium aluminide coatings increased with increasing normal load as shown in Fig. 7.18. Such behaviour is consistent with Eq. (7.22) given above.

The laser processing conditions employed are known to influence the wear performance of the modified surfaces. At a given normal load of 2 N, the wear volume of Ti_3Al coating and TiAl coating first increased and then decreased, while the wear volume of $TiAl_3$ coating first decreased and then increased with increasing sliding speed as shown in Fig. 7.19. This behaviour may be related to change of wear mechanism from mild oxidational wear to severe oxidational wear. Another illustration of how the speed at which the modified surface is scanned alters the

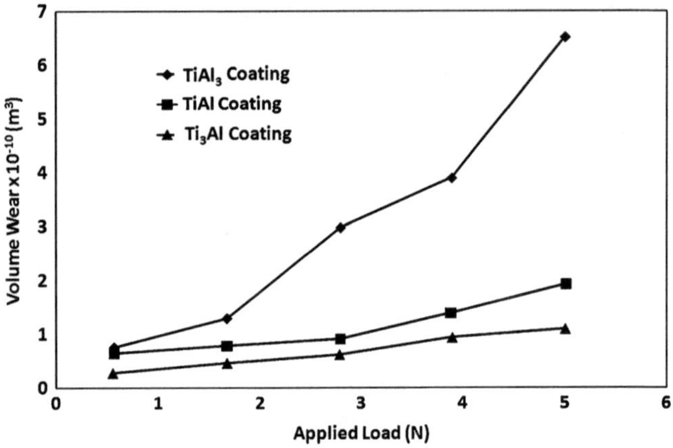

Fig. 7.18 Influence of applied load on the wear rate of laser-treated titanium aluminide coatings [48]

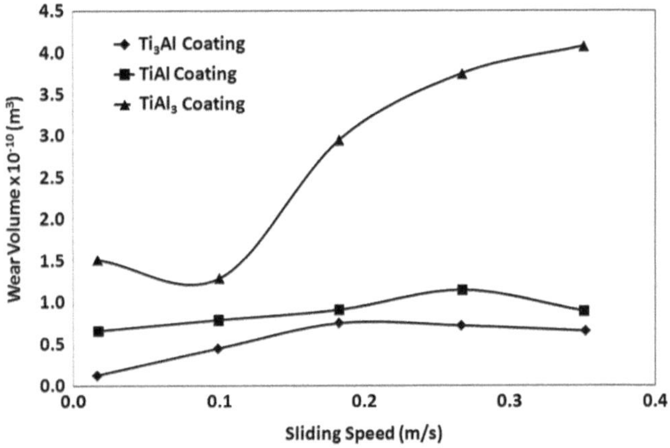

Fig. 7.19 Influence of sliding speed on the wear rate of laser-treated titanium aluminide coatings [48]

wear response of modified surfaces is presented in Fig. 7.20. The wear rate decreases with increase of scan speed. As the scan speed increases, the solidification rate increases thereby resulting in finer microstructure which is often accompanied by superior mechanical properties of modified surface. The improved mechanical properties, particularly increase in hardness, result in reduction of wear rate as per Eq. (7.22) described above.

Radek and Bartkowiak [59] have evaluated the tribological properties of laser-treated and electro-spark-deposited coatings. The tests were conducted for Mo and Cu coatings (the anode) which were electro-spark deposited over a C45 steel substrate (the cathode) and melted with a Nd:YAG laser beam. The coatings after

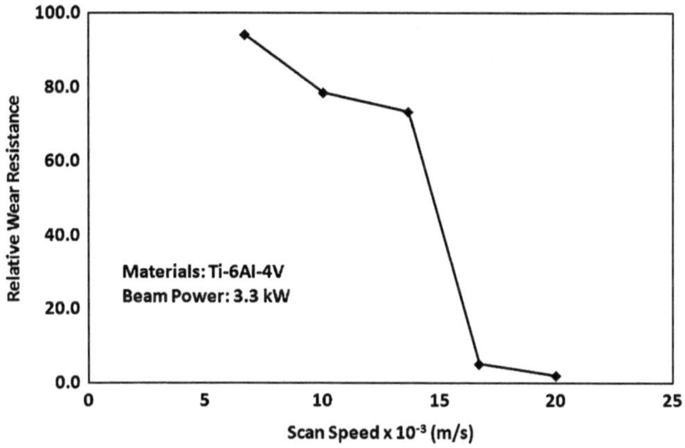

Fig. 7.20 Influence of scan speed at beam power of 3.3 kW on the wear resistance of Ti–6Al–4V alloy [55]

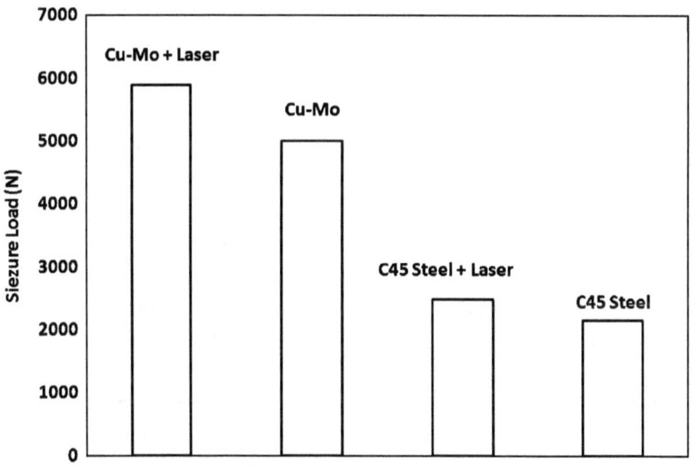

Fig. 7.21 Improvement of seizure load of C 45 steel by laser treatment of Cu–Mo coating deposited by electro spark technique [59]

laser processing are still distinguished by very good seizure load as shown in Fig. 7.21.

Persson et al. [54] examined the frictional behaviour and galling resistance of the laser-processed Fe-based alloy Norem 02 at temperatures ranging from room temperature (RT) to 527 K. Friction and galling testing was performed by sliding two crossed test rods in a load-scanning test rig. The results were compared with those from laser-processed Stellite 21, one of the best low frictions, galling resistant, Co-based alloys known. At room temperature, Norem 02 exhibited excellent sliding performance, comparable to that of Stellite 21, showing a low and stable friction coefficient of 0.25 and revealed no tendency to gall. However, at temperatures of 427 K or above, the friction coefficient rose to 0.5–0.7 and massive galling was observed. Stellite 21, however, maintained its excellent friction performance even at higher temperatures.

At an intermediate sliding speed and applied load, mild oxidational wear has been reported to become prevalent. The wear rate (W) for mild oxidation is given as [58]

$$W = \frac{FC^2 A_o}{vZ_c H_o} \exp\left(-\frac{Q}{RT_f}\right) \tag{7.23}$$

where A_o is Arrhenius constant, Q is the activation energy for oxidation, Z_c is the critical thickness of spalling, T_f is the temperature at the point of contact of the asperities, v is the sliding velocity, C is a constant which depends on the composition of the oxide scale and R is the molar gas constant.

The contact temperature at the asperities, known as flash heating temperature, is given by [58]

$$T_f = T_b + \frac{\mu v r}{2K_e}\left(\frac{H_o F}{NA_n}\right)^{\frac{1}{2}} \tag{7.24}$$

where H_o is the hardness of the oxide scale, N is the total no of the asperities, r is the radius of the contacting area and K_e is the equivalent thermal conductivity.

The oxidation under sliding wear takes place at a rate higher than that observed under static conditions [60–62]. The observed difference is due to change of Arrhenius constant A_o, which depends on material and wearing condition [63]. However, the activation energy for oxidation Q remains unchanged [64]. Hence, further discussion on lowering the wear rate will be confined to the activation energy Q and the critical thickness of spalling (Z_c). Examination of the activation energies for oxidation (Q) for a large number of oxide scales [58, 64] indicates that the activation energies for oxidation are high for Al_2O_3 and Cr_2O_3 scale formation. Hence, many surface engineering solutions to improve wear resistance in this regime involve forming of Al-or Cr-rich layers. This can be achieved by laser surface alloying.

Hirose et al. [65] have reported improved wear performance on alloying Ti with Al using a laser. When Ti undergoes mild oxidational wear, it results in the formation of TiO_2 and wears through a spallation mechanism. When alloyed with Al by laser surface alloying, a protective Al_2O_3 scale forms. Since the activation energy for formation of Al_2O_3 is much higher compared to TiO_2, the growth of this oxide scale is much slower. Thus, it takes a long time for the critical spalling thickness to be attained leading to reduced wear rate. Similarly, Tau and Doong [66] found that the chromium-alloyed surface exhibited higher wear resistance compared to untreated steel for a given hardness and structure. This behaviour can also be attributed to the fact that the chromium-rich layer formed a Cr_2O_3 scale, which exhibited better adhesion to the substrate than the Fe_2O_3 scale formed on the untreated steel. Isawane and Ma [67] also reported an improved wear performance of stainless steel when coated with Cr and Ni.

The third method of reducing the wear rate is to increase the critical thickness of spalling. For a given oxide scale substrate system, the critical thickness for spalling (Z_c) can be increased by the following means:

1. Adding *Y*, Hf or rare earth elements on the surface [68, 69].
2. Ensuring fine dispersoids near the surface, either by internal oxidation or by using mechanically alloyed material [70].
3. Aluminising with noble alloying elements, e.g. Pt, Au, Ag, etc. as such addition results in cleaner oxide formation [71–73].

Addition of *Y*, Hf and rare earth elements is known to increase the adhesion of the oxide scale, and this has been attributed to the following:

1. Scale grows into the substrate at scale–metal interface resulting in mechanical pegging [74].
2. Voids generation at the scale–metal interface is prevented by vacancy accommodation [75].
3. Grains become finer and grain boundary diffusion coefficient increases, thereby promoting stress relief by cobble creep [76].
4. Acting as nucleation sites and producing a fine grained scale, the dispersoids block the boundaries and sulphur segregation at the scale–metal interface is prevented. This prevents embrittlement at the scale–metal interface. This acts as oxygen active dopants [77].

The increase in critical thickness of spalling, resulting in increase of wear resistance, is reflected in the work of Wang et al. [38] who noted an increase in wear resistance with laser surface alloying of substrate, a low carbon steel by CeO_2. Their results illustrated in Fig. 7.22 depict wear rate as function of hardness. Clearly a lower wear rate can be seen with a CeO_2-alloyed surface. They attributed this observation to refinement of microstructure, strengthening and purification of the grain boundary and improvement in the morphology and distribution of eutectics. These refinements result in a more adherent oxide scale which, in turn, improves the wear resistance.

In order to improve the sliding wear resistance of stainless steel components, carbide-hardened surfaces were produced on AISI 316L austenitic stainless steel employing laser surface alloying by Tassin et al. [43]. Different powdered precursors (CrC, a mixture of CrC and Cr and a mixture of Ti and SiC) were applied to the steel surface and irradiated using a continuous wave 300 W Nd–YAG laser. The surface alloys resulting from the incorporation of CrC, alone or mixed with chromium, were composed of austenite (γ) dendrites surrounded by a γ-M_7C_3 eutectic (M = Fe or Cr) and their microhardness ranged from 380 to 450 HV. Superficial hardening by titanium carbide was obtained using a mixture of Ti and SiC powders as precursors. Ti and SiC particles dissolved in the melted pool, leading to the precipitation of fine TiC particles during solidification. The microhardness of the surface alloy was about 350 HV. For both chromium carbide

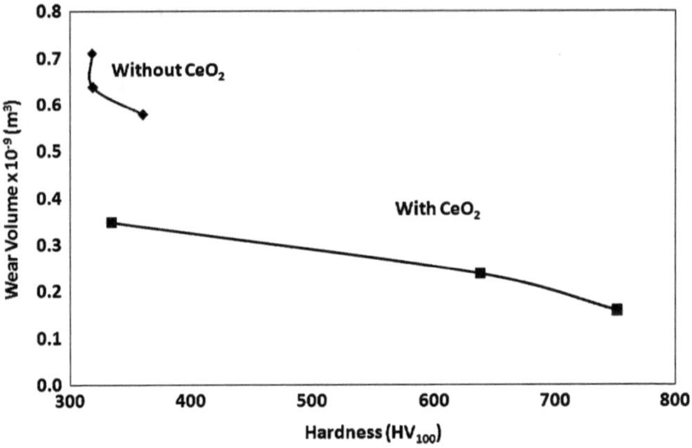

Fig. 7.22 Improvement of wear resistance of low carbon steel 1020 by addition of CeO_2 employing laser [38]

and titanium carbide surface-alloying treatments, a noticeable improvement in the sliding wear resistance of AISI 316L stainless steel was noted.

A significant amount of work has been reported on improvement of tribological properties on introduction of carbon on the ferrous materials by laser surface alloying, either using graphite or melting in presence of CO_2 or acetylene [78–80]. Laser-borided single-phase iron has been shown to be superior in terms of sliding wear resistance and rolling contact fatigue resistance to traditionally single-phase-borided samples. Laser alloying of high speed steel by BN and Ti/BN resulted in increased tool life [81]. Laser surface alloying of Cu–Ni alloy with boron powder using a defocused CO_2 laser has been reported to increase wear resistance as high as 40 times [82]. Gulan et al. [83] found improvement in wear by scratching of laser-alloyed Al80 as shown in Fig. 7.23. According to their result, laser alloying with Al results in marginal improvement of wear resistance. Alloying with Ni and Si led to limited enhancement, in spite of improved hardness. This fact is attributed to formation of brittle phases such as $Mg_{17}Al_{12}$, Mg_2Ni, Mg_2Si, etc. However, alloying with Cu was found to yield significant improvement in wear resistance.

The potential of producing rapidly solidified microstructures with enhanced hardness and wear resistance by laser cladding of Co, Fe and Ni base alloys on traditional substrate materials has also been demonstrated by several researchers [13, 84–86]. Belmondo and Castagna have successfully laser cladded a mixture of Cr–Ni–Mo and Cr–carbide on mild steel with increased wear performance [87]. A study by Singh and Majumder [88] has shown that laser cladding of Fe–Cr–Mn–C on steel yields a fine duplex microstructure of fine grain ferrite matrix with M_6C and M_7C_3 carbide precipitates which leads to superior wear properties compared to Stellite 6. Use of Cr_3C_2 and Mo_2C clad materials is found to yield a β-Ti + TiC microstructure which is desirable to improve wear characteristics [89].

A comparison between the wear rates of laser surface-hardened and laser surface-melted surfaces has been made by Roy and Manna [90]. In their study, an

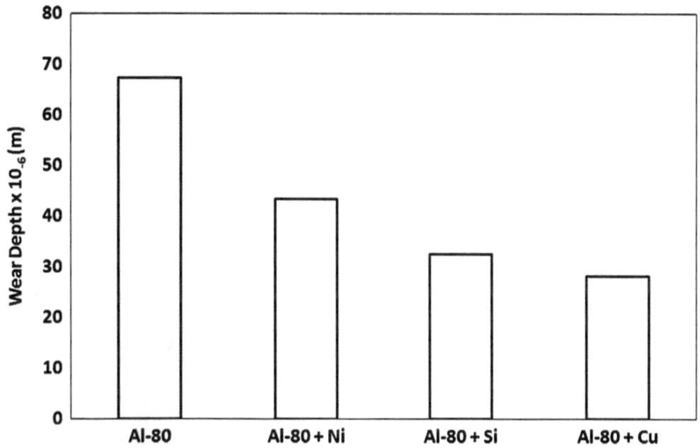

Fig. 7.23 Bar diagram showing improvement of wear resistance by laser surface alloying of aluminium alloy Al 80 [83]

attempt has been made to evaluate the scope of enhancing wear resistance of austempered ductile iron by laser surface melting (LSM) and laser surface hardening (LSH). A detailed study concerning microstructural evolution and mechanical properties following LSM and LSH indicates that LSM develops a relatively low microhardness in the near-surface region and a predominantly austenitic microstructure in the laser-melted zone. On the other hand, LSH, compared to LSM, results in a higher hardness and a more uniform microhardness profile due to a primarily fine martensitic microstructure in the laser-hardened zone. Careful X-ray diffraction analyses coupled with microstructural studies reveal that diffusion of carbon from graphite is responsible for a higher volume fraction of retained austenite and lower hardness in the laser-irradiated zone following LSM as compared to LSH. Furthermore, LSH leads to development of residual compressive stress, while LSM produces residual tensile stress on the surface. Finally, adhesive wear tests with a pin-on-disc machine and subsequent microstructural analyses show that LSH is more appropriate than LSM to enhance adhesive wear resistance of austempered ductile iron. Figure 7.24 presents a SEM image of a worn surface of a laser surface-melted sample showing delamination or removal of material during adhesive wear at a load of 5 kg. The high magnification scanning electron micrograph (Fig 7.24b) of the wear debris substantiates that loss of metal occurs primarily by the propagation of micro-cracks at the surface. The latter may be fatigue-cracks initiated at the sub-surface or may initiate from the austenite–martensite interface where deformation constraints are maximum. Improved wear performance of laser surface-hardened 52100 steel has also been demonstrated by Basu et al. [91]. Similarly, Padmavati et al. [92] found that laser surface melting of magnesium alloy improved wear resistance. Other studies have shown that laser surface alloying and laser composite processing [93, 94] result in improved wear resistance of aluminium alloys.

Fig. 7.24 SEM image of a worn surface of a laser surface-melted austempered ductile cast iron [90]

7.4.2 Abrasive Wear

Laser surface modification offers a promising route for improving the abrasive wear resistance of materials. Abrasive wear resistances of a variety of steels and other materials have been improved by laser surface modification. The abrasive wear rate of laser transformation hardened steels of varying carbon content, along with their corresponding hardness, is illustrated in Fig. 7.25. Clearly, the abrasive wear resistance increases with increase in hardness of the laser surface-modified steel layer in accordance with Eq. (1.16) of Chap. 1.

Jiang et al. [55] nitrided Ti–6Al–4V alloys using a laser in a gaseous nitrogen environment. A significant improvement in abrasive wear resistance upon laser nitriding was noted. The abrasive wear rate is also as expected governed by the processing parameters, for example, as shown in Fig. 7.26, the abrasive wear resistance decreases with increasing scan rate. This is attributed to the unique microstructure of the 'in situ' laser surface-alloyed composite coatings. The as-solidified dendritic TiN reinforcements strongly embedded in the matrix provide the laser surface-alloyed coatings excellent resistance. The volume fraction of the 'in situ'-solidified TiN reinforcement and thereby the wear resistance of the laser surface-alloyed composite coatings can be easily controlled by varying the laser surface alloying parameters.

The gas composition during laser surface gas alloying is also known to govern the abrasive wear behaviour. According to Grenier et al. [47], laser gas alloying of Ti result in increase in abrasive wear resistance as shown in Fig. 7.27. The best wear resistance was found when the surface was alloyed with a mixture of carbon monoxide and nitrogen gas. The best abrasive wear resistance was exhibited by the layer having highest hardness and it is plausible that the mixed carbides and

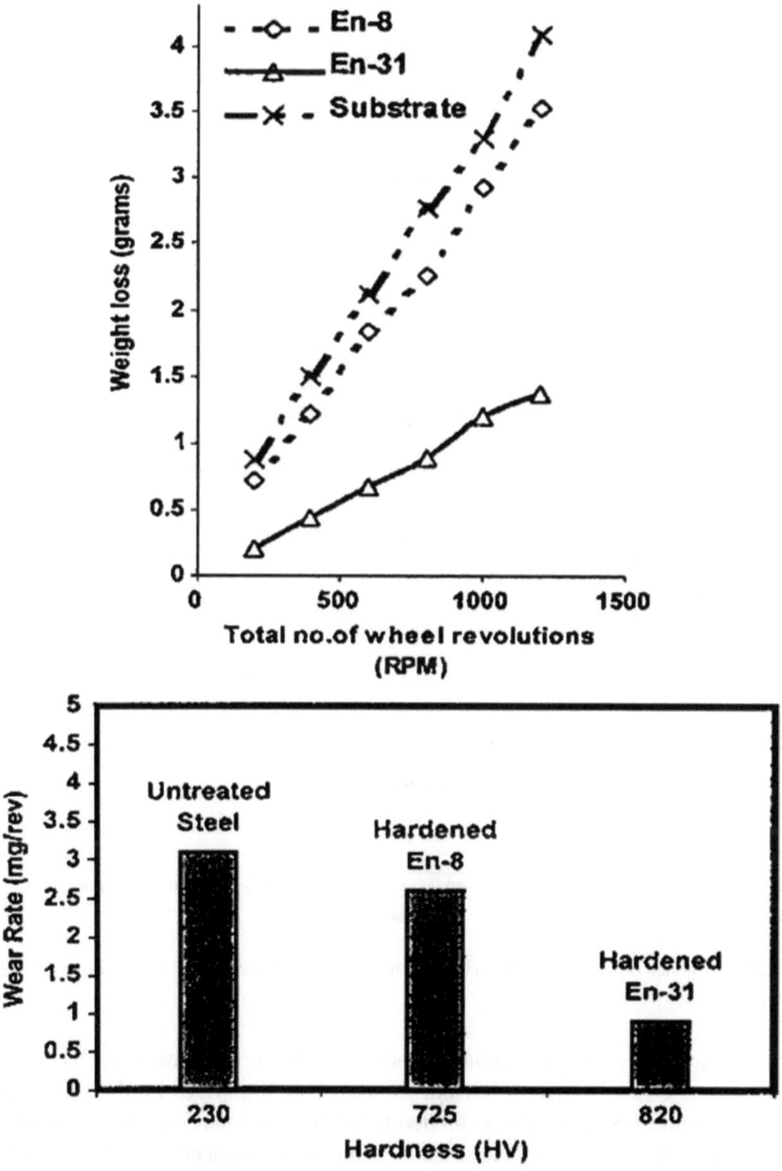

Fig. 7.25 Abrasive wear rate of laser transformation hardened steels of varying carbon content along with their corresponding hardness

nitrides that result in alloyed layer when the above gas mixture is used are responsible for the enhanced properties.

Hybrid processing by combining laser treatment with another coating process has great technological potential. For example, laser glazing of plasma-sprayed and

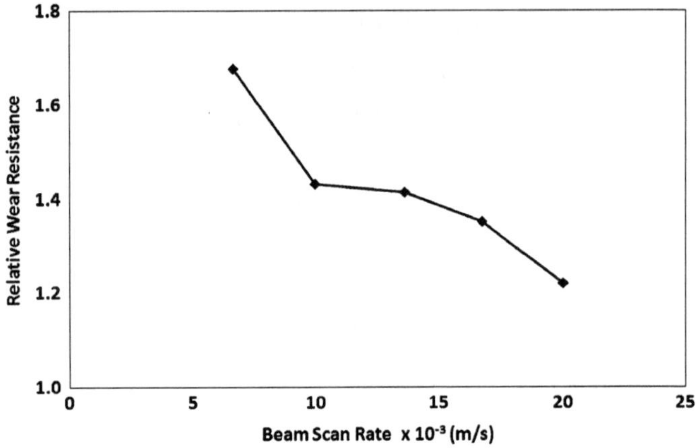

Fig. 7.26 Influence of beam scan speed on relative abrasive wear resistance of Ti–6Al–4V [55]

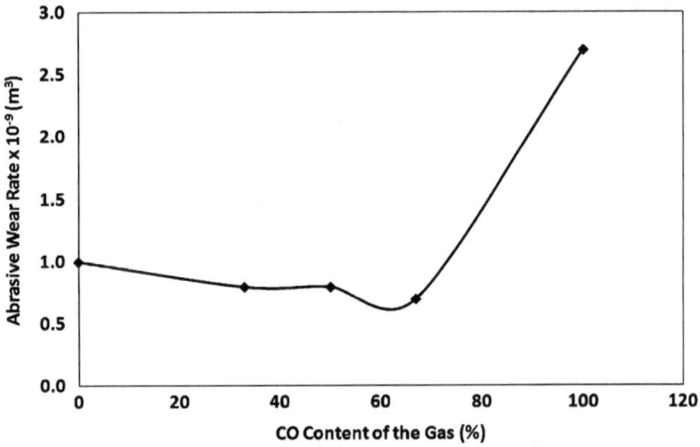

Fig. 7.27 Influence of gas composition (CO content) on the abrasive wear rate of laser gas-alloyed Ti [47]

detonation-sprayed coatings exhibited improved abrasive wear resistance as compared to as sprayed coating. Such behaviour is shown in Fig. 7.28. Further, laser-glazed detonation-sprayed coatings also exhibited better wear resistance than as-sprayed coating. However, the property enhancement was not as pronounced as in the case of laser post-treatment of plasma-sprayed coatings as the detonation-sprayed coatings are characterised by negligible porosity and provide little opportunity for further improvement. The result also highlights the importance of glazing power and speed in relation to final performance of the treated layer.

Rieker et al. [95] found enhancement in abrasive wear resistance of ferritic stainless steel by laser alloying with Mo as well as Mo + B. Laser-treated boronised surface exhibits improved abrasive wear rate at low laser power and higher wear rate at high laser power. This behaviour is illustrated in Fig. 7.29. Boas and

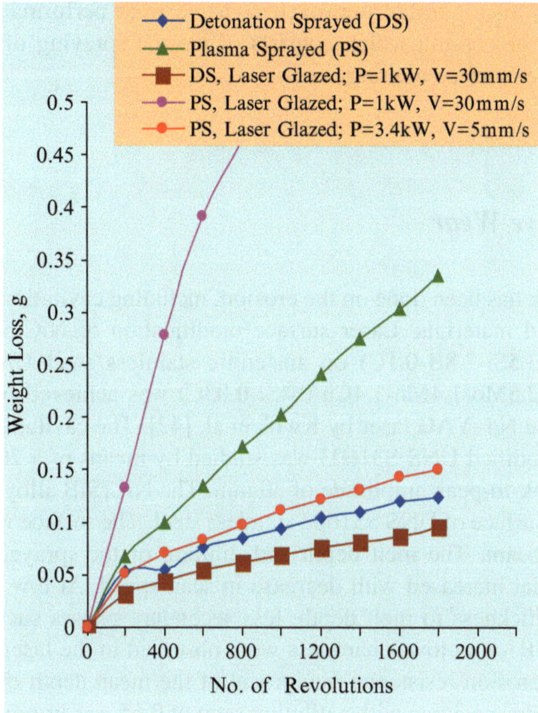

Fig. 7.28 Abrasive wear resistance of laser-glazed plasma-sprayed and detonation-sprayed coatings

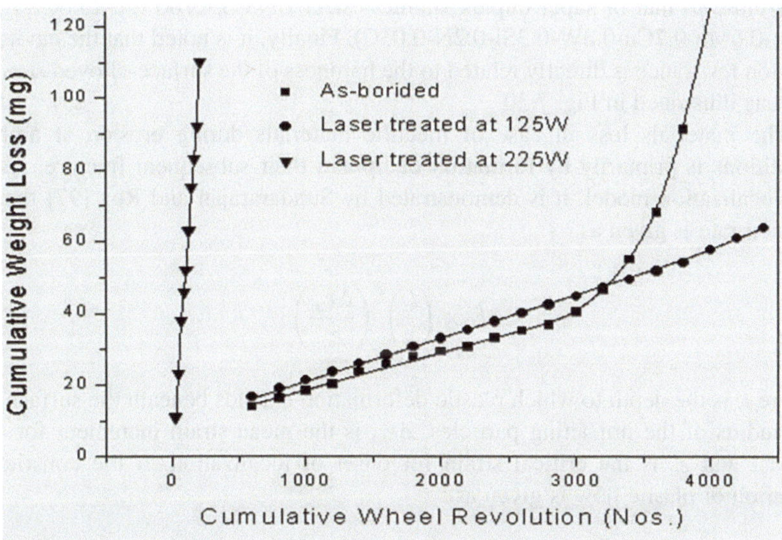

Fig. 7.29 Abrasive wear resistance of laser-treated boronised steel

Bamberger [96] also reported improved abrasive wear performance of low alloy steel by hybrid processing which consists of plasma spraying of triballoy T-400 followed by laser surface melting.

7.4.3 Erosive Wear

Significant work has been done on the erosion, including cavitation erosion of laser surface-modified materials. Laser surface modification of NiCrSiB coating (Ni-16.5Cr-15.5Fe-3.5Si-3.8B-0.1C) on austenitic stainless steel UNS S31603 (Fe-17.6Cr-11.2Ni-2.5Mo-1.4Mn-1.4Cu-0.4Si-0.03C) was achieved by using a 2-kW continuous wave Nd–YAG laser by Kwok et al. [42]. The cavitation erosion of the laser surface-modified UNS S31603 was studied by means of a 20-kHz ultrasonic vibrator at a peak-to-peak amplitude of 30 mm. The NiCrSiB alloy layer was flame sprayed on the surface of UNS S31603 stainless steel. The surface was then scanned with the laser beam. The melt depth and dilution of the sprayed layer with the substrate material increased with decrease in scan speed. At low dilution ratio of sprayed layer thickness to melt depth, less secondary phases such as CrB, CrB_2, Fe_2B and $M_7(CB)_3$ and lower hardness were observed in the laser-modified layer. The cavitation erosion resistance (reciprocal of the mean depth of penetration) of the laser-modified specimen with a dilution ratio of 0.65 was improved by 2.7 times as compared with that of the as-received UNS S31603 specimen. The cavitation erosion resistance of the laser surface-modified specimen with dilution ratio of 0.88 was improved fourfold as compared with that of the as-received UNS S31603 and very close to that of super duplex stainless steel UNS S32760 (Fe-25.6Cr-7.2Ni-4Mo-0.6Mn-0.7Cu-0.8W-0.3Si-0.2N-0.03C). Finally, it is noted that the cavitation erosion resistance is directly related to the hardness of the surface-alloyed stainless steel as illustrated in Fig. 7.30.

The materials loss in case of metallic materials during erosion at ambient conditions is primarily by formation of lip and their subsequent fracture. As per the localization model, it is demonstrated by Sundararajan and Roy [97] that the erosion rate is given as

$$E \propto \left(\frac{L}{r}\right)^3 \left(\frac{\Delta\varepsilon_{\rm m}}{\varepsilon_{\rm c}}\right) \tag{7.25}$$

where L is the depth to which plastic deformation extends beneath the surface, r is the radius of the impacting particles, $\Delta\varepsilon_{\rm m}$ is the mean strain increment for each impact and $\varepsilon_{\rm c}$ is the critical strain for onset of localization. If the constitutive equation of plastic flow is given as

$$\sigma = K\varepsilon^n(1 + CT) \tag{7.26}$$

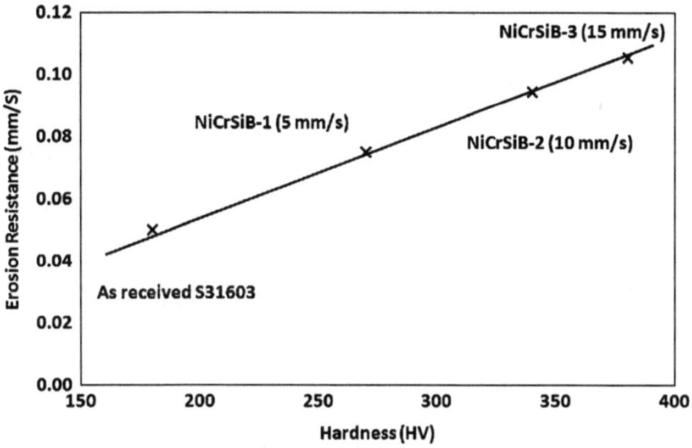

Fig. 7.30 Influence of hardness on cavitation erosion resistance of NiCrSiB coating [42]

where K is the strength coefficient, C is the temperature coefficient of flow stress and n is the strain hardening exponent, ε_c can be obtained as

$$\varepsilon_c = \left[\frac{n\rho C_P}{KC}\right]^{\frac{1}{n+1}} \tag{7.27}$$

where ρ is the density of the eroding material and C_P is the specific heat. K can be replaced by the melting point (T_m) of the eroding materials according to

$$C = \frac{4.5}{T_m} \tag{7.28}$$

Thus, the critical strain for localization can be written as

$$\varepsilon_c = \left[\frac{n\rho C_P T_m}{4.5K}\right]^{\frac{1}{n+1}} \tag{7.29}$$

Or

$$\varepsilon_c \cong \frac{n\rho C_P T_m}{4.5K} \tag{7.30}$$

since erosion takes place at high strain and n at high strain is normally very low thereby making $n + 1$ nearly equal to 1. From Eq. (7.25), it is clear that, for a given eroding condition, L, ρ and $\Delta\varepsilon$ are constant. Thus the erosion rate is inversely proportional to the critical strain for localization and this critical strain for flow localization is directly proportional to melting temperature. In other words, the

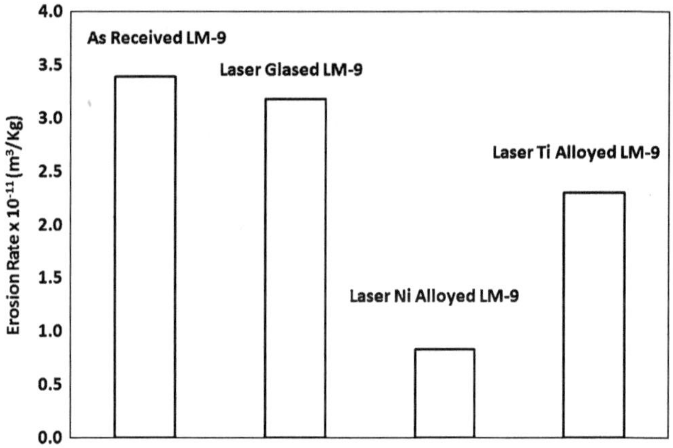

Fig. 7.31 Improvement of erosion resistance of laser-alloyed LM-9 with Ti and Ni [98]

erosion rate is inversely proportional to the melting temperature of the eroding materials as given below

$$E \propto \frac{4.5K}{n\rho C_P T_m} \tag{7.31}$$

Hence, higher the melting temperature of the surface, lower is the erosion rate. Roy et al. [98] employed this concept to laser surface alloying of Al alloy LM-9 by Ni and Ti and found improved erosion rate. Their result is presented in Fig. 7.31. Erosion resistance of laser-glazed, laser Ni-alloyed and laser Ti-alloyed LM-9 improved by 6.2 %, 76 % and 32 % respectively.

The study by Datta Majumder and Manna [51] concerns laser surface alloying (LSA) of AISI 304 stainless steel (304-SS) with pre-deposited Mo (by plasma spraying) to enhance erosion–corrosion resistance of the substrate. The optimum LSA conditions in terms of pre-deposit thickness, laser power density q and scan speed v have been identified to achieve the desired microstructure, composition and mechanical properties. LSA under such optimum conditions significantly improves the erosion–corrosion resistance in 20 wt.% sand in 3.56 wt.% NaCl solution as depicted in Fig. 7.32. Thus, it is concluded that LSA of 304-SS with Mo is an appropriate technique to enhance the resistance to erosion–corrosion in stainless steel. Datta Majumder and Manna [50] also explored the possibility of improving erosion resistance of Cu by laser surface alloying with Cr. As shown in Fig. 7.33, an improvement in room temperature and elevated temperature erosion resistance is noted. The above observation can be explained in terms of erosion behaviour at elevated temperature. As such high temperature, erosion will occur primarily by spallation of oxide scale and therefore the erosion rate will be determined by the steady state oxide scale thickness which is governed by the balance of growth of oxide scale by oxidation and erosion of oxide scale due to impact. If it is assumed

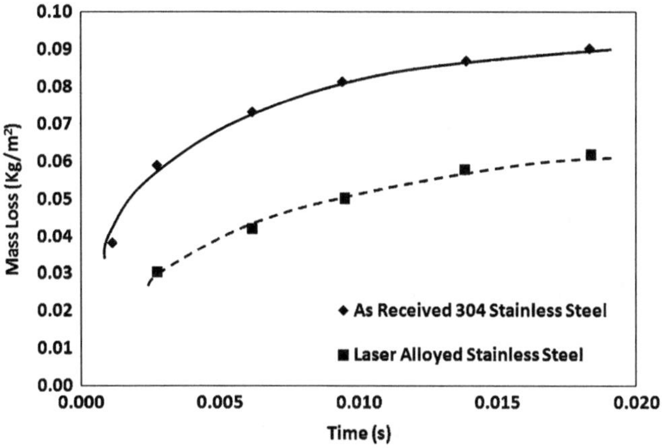

Fig. 7.32 Improvement in erosion–corrosion resistance of laser surface-alloyed 304 stainless steel with Mo in 20 wt.% sand in 3.56 wt.% NaCl solution [51]

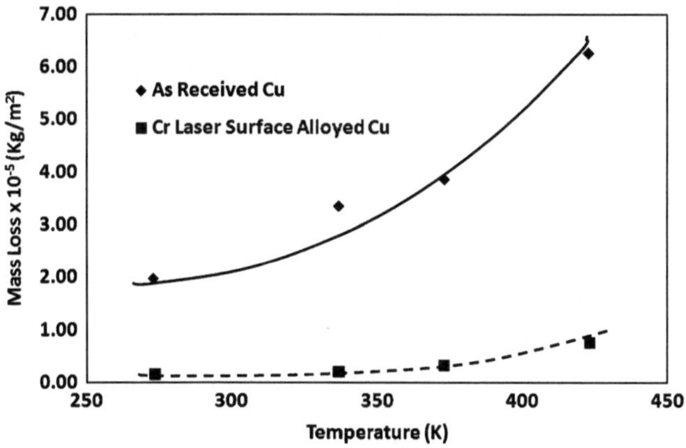

Fig. 7.33 Improvement of room temperature and elevated temperature erosion resistance of copper laser surface alloyed with Cr [50]

that the oxide scale which forms on the eroding material during erosion is adherent and sufficiently ductile to withstand repeated impact without developing cracks, steady-state oxide scale thickness can be defined. It can be assumed that the oxidation of the eroding material follows parabolic kinetics given by the Eq. (7.32)

$$\Delta m^2 = K_{\mathrm{p}}^{\mathrm{o}} t \tag{7.32}$$

where Δm is the mass gain experienced per unit area due to intake of oxygen to form the oxide scale, $K_{\mathrm{p}}^{\mathrm{o}}$ is the parabolic rate constant and t is the time of exposure. The parabolic rate constant is usually expressed in the form

$$K_p^o = A_o \exp\left(\frac{-Q}{RT}\right) \tag{7.33}$$

where A_o is the Arrhenius constant, Q is the activation energy for oxidation, R is the gas constant and T is the absolute temperature.

In order to represent the E–O interaction in mathematical terms, one needs the rate of growth of the oxide scale thickness with time rather than the weight gain given by Eq. (7.32). As noted by Lim and Ashby [57], once the composition of the oxide scale is known, Eq. (7.32) can be transformed to give

$$Z^2 = 2K_p t \tag{7.34}$$

$$K_p = 0.5C^2 K_p^o \tag{7.35}$$

where C is a constant for a given oxide composition (with units is m^3/kg). K_p is usually referred to as the scaling constant. A value appropriate to erosion conditions should be chosen for K_p, since Levy et al. [99] have clearly demonstrated that the oxide scales grow much more rapidly under erosion conditions as compared to static conditions. From Eq. (7.34), the rate of increase of oxide scale thickness with time is given as

$$\frac{dZ}{dt} = \frac{K_p}{Z} \tag{7.36}$$

If E_o is assumed to be the erosion rate of the oxide scale and F the particle flux rate given by the ratio of particle feed rate (f) to the eroded area, then the rate of decrease of the oxide scale thickness due to its erosion is obtained as

$$\frac{dZ}{dt} = -\frac{E_o F}{\rho_o} \tag{7.37}$$

where ρ_o is the density of the oxide.

Finally, a situation will arise when the oxide growth by oxidation [Eq. (7.36)] will be equal to the oxide removal by erosion [Eq. (7.37)]. Under such condition, the steady state oxide thickness (Z_{ss}) can be obtained as

$$Z_{ss} = \frac{K_p \rho_o}{E_o F} \tag{7.38}$$

Hence, the steady state oxide thickness increases with increasing temperature (through K_p), decreasing oxide erosion rate and decreasing particle flux rate. It should be mentioned that E_o in Eq. (7.38) represents the erosion rate of pure oxide. Now K_p can be expressed as

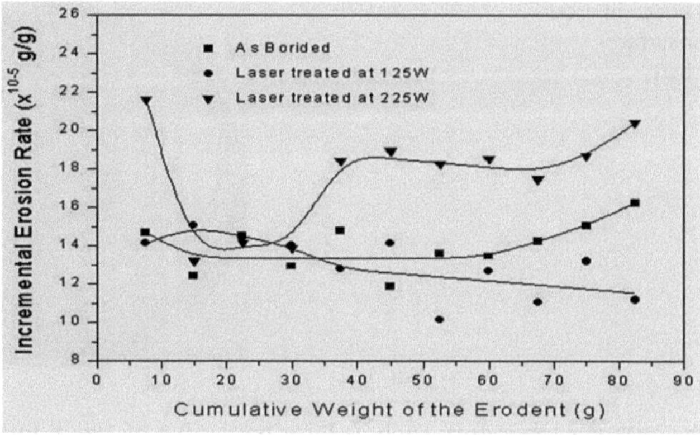

Fig. 7.34 Effect of hybrid processing such as boronising followed by laser treatment on erosion resistance of steel

$$K_{\mathrm{p}} = A_o \mathrm{Exp}\left(-\frac{Q}{RT}\right) \tag{7.39}$$

where A_o is Arrhenius constant and Q is the activation energy for oxidation. From Eq. (7.39) the higher the value of Q, lower is the value of K_{p} and hence lower the Z_{ss} or erosion rate. The substrate Cu will form a CuO scale and Cr-alloyed surface will form a Cr_2O_3 scale. The activation energy for anion diffusion for formation of CuO and Cr_2O_3 scales are 36,100 and 61,100 cal/mol respectively. Hence, the steady state thickness of the oxide scale and, in turn, the erosion rate is significantly higher for Cu-alloyed surface than for a Cr-alloyed surface.

The influence of laser processing conditions on the erosion rate can be seen from Fig. 7.34 in case of a hybrid laser process in which the steel was first boronised and the boronised surface was then modified with a laser. Subsequently the modified surface was subjected to solid particle erosion. The surface modified at a lower laser power was found to exhibit an erosion rate lower than the as-boronised surface. However, when the boronised surface is laser processed at a higher power, the erosion rate is found to be higher than the as-boronised surface.

7.5 Applications of Laser Surface-Modified Components

Laser surface modification is suitable for a wide range of materials and the process can address diverse degradation modes. The earliest industrially accepted applications have involved laser transformation hardening or laser remelting. For example, an image of a laser surface-remelted camshaft is presented in Fig. 7.35. Laser surface melting of a cast iron camshaft has already been commercialised [100]. Similarly there are numerous other applications that have been successfully

Fig. 7.35 An image of cross section of laser surface-remelted camshaft

realised in the author's laboratory such as laser hardening of steam turbine blades, laser remelting of camshafts, etc.

The materials that can successfully be modified using a laser include metals, intermetallics and ceramics. Although the generic applications of laser cladding and laser surface alloying are similar, the fact that there is negligible dilution of clad materials from the substrate makes laser cladding better suited for most tribological applications. However, as laser surface alloying involves addition of relatively lower amounts of the alloying element into the substrate, it can also play crucial role in conserving expensive scarce material. Stellite cladding has been successfully carried out on the surface of valve seat rings and steam generator turbine blades. Silver and copper cladding is also routinely done for electrical contactor industry. For aircraft industries, wear facing using a laser has been performed for tracks of aircraft carrier launch rails and naval aircraft skid plates. The shroud interlocks between turbine blades are laser cladded to reduce wear that may occur when the cold engine is heating up to fill the expansion gap between the blades. Rolls Royce has introduced laser cladding for hardfacing turbine blades of aero engines [100]. Laser cladding is also employed for repairing damaged zones of high temperature turbine blades [101]. In the nuclear industry, laser surface modification is done for nuclear support plates, nuclear displacement rod drive mechanisms, nuclear pump seal rings, etc. Laser cladding of municipal combustor waste components and submarine shaft seals has been successfully developed. Toyota R&D has developed laser cladding of copper base alloys on the valve seat with considerable improvement of adhesive and abrasive wear resistance over a range of temperatures [102]. Laser cladded Stellite 6 on valves of diesel engines for ships made of Nimonic80 is shown in Fig. 7.36.

NiTi is an ideal biomaterial for stents applications. In order to reduce the danger of Ni release, this material can be laser gas nitrided to improve biocompatibility [103]. Hydroxyapatite has also been cladded using a laser onto Ti prosthesis to create a coating which allows tissue growth between the prosthesis and the bone for better fixation [104]. Laser alloying is routinely carried out for making machine readable coinage and other metallic objects [105, 106].

Laser cladding is also increasingly used for reclamation and repair. Laser cladding has been used to rebuild the spline on drive shafts [107]. Similarly,

Fig. 7.36 Laser-cladded satellite 6 on valves of diesel engine for ship made of nimonic80

bearings with oil seals used in mining industry are resurfaced by laser cladding. Turbines blades can also be rebuilt at their tips and leading edges. An example of a reclaimed blower shaft of a refinery by laser coating of Stellite is shown in Fig. 7.37 along with a coated and ground portion of the component and the microstructure of the coating.

7.6 Direction of Future Work

The field of hybrid laser processing has attracted growing attention in recent times. As laser is used not only for enhancing wear properties but also for improving corrosion resistance [108, 109], further work on hybrid processing is most relevant. Hybrid processing, such as boronising or thermal spraying followed by laser surface melting has already yielded promising results. Further development and establishment of their applications in the engineering industry are expected in near future as described in several review papers [110, 111]. Other hybrid processes, laser-assisted physical vapour deposition [112, 113], laser-assisted chemical vapour deposition [114, 115], laser-assisted electrodeposition [116, 117], laser-assisted machining [118, 119], laser-assisted drilling [120, 121] and laser-assisted electromachining [122, 123] also provide unique advantages and are worthy of being explored to realise various techno-commercially viable applications.

Advanced processes such as rapid prototyping or low volume manufacturing are another important process that is receiving attention. The rapid prototyping process enables thin-walled, precisely designed metal structures with a density close to 100 % [124] to be obtained. The structures can be produced with the use of a precisely delivered powder introduced through a concentric nozzle into a laser-melted pool (Fig. 7.14). The thickness of the walls produced by rapid prototyping could be as small as the laser spot (from 0.1 mm to a few millimetres). The microstructures of these three-dimensional products are similar to that obtained by powder sintering. The process is applied to form shapes for stamping machines and to fill cracks in damaged bearings, crankshafts and cylinders in automotive engines [125]. Significant development is also taking place in this direction.

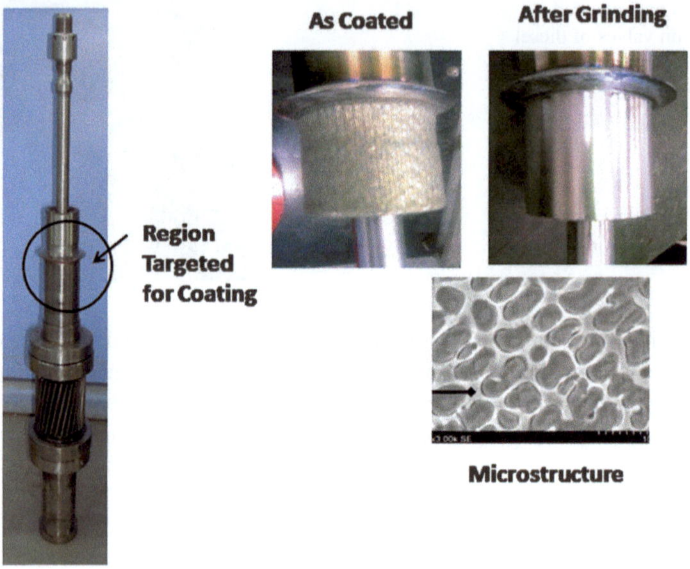

Fig. 7.37 Reclaimed blower shaft of a refinery by laser cladding of Stellite clad and ground portion of the shaft and the microstructure of the clad are also shown

Process automation during laser surface modification is another important area which is getting updated at a rapid pace. The laser beam can be controlled by controlling the power, diameter, structure and location of the beam. Similarly, the traverse speed, vibration, focal point, shroud velocity of the work station can be monitored. The surface absorptivity, seam location, temperature of the work piece can be changed as per requirement. Acoustic signal-based sensors such as mirror, probe, nozzle, workpiece, etc. are currently under investigation. Investigations are also in progress for radiation emission signal-based sensors such as photoelectric, pyroelectric, CCD camera and infrared camera, etc. Other interesting areas are space charge signal-based plasma charge sensor nozzle or ring, Langmuir probe with or without external voltage, etc.

Acknowledgement Various colleagues of one of the authors (SJ) who have been actively involved in R&D as well as application development efforts highlighted in this chapter from Center for Laser Processing of Materials at International Advanced Research Center for Powder Metallurgy and New Materials are sincerely acknowledged.

References

1. Maiman TH (1960) Nature, 6 August
2. Basov NG, Kroklin ON, Popov YM (1961) Sov Phys JETP 13:1320
3. Hall RN, Fenner GE, Kingsley JD, Soltys TJ, Carlson RO (1962) Phys Rev Lett 9:366

4. Nathan MI et al (1962) Appl Phys Lett 1:62
5. Quist TM et al (1962) Appl Phys Lett 1:91
6. Patel CKN (1964) Phys Rev A 136:1187
7. Ewing JJ, Brau CA (1975) Appl Phys Lett 27:330
8. Chande T, Majumder J (1983) Metall Trans B14:181
9. Roy M, Bharti A (2000) In: Proceedings of 3rd workshop on application of laser in mechanical industry, February, Jadavpur, Calcutta
10. Bharti A, Sivakumar R, Goel DB (1989) J Laser Appl 5:43
11. Kovalemko VS, Volgin VI, Mixhailov VV (1977) Tech Organ Proizvod 3:50
12. Xue L, Islam MU, Koul AK, Wallace W, Bibby M (1997) Mater Manuf Process 12:799
13. Komvopoulos K, Nagarathnam K (1990) J Eng Mater Technol 112:131
14. Morisige N (1997) In: Khanna AS, Totlani MK, Singh SK (eds) Corrosion and it's control. Elsevier, Bombay
15. Ganamuthu DS (1980) Opt Eng 19:783
16. Jain AK, Kulkarni VK, Sood DK (1981) Appl Phys 25:127
17. Bergman HW, Mordike BL (1983) Z Werkstofftech 14:228
18. Kear BH, Breinan EM (1977) In: Proceedings of Sheffield international conference on solidification and casting, Sheffield
19. Singh J, Mazumder J (1986) Mater Sci Technol 2(7):709
20. Steen WM, Courtney CGH (1980) Met Technol June
21. Draper CW, Ewing CA (1984) J Mater Sci 19:3815
22. Draper CW, Poate JM (1985) J Int Met Rev 30:85
23. Jouvard JM, Grevey D, Lemoine F, Vannes AB (1997) J Phys (France) III 7:2265
24. Frenk A, Vandyoussefi M, Wagnière J-D, Zryd A, Kurz W (1997) Metall Mater Trans 28B:507
25. Li Y, Ma J (1997) Surf Coat Technol 90:1
26. Lin J (1996) J Laser Appl 12:28
27. Hayhurst P, Tuominen J, Mantyla J, Vuoristo P (2002) Proceedings of ICALEO 2002, Phoenix, AZ, October, LIA, paper 1201
28. Fellowes FC, Steen WM, Coley KS (1989) Proc. conf., IFHT, Lisbon, Portugal, September, p 435
29. Ayers JD, Schaefer RJ, Robey WP (1980) J Met Aug:19
30. Kloosterman AB, De Hosson JThM (1997) In: Aliabadi MH, Berbbia CA (eds) Surface treatment '97. Computational Mechanics Publications, Southampton, p 286
31. Hlawka F, Song GQ, Cornet A (1996) J Phys (France) IV Colloque C4 6:123
32. Hu C, Xin H, Baker TN (1995) J Mater Sci 30:5985
33. Ashby MF, Easterling KE (1984) Acta Metall 32:1935
34. Cline J, Anthony H (1977) J Appl Phys 48:3895
35. Davis M, Kapadia P, Dowden J, Steen WM, Courtney CHG (1986) J Phys D Appl Phys 19:1981
36. Bradley JR (1988) J Phys D Appl Phys 21:834
37. Das DK, Roy M, Singh AK, Sundararajan G (1996) Mater Sci Technol 12:295
38. Wang Y, Kovacevic R, Liu J (1998) Wear 221:47
39. Bharti A, Roy M, Sundararajan G (1992) In: Proceedings of 4th European conference on laser application, October, p 299
40. Staiaa MH, Cruza UM, Dahotre NB (2000) Thin Solid Films 377–378:665
41. Man HC, Zhang S, Cheng FT, Yue TM (2002) Scr Mater 46:229
42. Kwok CT, Cheng FT, Man HC (1998) Surf Coat Technol 107:31
43. Tassin C, Laroudie F, Pons M, Lelait L (1996) Surf Coat Technol 80:207
44. Fu Y, Batchelor AW (1998) Surf Coat Technol 102:119
45. Khedkar J, Khanna AS, Gupt KM (1997) Wear 205:220
46. Shamanian M, Mousavi Abarghouie SMR, Mousavi Pour SR (2010) Mater Des 31:2760
47. Grenier M, Dube D, Adnot A, Fishet M (1997) Wear 210:127

48. Guo B, Zhou J, Zhang S, Zhou H, Pu Y, Chen J (2008) Surf Coat Technol 202:4121
49. Staia MH, Cruz M, Dahotre NB (2001) Wear 251:1459
50. Dutta Majumdar J, Manna I (1999) Mater Sci Eng A268:227
51. Dutta Majumdar J, Manna I (1999) Mater Sci Eng A267:50
52. Dutta Majumdar J, Mordike BL, Manna I (2000) Wear 242:18
53. Manna I, Dutta Majumdar J, Ramesh Chandra B, Nayak S, Dahotre NB (2006) Surf Coat Technol 201:434
54. Persson DHE, Jacobson S, Hogmark S (2003) Wear 255:498
55. Jiang P, He XL, Li XX, Yu LG, Wang HM (2000) Surf Coat Technol 130:24
56. Hutchings IM (1992) Tribology. Edward Arnold, London
57. Rabinowicz E (1965) Friction and wear of materials. Wiley, New York
58. Lim SC, Ashby MF (1987) Acta Metall 35:11
59. Radek N, Bartkowiak K (2011) Phys Procedia 12:499
60. Quin TFJ, Sullivan JL, Rowson DM (1984) Wear 94:175
61. Rowson DM, Quinn TFJ (1980) J Phys D Appl Phys 13:209
62. Quin TFJ (1998) Wear 216:262–275
63. Malgaord J, Srivastaba VK (1977) Wear 41:263
64. Roy M, Ray KK, Sundararajan G (1999) Oxid Met 50
65. Hirose A, Ueda T, Kobayashi KF (1993) Mater Sci Eng 160A:143
66. Tau YH, Doong JL (1989) Wear 132:9
67. Isawane DT, Ma U (1989) Wear 129:123
68. Hou PY, Stringer J (1992) Oxid Met 38:323
69. Sigler DR (1993) Oxid Met 40:555
70. Bennett MJ, Haulton MR (1990) In: Bachelet E (ed) High temperature materials for power engineering. Kluwer Academic, Dordrecht, p 189
71. Tatlock GJ, Hard TJ (1984) Oxid Met 22:201
72. Gobel M, Rahmel A, Schertzc M (1993) Oxid Met 3–4:231
73. Sun JH, Jang HC, Chang E (1994) Surf Coat Technol 64:195
74. Vernon-Perry KD, Crovenor CRM, Needhan N, English T (1988) Mater Sci Technol 4:461
75. Pint BA (1997) Oxid Met 48:303
76. Ramnarayanan TA, Raghavan M, Petkvichuton R (1984) J Electrochem Soc 131:923
77. Forest C, Davidson JH (1995) Oxid Met 43:479
78. Walker A, Flower HM, West DRF (1985) J Mater Sci 20:989
79. Walker A, West DRF, Steen WM (1984) Mater Technol 11:399
80. Mordike BL, Bergmann HW, Grob N (1983) Z Werkstofftech 14:253
81. Shehata GH, Moussa AMA, Molian PA (1993) Wear 171:199
82. Nakata K, Tomoto K, Matsuda F (1996) Trans Jpn Weld Res Inst 25:37
83. Galun R, Weisheit A, Mordike BL (1996) J Laser Appl 8:299
84. Li R, Ferreira MJS, Anjos M, Vilar R (1996) Surf Coat Technol 88:90
85. Li R, Ferreira MJS, Anjos M, Vilar R (1996) Surf Coat Technol 88:96
86. Anjos M, Vilar R, Qui YY (1997) Surf Coat Technol 92:142
87. Belmondo A, Castagna M (1979) Thin Solid Films 64:249
88. Singh J, Majumder J (1987) Metall Trans 18:313
89. Folkes JA, Shibata K (1994) J Laser Appl 6:88
90. Roy A, Manna I (2001) Mater Sci Eng A297:85
91. Basu A, Chakraborty J, Shariff SM, Padmanabham G, Joshi SV, Sundararajan G, Dutta Majumdara J, Manna I (2007) Scr Mater 56:887
92. Padmavathi C, Sarin Sundar JK, Joshi SV, Prasad Rao K (2006) Trans Indian Inst Met 59:99
93. Staia MH, Cruz M, Dahotre NB (2000) Thin Solid Films 377–378:665
94. Ghosh K, McCay MH, Dahotre NB (1999) J Mater Process Technol 88:169
95. Rieker C, Morris DF, Steffen J (1989) Mater Sci Technol 5:590
96. Boas M, Bamberger M (1988) Wear 126:197
97. Sundararajan G, Schewmon PG (1983) Wear 84:237

98. Roy M, Das DK, Sivakumar R, Sundararajan G (1991) In: Darakadasa ES, Seshan S, Abraham KP (eds) Proceedings of the 2nd international conference on aluminium, July, Bangalore, p 947
99. Levy AV, Slamovich E, Jee N (1986) Wear 110:117
100. McIntyre M (1983) In: Metzbower EA (ed) Applications of lasers in material processing. ASM, Materials Park, OH
101. Regis V, Bracchetti M, Cerri W, Angelo DD, Mor GP (1990) In: Proceedings of high temperature materials for power engineering, September, Liege. Kluwer Academic, Dordrecht
102. Tanaka K, Saito T, Shimura Y, Mori K, Kawasaki M, Koyama M, Murase H (1993) J Jpn Inst Met 57:1114
103. Man HC, Cui ZD, Yue TM (2002) J Laser Appl 14:242
104. Lusquinos F, Pou J, Arias JL, Boutinguiza M, Leon B, Perez-Armor M (2001) In: Proceedings of ICALEO 2001, October, Jacksonville, FA, LIA, paper 1003
105. Liu Z, Watkins KG, Steen WM (1999) J Laser Appl 11:136
106. Liu Z, Pirch N, Gasser A, Watkins KG, Hatherley PG (2001) J Laser Appl 13:231
107. Anderson T (2002) Proceedings of ICALEO 2002, October, Pheonix, AZ, LIA, paper 1505
108. Padmavathi C, Sarin Sundar JK, Joshi SV, Prasad Rao K (2006) Mater Sci Technol 22:1
109. Zaplatynski I (1982) Thin Solid Film 95:275
110. Joshi SV, Sundararajan G (1998) In: Dahotre NB (ed) Lasers in surface engineering. ASM, Materials Park, OH, p 121
111. Pawlowski L (1999) J Therm Spray Technol 8(2):279
112. Voss A, Funken J, Alunovic M, Sung H, Kreutz EW (1992) Thin Solid Film 220:116
113. Funken J, Kreutz EW, Krische M, Sung H, Voss A, Erkens G, Lemmer O, Leyendeker T (1992) Surf Coat Technol 52:221
114. Oliveira JC, Paiva P, Oliveira MN, Conde O (1999) Appl Surf Sci 138–139:159
115. Wang Q, Schliesing R, Zacharias H, Buck V (1999) Appl Surf Sci 138–139:429
116. Zouari I, Lapicque F, Calvo M, Cabrera M (1990) Chem Eng Sci 45:2467
117. Grishko VI, Duley WW, Gu ZH, Fahidy TZ (2001) Electrochim Acta 47(4):643
118. Rebro PA, Shin YC, Incropera FP (2004) Int J Mach Tool Manuf 44(7–8):677
119. Ding H, Shin YC (2010) Int J Mach Tool Manuf 50(1):106
120. Bandyopadhyay S, Sarin Sundar JK, Sundararajan G, Joshi SV (2002) J Mater Process Technol 127:83
121. Tam SC, Willams R, Yang LJ, Jana S, Lim LEN, Lau MWS (1990) J Mater Process Technol 23:177
122. Pajak PT, De Silva AKM, McGough JA, Harrison DK (2004) J Mater Process Technol 149:512
123. Thomson G, Pridhan M (1998) Opt Laser Technol 30:191
124. Hoffmann E, Backes G, Gasser A, Kreutz EW, Stromeyer R, Wissenbach K (1996) Laser und Optoelektronik 28(3):59
125. Graydon O (1998) Opt Lasers Eng Feb:33

[illegible]

Chapter 8
Surface Engineering for Biotribological Application

D.V. Shtansky and Manish Roy

8.1 Introduction

Advances in medical science and development in biomedical materials have led to a remarkable increase in ageing population and improvement of quality of life. Monitoring of friction and wear in human body has played a very significant role in enhancing human lifespan. Several organs, joints and critical parts of human body wear out and need to be replaced. Friction and wear play important role in several cardiovascular devices. Heart disease is one of the most common diseases requiring replacement of heart valve. Artificial heart valves are used to replace damaged or diseased natural valves resulting in significant improvement of life span and quality of life. These valves are generally made of pyrolytic carbon (PyC). However, this material is brittle and has low blood compatibility. Consequently, the patients suffer from thrombosis and require taking anti-coagulation medicine which has side effects. Synthetic vascular grafts are used to repair weakened blood vessels to bypass blockages has resulted in enhanced blood flow to severely ischemic organs and limbs. In order to improve the pumping function of the heart, intra-aortic balloon pump, ventricular assist device, total implantable artificial heart, etc. are being used. Pacemakers and automatic internal defibrillators are widely used to over ride or correct aberrant, life-threatening cardiac arrhythmias. All these devices are subjected to wear and tear during implanting or during operation. In addition, life threatening occlusive diseases of the arteries is treated with implantation of stents. It is a device which is inserted into the body passages using catheter to maintain the flow of blood, urine, bile or air. It is important to minimise the friction

D.V. Shtansky (✉)
National University of Science and Technology "MISIS", Leninsky pr. 4, Moscow 119049, Russia
e-mail: Shtansky@shs.misis.ru

M. Roy
Defence Metallurgical Research Laboratory, Kanchanbagh, Hyderabad 500058, Andhra Pradesh, India

M. Roy (ed.), *Surface Engineering for Enhanced Performance against Wear*,
DOI 10.1007/978-3-7091-0101-8_8,

during insertion and removal of stents in order to minimise damages of tissue. There are several approaches to minimise such friction. Surface coating with self-lubrication layer is one measure which has become very popular in recent times. However, such treatment suffers from the drawbacks of thrombotic occlusion/ stenosis and restenosis [1–3] due to platelet activation, etc. [4]. Similarly, thromboembolism is another problem for management of implanted mechanical heart valve prosthesis [5]. Further, medical implants can fail due to wear, fatigue, chemical degradation, infection, etc. leading to osteolysis, loosening.

Two well-known fields for biomedical implants are orthopaedics and dentistry. Among clinical orthopaedics implant, total joint replacements of hip, knee, shoulder and ankle have received maximum attention in recent time. These joints operate by low-friction articular cartilage-bearing surfaces, which are conforming and self-regenerating [6, 7]. When natural joints are severely damaged, e.g. due to osteoarthritis, they are replaced by implants. In total joint replacement, a cup is fixed in the acetabulum against which is articulated a femoral head. The implant surface must be tolerant to dynamic loading to which it is subjected during usage and should have ability for long-term exposure to biological interaction with surrounding tissue. Similarly, dental implant abutment screws joint tend to loosen and undergo wear and friction. However, prolong usage of implanted joints generates wear debris which results in formation of granulomatous inflammatory response of the periprosthetic tissues (osteolysis). The wear debris simulates formation of multinucleated giant cells, whereas small particles are phagocytosed by macrophases. These small particles produce several osteolytic cytokines and mediators that change the osteoclast and osteoblast activities. This causes bone resorption and subsequent implant loosening. Thus further development of joint replacement is concerned with durability of the implant.

While metals and alloys meet many requirements of biomaterials, the interfacial bonding between the metallic surface and the surrounding bone is poor or does not exist at all [8]. Ceramics are potential bioactive implant materials. However, it should be noted that the clinical application of bioactive ceramics in high-load bone implants is limited because of its low fracture toughness. An effective way to promote the formation of bone-like layer on the implant surface is the deposition of bioactive or biotolerant coating. Further, it is noted that application of various coating enhances wear and friction behaviour, biocompatibility, hemocompatibility, etc. of metallic implant. Thus, the aim of this chapter is to examine the present state of art of different surface modification techniques available to improve the performances of physiological implants against tribological degradations.

8.2 Wear of Implants

Knee, shoulder and hip joints involve a sliding contact between the femoral component and the tibial or acetabular component during the motion of the human body. As a result, the metallic components of the artificial joint are

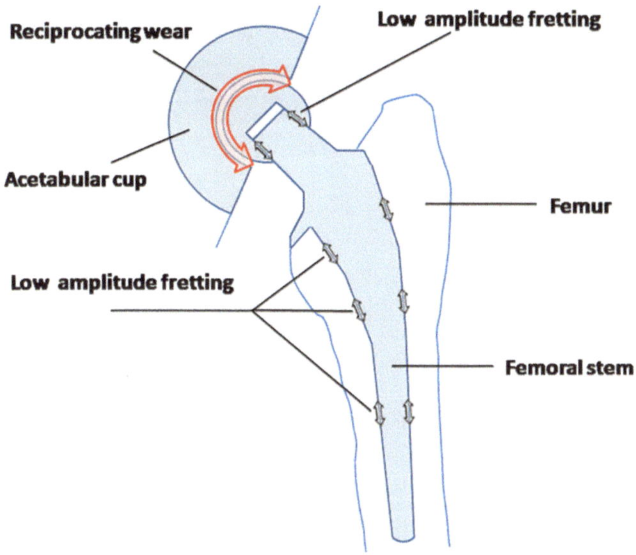

Fig. 8.1 A schematic representation of a total hip joint replacement prosthesis showing the types of motion and surface degradation mechanisms of the implant's metallic components

susceptible to sliding wear as shown in Fig. 8.1. Fretting wear is a particular form of wear involving a small amplitude relative displacement [9]. In the particular case of orthopaedic implants, micro-motions are known to occur at points of fixation between the different components [10], while corrosion is caused by the body fluid which contains various inorganic and organic ions and molecules [11] (see Fig. 8.1). Fretting has been identified at the neck/head contact and at the stem/bone interface of modular hip joints and at the screw head/plate junction of fixation plates [12]. In contrast to sliding, during fretting, a considerable part of the displacement may be accommodated in the contact by elastic deformation and thus the elastic properties of protective coatings can affect the implants behaviour and functionality. Furthermore, due to the closed geometry of the contact, debris is easily trapped and thus the behaviour of the third body is critical.

As mentioned earlier, stents is a device which is inserted into the body passages using catheter to maintain the flow of blood, urine, bile or air. It is important to control the friction during insertion and removal of stents in order to minimise damages of tissue. This problem is counter either by using hydrophobic polymer or by introducing special lubricant known as RotaGlide™ in to the guiding catheter [13]. Several cardiovascular devices, such as heart valves, vascular grafts, intra-aortic balloon pump, etc., are subjected to wear and tear during implanting or during operation.

Methods of joining prosthesis components in dental implant systems are very important to prevent implant/abutment rotation. Figure 8.2 shows a dental implant system. If the abutment screw does not remain properly affixed during masticator loading, it is necessary to retighten the screw to prevent prosthesis rotation, thus increasing the cost of the treatment and the patient discomfort. Although bolt joints have been extensively studied, they are a very complex structure and their

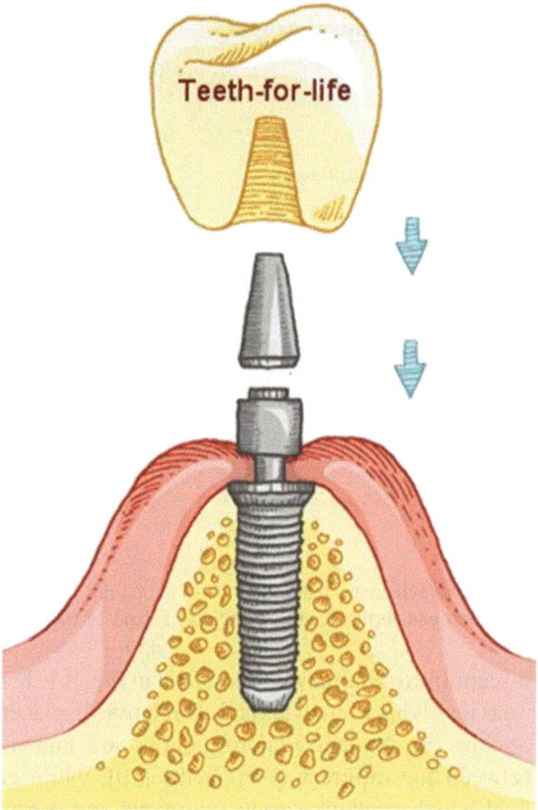

Fig. 8.2 The dental implant system

behaviour is not well known. They are not very stable and can flex, shift and relax under environmental changes and/or variations in external load. The torque applied to the screw joint during tightening is not entirely translated into screw preload since part of this torque is expended to overcome friction. About 50 % of the energy transmitted by torque tightening is expended to overcome the friction between the abutment screw head and the abutment-seating surface. About 40 % of the applied torque is used to overcome thread friction and only 10 % produces the screw tension. As a result the effective joint preload is smaller than the applied torque. In addition, even in properly joined implant various parts of dental implants are subjected to frictional degradation during mastication.

8.3 Classification of Coatings for Biomedical Application

Coating for biomedical application can be classified depending on their biocompatibility and tissue response. They can be described as biotolerant, bioinert and bioactive. Biotolerant coatings release materials which are non-toxic and they may cause benign tissue reactions such as formation of a fibrous connective tissue capsules or weak immune reaction leading to formation of giant cells or phagocytes. These coatings are characterised by development of fibrous tissues around the coatings [14, 15]. Stainless steel coatings deposited by thermal spraying technique or bone cement coatings containing polymethylmethacrylate (PMMA) can be categorised under this head. The interaction between the coating and the surrounding tissue is minimum for the bioinert coating. These coatings do not show any positive interaction with living tissues [16]. Various titanium-based coatings, ceramic coatings such as alumina, zirconia, titania, etc. and some polymeric coating are bioinert coatings. Those coatings which develop positive bone response and cause bone growth and regeneration in vivo site of application are known as bioactive coatings. They can bond to bone by formation of bone-like apatite layer on the surface of implants through chemical bonding along the bone–implant interface. These layers are essentially bone-like carbonate appetite layer that is chemically and crystallographically equivalent to the mineral phase in nature bone [17]. These coatings can act as scaffolding for bone cell migration and proliferation leading to new bone growth. Therefore, it is believed that a coating should contain HAP ($Ca_{10}(PO_4)_6(OH)_2$) components, such as Ca, P and O for it to have apatite nucleation ability. It has been shown that bioactive glasses and certain ceramics release Ca^{2+} ions through an exchange with protons of the body fluids when implanted into bone [18]. Nevertheless, evaluation of various bioceramics has shown that even P_2O_5-free $CaO–SiO_2$ and CaO- and P_2O_5-free $Na_2O–SiO_2$ glasses can form apatite in simulated body fluid [19]. It has been also shown that apatite nucleation is induced by hydroxyl functional groups on the bioactive coatings surface [19, 20]. The body environment of these functional groups assumes a negative surface charge [19] that stimulates apatite growth. There interfaces permit transmission of compressive, tensile and shear stresses. Synthetic hydroxyapatite (HA), calcium phosphate and bioglass coatings can be considered as bioactive coatings.

8.3.1 Bioactive Coatings

As discussed earlier that the effective way to promote the formation of bone-like layer on the implant surface is to modify the surface with bioactive layer. One way to do that is to deposit multifunctional bioactive coating. Ceramics coatings, such as ZrO_2, TiN, TiO_2, SiO_2 and SiC are potential coatings on implant surface. Zirconia (ZrO_2) ceramics and its composites possess excellent mechanical properties. These ceramics were evaluated as biomaterials in the late 1960s and are used for orthopaedic implants as ball heads in artificial hip joints [21, 22]. However, ZrO_2 does not bond directly to bone. TiN coatings because of its excellent combination of useful properties such as high hardness, wear and corrosion resistance and biocompatibility have a wide range of applications such as dental prosthesis, materials for hip joint and heart valve replacements [23, 24]. TiN [25, 26] and TiO_2 [27] films have shown to possess excellent hemocompatibility. Further discussion on biotribology of TiN, ZrO_2 and TiO_2 coatings is carried out in subsequent section. SiO_2 and TiO_2 ceramics induce apatite formation due to a negative surface charge under physiological pH [28]. Bioactive silicate glasses can form a bond with bone tissue [29] and have been used in a number of clinical applications [30]. Amorphous SiC alloys have been employed to enhance the biocompatibility of artificial heart valves [31].

Shtansky et al. developed multifunctional bioactive nanostructured films (MuBiNaFs) and evaluated them as perspective biomaterials to be used in load-bearing medical applications [32, 33]. A comparative investigation of MuBiNaFs based on the systems Ti–Ca–C–O–(N), Ti–Zr–C–O–(N), Ti–Si–Zr–O–(N) and Ti–Nb–C–(N) is carried out by them [34]. $TiC_{0.5}$ + 10 % CaO, $TiC_{0.5}$ + 20 % CaO, $TiC_{0.5}$ + 10 % ZrO_2, $TiC_{0.5}$ + 20 % ZrO_2, Ti_5Si_3 + 10 % ZrO_2, $TiC_{0.5}$ + 10 % Nb_2C and $TiC_{0.5}$ + 30 % Nb_2C composite targets were manufactured by means of self-propagating high-temperature synthesis. Subsequently above-mentioned films were deposited by DC magnetron sputtering in an atmosphere of argon or in a gaseous mixture of argon and nitrogen. The films were characterised in terms of their structure, chemical composition, surface topography, hardness, elastic modulus, elastic recovery, surface charge, friction coefficient and wear rate. The biocompatibility of the films was evaluated by both in vitro and in vivo experiments. The transmission electron microscopy image of Ti–Si–Zr–O–N film and Ti–Nb–C–N film along with their corresponding diffraction pattern is presented in Fig. 8.3. Ti–Si–Zr–O–N films deposited in a gaseous mixture of Ar + N_2 exhibits a cubic structure (lattice type B1) having nanocrystalline grains. Ti–Nb–C–N film also displays nanocrystalline structure with the grain size $d < 50$ nm. The films deposited under optimal conditions showed high hardness in the range of 30–37 GPa, significant reduced Young's modulus, low friction coefficient down to 0.1–0.2 and low wear rate in comparison with conventional magnetron-sputtered TiC and TiN films. The friction coefficient of a series of films deposited on Ti base alloy is presented in Fig. 8.4 [32]. These films were tested against WC–Co ball. All films have very low friction coefficient, although addition of nitrogen does not alter it significantly. Similarly wear rate of a series of MuBiNaFs deposited on hard metal is illustrated in the form of bar diagram in Fig. 8.5 [32]. Except Ti–Si–Zr–O–N film, other films have reasonably low wear rate. High wear rate of Ti–Si–Zr–O–N film is related to its high friction coefficient. The predominant wear mechanism is found to be abrasion. The morphology of worn surfaces of Ti–Ca–P–C–O film and Ti–Ca–C–O film after approximately 14,000 cycles is presented in Fig. 8.6a, b. Although abrasive wear proceeded mainly by plastic deformation mechanism, it also occurred through brittle fracture when a critical load was exceeded. It can be seen that the film failure started from the generation of a few chevron cracks within the wear track leading to changes in stress distribution and increase in local sliding friction force. This resulted in fast ductile perforation of the film due to the cyclic loading. Although abrasion was a dominant wear mechanism, contribution of tribochemical reaction and corrosive wear was also evident.

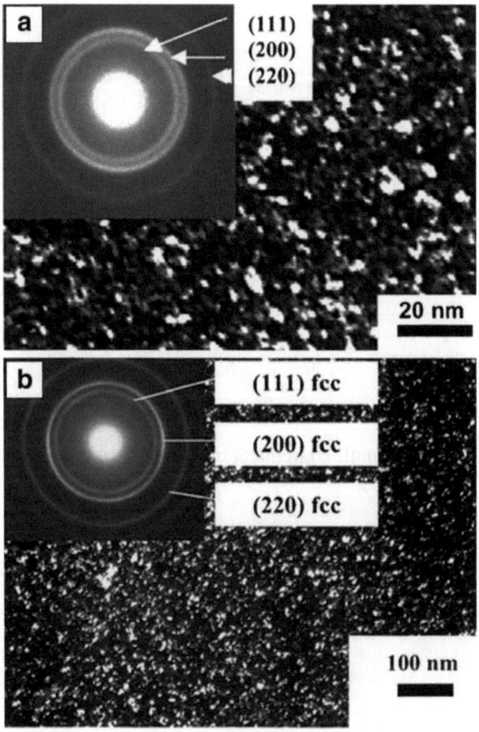

Fig. 8.3 The transmission electron microscopy images of (**a**) Ti–Si–Zr–O–N film and (**b**) Ti–Nb–C–N film along with their corresponding diffraction patterns [32]

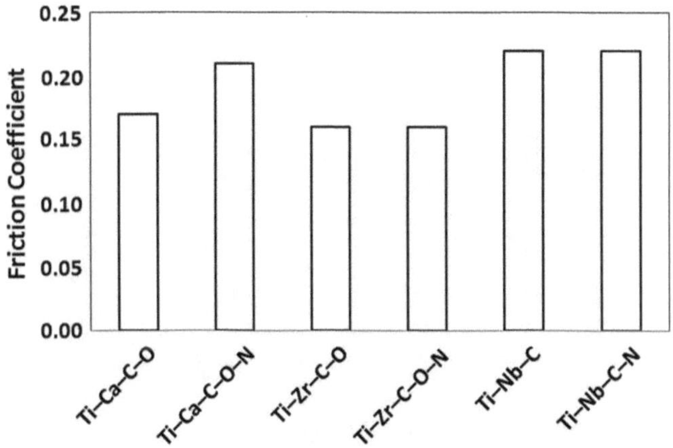

Fig. 8.4 The friction coefficient of a series of multicomponent bioactive Ti-based film deposited on Ti base alloy using dc magnetron sputtering [32]

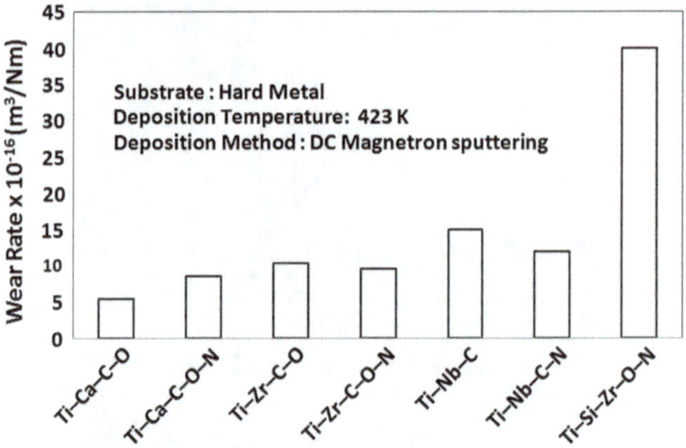

Fig. 8.5 Wear rate of a series of multicomponent bioactive Ti-based films deposited on hard metal substrates [32]

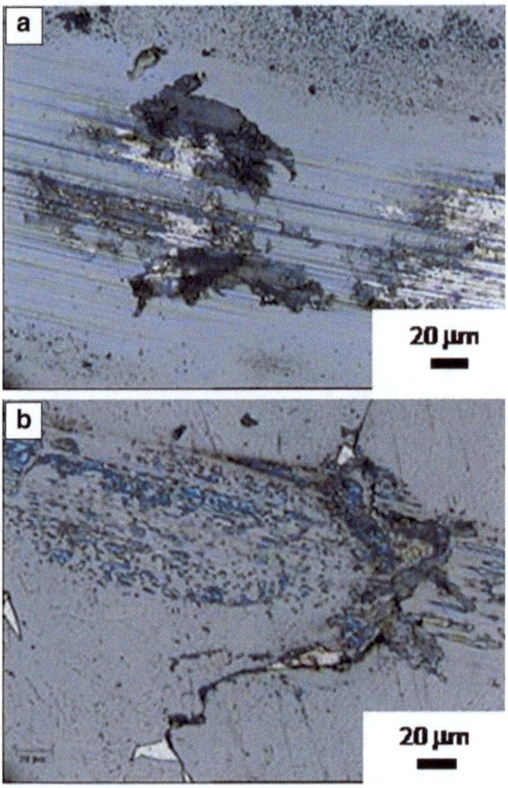

Fig. 8.6 SEM images showing the morphology of worn surfaces of (a) Ti–Ca–P–C–O film and (b) Ti–Ca–C–O films after approximately 14,000 cycles [39]

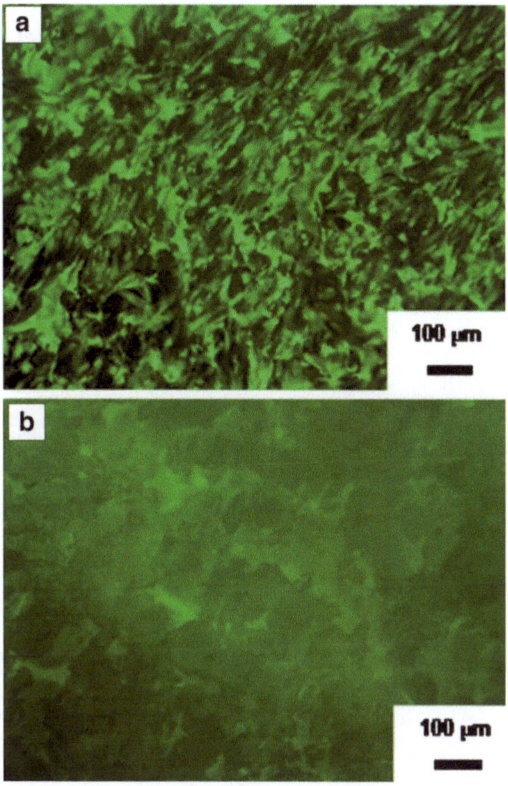

Fig. 8.7 The fluorescence microscopy images showing the high population of fibroblast cells on the surface of PTFE coated with the TiCaPCON film and a low cell density on the uncoated PTFE surface [38]

More recently, Ta- and Si-doped Ti–Ca–P–C–O–N films have been developed [35, 36]. The films possessed a nanocomposite structure with various functional groups on the film surface that stimulate accelerated osseointegration. For instance, the Si-doped Ti–Ca–P–C–O–(N) films sputter-deposited using composite $TiC_{0.5}$ + CaO + Si and $TiC_{0.5}$ + CaO + Si_3N_4 targets consisted of TiC(N) as a main phase with a minor amount of TiO_x, SiN_x, SiO_x, SiC and CaO phases which were probably mainly in an amorphous state at the grain boundaries and COO– groups on the film surface. Excess carbon atoms precipitated in the Ti–Si–Ca–P–C–O–N film (N-free target) in a DLC form. The excellent biocompatibility of these films were characterised by investigation using fluorescence microscopy which identified the high population of fibroblast cells on the surface of PTFE coated with the TiCaPCON film, while a low cell density was observed on the uncoated PTFE surface as shown in Fig. 8.7a, b. The Ta-doped Ti–Ca–(P)–C–O–(N) films deposited in argon consisted of (Ti,Ta)C, Ti_xO_y and CaO phases in an amorphous matrix with P–O, C–O and O–H bonding. The films reactively sputtered in Ar + N_2 atmosphere exhibited (Ti,Ta)(C, N), Ti_xO_y and CaO phases, diamond-like carbon, *bcc* Ta and indications of P–O bonding. A typical XPS spectra obtained from Ti–Ta–Ca–P–C–O film is presented in Fig. 8.8. Main peak is at 454.6 and 460.7 eV confirms presence of TiC. Existence of TiC peak is also confirmed by characteristics band at 281.9 eV. A weak peak at

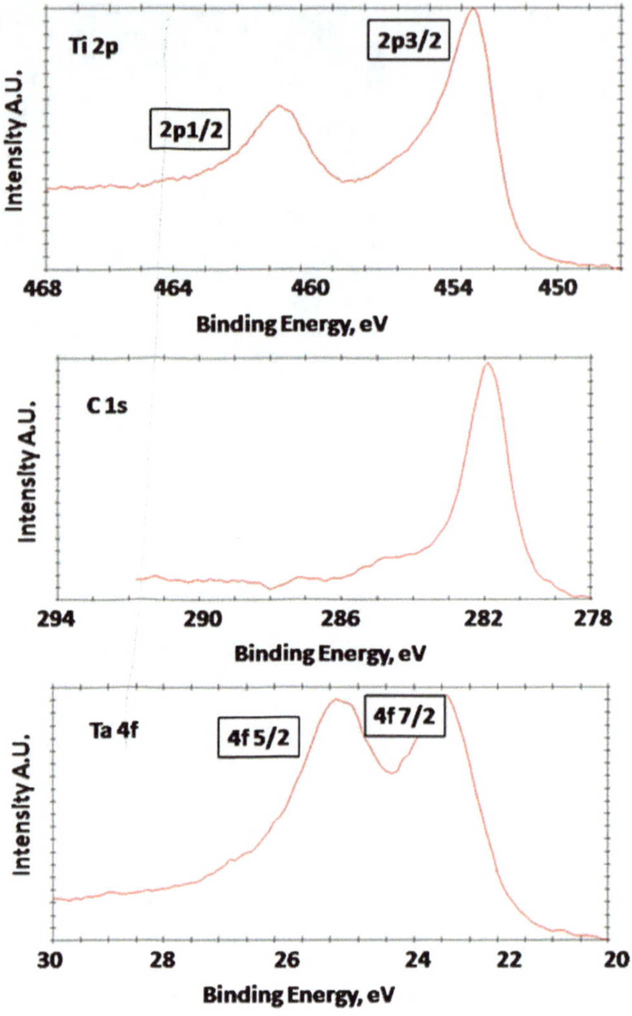

Fig. 8.8 The typical XPS spectra obtained from Ti–Ta–Ca–P–C–O film [35]

286 eV suggests presence of C–C. The tantalum peak at 22.75 eV can be attributed to Ta carbide.

A comparison of properties of the MuBiNaFs with the properties of bulk materials (Ti-, Ni- and Co-based alloys, stainless steel, ceramics) and thin films produced by alternative methods shows a noticeable advantage from the viewpoint of the whole combination of physical, mechanical, tribological, and biological properties. The MuBiNaFs are characterised by high hardness 25–40 GPa, combined with a high percentage of elastic recovery (up to 75 %) and reduced Young's modulus of 230–350 GPa, which is lower than those of bulk ceramics (TiN—440 GPa, TiC—480 GPa, SiC—450 GPa and Al_2O_3—390 GPa) and close to the

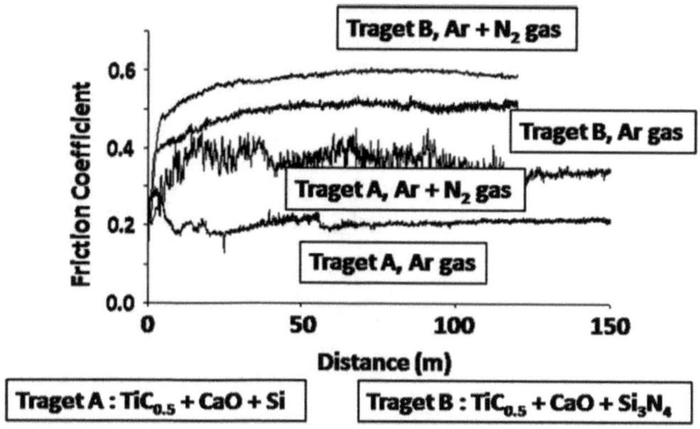

Fig. 8.9 Typical friction curve of MuBiNaF coatings [36]

modulus of stainless steel (200 GPa) and Ti (120 GPa). The benefits of a moderately low Young's modulus in bioimplant applications are well known: better transfer of functional loads to bone and reduced interfacial stresses between the film and substrate materials. The MuBiNaFs also showed high resistance to plastic deformation (up to 0.9 GPa) and long elastic strain to failure, which was previously reported as a good indicator of high film durability and wear resistance [37]. Other mechanical characteristics include a high fatigue limit of 350 MPa (MuBiNaFs on a Grade 4 Ti substrate), high adhesion strength (up to 50 N) and excellent impact resistance. The MuBiNaFs demonstrated hydrophilic nature, negative surface charge at pH 5–8.5, positive values of corrosion potential with low current density under various biological solutions, bioactivity, biocompatibility and non-toxicity.

The Ta-doped MuBiNaFs tested against Al_2O_3 ball showed low friction coefficient down to 0.2–0.25 and wear rate of $(0.7–6.8) \times 10^{-6}\ mm^3\ N^{-1}\ m^{-1}$ both in air and under physiological solution and this is two orders of magnitude lower than that of Ti [35]. Depending on the target used, a large difference in the friction coefficient of Ti–Si–Ca–C–O–N films was reported [36]. Under physiological solution (normal saline), the films deposited using the $TiC_{0.5}$ + CaO + Si target displayed a friction coefficient within the range of 0.2–0.33, whereas the films deposited using the $TiC_{0.5}$ + CaO + Si_3N_4 target had a higher friction coefficient of about 0.5–0.55. Typical friction curve of these coatings is illustrated in Fig. 8.9. A transfer layer observed in the bottom of wear track (due to the wear of the ball) indicated adhesion wear as the dominant wear mechanism. The Ti–Si–Ca–C–O–N films tested in a Dulbecko modified Eagle medium with Fetal calf serum (DMEM + FCS) showed low values of the friction coefficient and a wear rate down to 0.15 and $(0.9–1.5) \times 10^{-6}\ mm^3\ N^{-1}\ m^{-1}$, respectively. This finding can be explained by an increase of the solution viscosity which results in an increase of the hydrodynamic component of the lubrication. The wear track was relatively smooth and free from the adherent or transferred material. Thus the dominant wear mechanism changed from adhesion to abrasive wear when the samples were transferred from

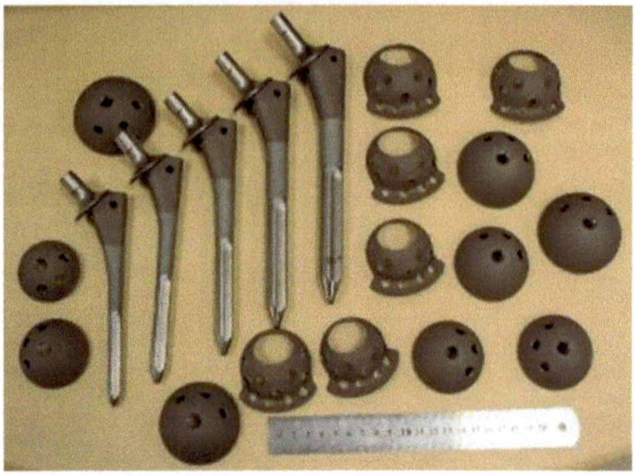

Fig. 8.10 Images of hip implants (CONMET, Russia) coated with MuBiNaF (MISIS, Russia)

a physiological solution to a DMEM + FCS medium. Image of MuBiNaFs-coated hip implants is presented in Fig. 8.10.

Shtansky et al. [38] showed that the modification of the PTFE surface by the deposition of TiCaPCON films with and without stem cells was an effective way to improve the chemical and mechanical characteristics of polymer implants and provided them with a high osseointegration potential. The friction behaviour of the TiCaPCON films was examined in different environments. The friction coefficient μ of PTFE substrate tested under physiological solution displayed initial maxima followed by a drop to a value about 0.035 after a short run-in period about 10 m. The somewhat high value of μ at the beginning of the test can be ascribed to a high PTFE surface roughness. The steady state friction coefficient of the TiCaPCON film deposited on PTFE was 0.13. This value was lower compared with 0.2–0.24 reported for the TiCaPCON films deposited on the Al_2O_3 substrate and tested under similar conditions [39]. The use of a DMEM + FCS medium did not change the steady-state friction coefficient of the TiCaPCON films which reached the value of 0.13 after the sliding distance of 25 m. The friction profile of the TiCaPCON films in DMEM + FCS was smoother compared with that in normal saline but exhibited small fluctuations (± 0.01).

Park et al. [40] examined the characteristics of calcium phosphate coatings deposited by electrodeposition on as-received Ti and titanium substrate treated with H_2O_2 in a modified simulated body fluid (SBF). A porous coating comprising of mainly hydroxyapatite (HA) was formed on the H_2O_2-treated titanium substrate. During immersion in the SBF, this coating was transformed into carbonate and calcium-deficient HA layers with a bone-like crystallinity. Interestingly, uniformed coating containing amorphous calcium phosphate formed on the untreated titanium substrate was transformed into poorly crystalline HA during immersion in the SBF. This difference was attributed to the increased surface area of the titanium substrate due to the H_2O_2 treatment and to the release of the OH_ ions from this modified surface during electrodeposition. Thus, H_2O_2 treatment prior to electrodeposition is an effective method for preparing a potentially bioactive calcium phosphate coating by electrodeposition.

The widely studied bioactive coating is a calcium-phosphate-based material hydroxyapatite (HA) having formula $Ca_{10}(PO_4)_6(OH)_2$ [41]. HA is one of the most attractive materials for human hard tissue implants because of its close resemblance to chemical composition (Ca/P ratio) of teeth and bones [42, 43]. This coating accelerates the bone growth process in the vicinity of the prosthesis [44–49]. This material has been clinically applied as a coating on bioinert metallic implants [50]. The excellent biocompatibility and bioactivity for skeletal and dental Ti implants coated with HA may be due to the OH, and O concentrations decrease with the distance from the surface of HA coatings reaching the minimum at the interface between HA and Ti alloys [51]. HA coatings on a Ti–6Al–4V substrate obtained by plasma spraying are found to increase both resistances to elastic and plastic deformation especially after a heat treatment [52]. However, such coatings have no effects on strain hardening behaviour. The coating also has improved resistance to coating/substrate separation [53]. The fracture toughness of the coating is about 1 MPa $m^{1/2}$and it has low fatigue resistance [54, 55]. In order to achieve the mechanical characteristics needed for biomedical applications, blending with a tough phase is essential. β-Ti alloy, yttria-stabilized zirconia (YSZ), Ni_3Al and Al_2O_3 ceramics have been considered good candidates as the reinforcing phases [56–60]. These coatings can be applied by a wide range of surface deposition techniques, such as high velocity oxy-fuel spraying (HVOF), plasma spraying, pulsed laser ablation, sputtering, electrophoretic deposition, sol–gel and conventional ceramic processes that involve pressing and sintering [61–66]. The thickness of HA coating can be controlled between 40 and 400 μm [66]. Different phases in HA coatings exhibit different solubilities and dissolution behaviours. Amorphous HA has been found to be more soluble than crystalline HA [67–69] in aqueous solutions. HA is thermodynamically unstable at high temperatures and when used in plasma spraying, it promotes the formation of CaO which reacts with water and has a high solubility in body fluids. High temperature deposition processes are also responsible for the formation of amorphous phases that reduce the coating–metal interfacial strength. Therefore recent developments have concentrated on improving coating stability and adhesion. A controlled atmosphere plasma spraying (CAPS) system is found to be useful for controlling the degree of melting of the HA powder, enabling coatings of tailored microstructure to be produced [70]. Yttria-stabilized zirconia incorporated into HA coating results in composite coatings containing more unmelted particles and greater porosity [71]. ZrO_2 reacts with CaO to form $CaZrO_3$. This has been achieved by thermal treatment of the coating after deposition. A post-heat treatment promotes the complete crystallisation of the amorphous calcium phosphate and this results in much improved adhesion [72]. A recent study [73] addressed the poor bond strength between HA coating and the metal substrate by considering the mismatch in physical and chemical properties between the ceramic coating and metal. The large difference in thermal expansion coefficient and rapid cooling of the plasma-sprayed droplets results in residual stress in the coating that is responsible for compromise of adhesion. To eliminate the delamination problem of HA coatings from their substrate [74], a bond coating, dicalcium silicate, is applied using the plasma coating technique. The plasma-sprayed dicalcium silicate, Ca_2SiO_4, is a bond coat for improving the bonding of HA coatings on Ti alloys substrates [75]. Overall, HA coatings can improve the adhesion and fixation of an implant device. A

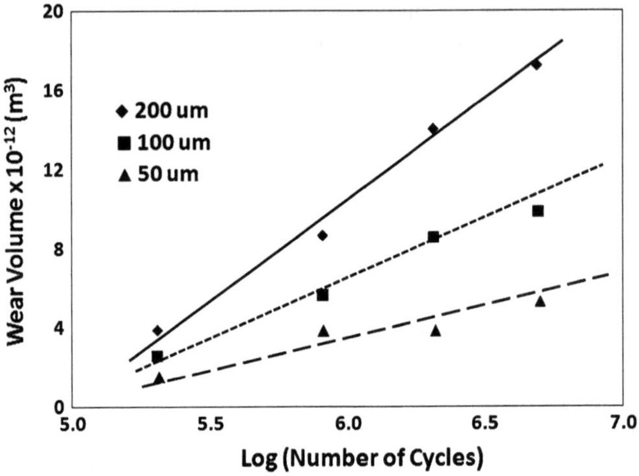

Fig. 8.11 The fretting wear of HA coating under lubricated conditions at the amplitudes of 50, 100, and 200 μm as a function of fretting cycles and normal loads of 10 N [76]

mechanical interlocking of implant with bone can eliminate problems associated with micromotion, i.e. loosening. Consequently coated implants can be recommended for applications involving younger or more active patients.

Fu et al. [76] studied the fretting wear behaviour of plasma-sprayed hydroxyapatite HA coatings lubricated with bovine albumin solution as a function of normal load, oscillatory cycle and slip amplitude. Wear mechanisms, wear volume and coefficient of friction of HA coating under lubricated fretting conditions were investigated. Fretting wear mechanisms of plasma-sprayed HA coatings under lubricated conditions were found to be mainly delamination, pitting and abrasive wear. Figure 8.11 shows the fretting wear volume of HA coating under lubricated conditions at the amplitudes of 50, 100 and 200 μm as a function of fretting cycles and normal loads of 10 N. The wear volume decreases with decrease in amplitude of cycles. Bovine albumin is an effective lubrication during fretting as shown in Fig. 8.12. It is also clear from Fig. 8.12 that hot isostatic pressing (HIP) treatment densified the coating structure, decreased the generation of fretting wear debris and improved the fretting wear resistance and hence, it could be used as a post-treatment method for as-sprayed HA coating. Although significant work is done on tribocorrosion behaviour of Ti base alloys under fretting condition [77, 78], fretting corrosion of bioactive coating has not been reported so far.

8.3.2 *Bioinert Coating*

Ti-based surface can be considered to be bioinert. These alloys exhibit excellent corrosion resistance in physiological solution [79]. Surface modification of Ti-

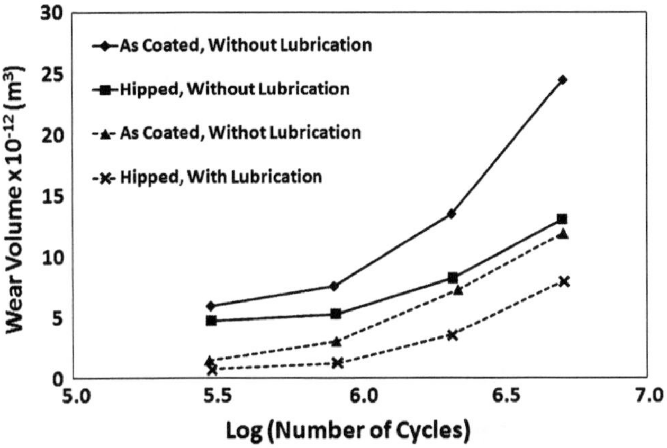

Fig. 8.12 The fretting wear of HA coating under lubricated conditions showing hot isostatic pressing (HIP) treatment improves wear performance [76]

based joint implant is carried out by employing laser-engineered net shape processing [80, 81]. In general, Ti-based alloys have poor tribological properties [82, 83]. The tribological properties of TiB/Ti–6Al–4V metal matrix layer fabricated by laser cladding and laser melting injection exhibited markedly improve wear performances [84]. Wear behaviour was studied for the composite layer laser processes using a mixture of 51 wt.% Ti + 35 wt.% b + 7 wt.% Zr + 5 wt.% Ta and 2 wt.% B. Wear studies indicate when Si_3N_4 is used as counterbody, TiB precipitates are pulled out resulting in higher friction coefficient and higher wear rate. If the counter body is 440C stainless steel ball, a drastic improvement of friction coefficient and wear rate is noted.

Nanocrystalline diamond coatings (NCD) are gaining increasing interest as serious candidate materials for high performance engineering surfaces in mechanical and tribological applications, due to their intrinsic smoothness combined with most of the outstanding properties of natural single-crystal diamond [85, 86]. High resolution transmission electron micrograph of NCD film is presented in Fig. 8.13 [86]. Small grains of the order of 10–15 nm can be observed. A typical diffraction pattern obtained from EELS is given in Fig. 8.14. The deposition of nanocrystalline diamond (NCD) coatings on load bearing surfaces can improve their wear resistance and life span. NCD films possess outstanding physical and tribomechanical properties such as very low friction response and high wear resistance. In addition, their biocompatible character makes them an ideal choice for biotribological purposes. Reports of works on NCD-coated metal alloys for implants such as hip, knee and temporomandibular joints, with the purpose of wear resistance improvement are found in the literature [87–89].

A nanocrystalline diamond (NCD)-coated Si_3N_4–bioglass composite, with potential use for hip and knee joint implants, was tribologically tested in simulated physiological fluids by Amaral et al. [90]. NCD was deposited using a hot-filament chemical vapour deposition (HFCVD) apparatus in an $Ar–H_2–CH_4$ gas mixture.

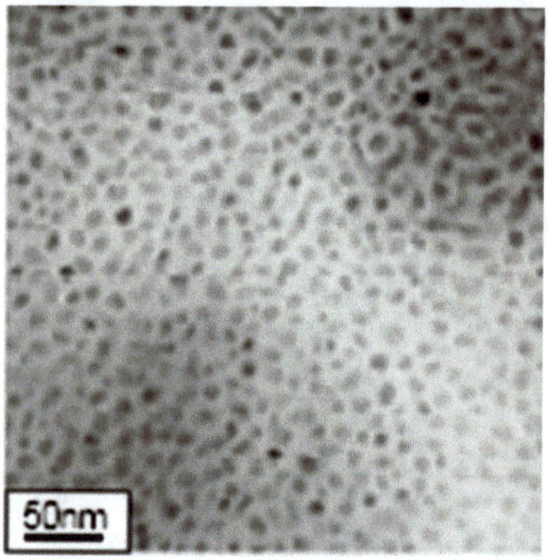

Fig. 8.13 TEM image of nanocrystalline diamond film [86]

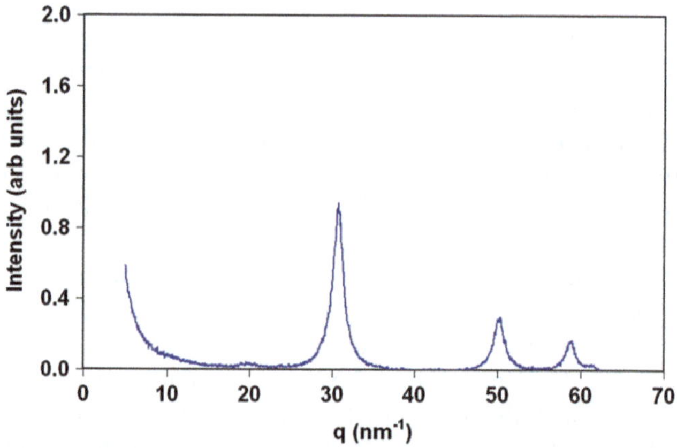

Fig. 8.14 A typical diffraction pattern obtained from electron energy loss spectroscopy of nanocrystalline diamond film [86]

Self-mated reciprocating experiments were performed using a pin-on-flat geometry in Hanks' balanced salt solution (HBSS) and dilute foetal bovine serum (FBS). A nominal contact pressure of 25 MPa was applied for up to 500,000 cycles. Very low friction coefficients of 0.01–0.02 were measured using HBSS, while for FBS lubricated tests the values are slightly higher (0.06–0.09) due to a protein attaching effect. AFM assessed wear rates by an approach using the bearing function for volume loss quantification, yielding wear rates of $k \sim 10^{-10}$ mm^3 N^{-1} m^{-1} in HBSS and $k \sim 10^{-9}$–10^{-8} mm^3 N^{-1} m^{-1} for FBS, characteristic of very mild wear regimes. Similar study was conducted by Amaral et al. [91] using the reciprocating tests under an applied load of 45 N during 500,000 cycles with a NCD-coated Si_3N_4 biocompatible ceramic substrates having two different surface preparations, polished and plasma etched (PE). Friction coefficient values of 0.02

and 0.12 were measured for the polished samples under HBSS and FBS lubrication, respectively. Polished and plasma etched samples showed increased adhesion relative to polished ones and withstood 6 km of sliding distance without any evidence of film fracture but with friction coefficients of 0.06 for HBSS and 0.10 for FBS experiments. Evidences of protein attachment and salt deposition were found, being the responsible for the enhancement of friction under FBS relatively to HBSS. The wear rates measured for the NCD films were in the range of $\sim 10^{-9}$–10^{-8} $mm^3\ N^{-1}\ m^{-1}$, values that are similar to the best values found for ceramic-on-ceramic combinations.

Shenhar et al. [92] employed a simple and original power immersed reaction-assisted coating (PIRAC) nitriding method suitable for surface modification of large complex shape orthopaedic implants of CP Ti and Ti–6Al–4V alloy. CP Ti and Ti–6Al–4V alloy samples were annealed at 1,123–1,373 K in sealed stainless steel containers that allow selective diffusion of nitrogen atoms from the atmosphere. PIRAC-nitrided surfaces were found to have a layered structure with a TiN: Ti_2N coating followed by nitrogen-stabilized α-Ti. In contrast to the coatings obtained by plasma nitriding [93], the compound layer on PIRAC-nitrided Ti–6Al–4V is always thicker than that on CP Ti after similar treatments. In contrast to surface-nitrided Ti–6Al–4V alloy, a Ti_3Al intermetallic phase was detected at the Ti_2N:α–Ti interface acting as a barrier for nitrogen diffusion. Importantly for biomedical applications, no toxic Al or V was detected in the surface layer of PIRAC-nitrided Ti–6Al–4V alloy.

Titania, zirconia, etc., are also considered as bioinert materials. Titanium dioxide (TiO_2) is widely used in biomedical applications because of its excellent biocompatibility. In biomedical implant applications, TiO_2 possesses three major biomedical advantages: corrosion resistance, biocompatibility and blood compatibility. The formation of TiO_2 coatings on Ti–6Al–4V alloys was proposed to ameliorate the biocompatibility of load bearing prostheses [94]. The titania coatings also present a barrier function which would avoid the negative effects of Al and V ions released by wear processes over the Ti–6Al–4V prostheses [95]. Both effects, i.e. apatite nucleation induction and diffusion barrier properties, permit to envisage applications for the improvement of metallic load bearing prostheses, even more when a high bonding strength to the commonly used Ti–6Al–4V alloy is ensured. There are several techniques for preparing titanium dioxide films, such as anodic oxidation [96], thermal oxidation [97], cathodic vacuum arc deposition [98], magnetron sputtering [99], plasma immersion ion implantation (PIII) [100], sol–gel [101], electrosynthesis [102], sol–gel [103], ion beam-enhanced deposition (IBED) [104] and plasma spraying [105]. Plasma-sprayed titanium coatings with porous structure have been used in teeth root, hip, knee and shoulder implants. The porous surface improves fixation via the growth of bone into the coating forming a mechanical interlock. Also TiO_2 thin films fabricated by PIII exhibit superior thromboresistant properties in long term tests [105].

Zirconium is widely used to build prosthetic devices because of its good mechanical and chemical properties. When exposed to oxygen, zirconium becomes zirconium oxide (ZrO_2) which is biocompatible [106]. Recent studies by Ferranis et al. [107] and Piconi et al. [108] proved the suitability of zirconia as the best implant material for hard tissue repair due its higher bending strength and fracture toughness. ZrO_2 is a bioinert non-resorbable metal oxide which has been used in dental implants [109] and in femoral heads of total hip replacements. In this case, ZrO_2 ball heads have shown an ultimate compressive load that is 2–2.5 times higher than that of aluminium oxide heads [108]. ZrO_2 is used in prosthetic surgery of the hip giving a prolonged life of the implant because of its low friction surface and low debris products. ZrO_2 implants have an excellent resistance to corrosion and a high wear resistance [110]. ZrO_2 has high affinity for bone tissue [109] and the bone–implant interface is similar to that seen around titanium implants [110]. It was shown that ZrO_2 did not present any signs of toxic, immune or carcinogenetic effects [111] and had no oncogenic effects in vitro [112]. ZrO_2, in fact, is mainly used to enhance bone growth, to minimise friction and corrosion and to improve biocompatibility of total joint prostheses [113]. A good biocompatibility of this material was found in animal studies with direct bone apposition to the implants [114]. ZrO_2 can also be prepared as a colloidal suspension and then used to coat surfaces for improving their characteristics [115]. In another study, Richard et al. [116] investigated the behaviour of new nano ZrO_2 and Al_2O_3-13 wt.% TiO_2 thermal sprayed coatings on commercially pure (CP)-Ti (grade 4) and titanium alloy substrates. Friction and wear tests against Al_2O_3 balls showed that the wear resistance of Al_2O_3-13 wt.% TiO_2 is better than that of ZrO_2 coating. Both plasma sprayings have similar abrasive wear behaviour; however, the average friction coefficient is higher for alumina–titania coating. The major wear mechanism was found to be soft abrasion.

Carbon nitride CN is an excellent candidate for use as biocompatible coatings on biomedical implants [117]. Despite few reports about biocompatibility of CN coating, its novel properties such as extreme hardness and wear resistance [118] are already established. The properties of these coatings are attractive to such medical devices as stents, dental roots, catheters, internal leads and electrodes, guidewires, prostheses, inferior vena cava filters, heart valves, blood tubing, cannulae, dental instrument tips and scalpels. It is also applied as bio-sensors, anti-biofouling coatings for urinary-and blood-based implants, needles, orthopaedic pins, reducing bone cement wear, coatings on artificial heart organs, intraocular lens, etc. [119].

8.3.3 Biotolerant Coating

Diamond-like carbon or DLC coatings are considered to be biotolerant coatings. It is known that DLC shows a low wear and also low friction in atmosphere against most materials except some polymers. DLC has, when sliding against certain materials, the ability to form a transfer layer on the softer counterpart, protecting it from wear. However, it is not clear whether this situation is also present in an in vivo joint where body fluids are capable of reacting with wear products and of

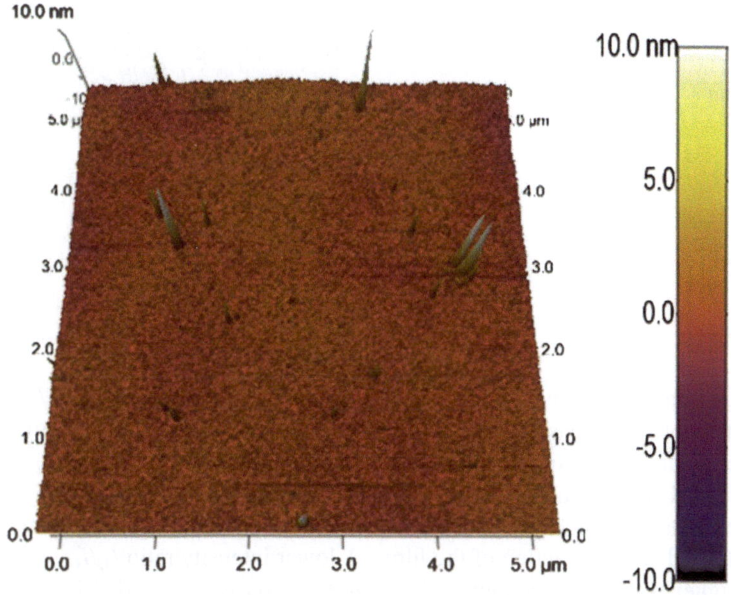

Fig. 8.15 An AFM image showing the topography of ultra smooth carbon film obtained by direct current sputtering

removing wear products out of the tribological contact area. Due to its excellent biocompatibility and good tribological properties, DLC is a promising candidate, as a coating in orthopaedic applications. There are many literatures reporting on experiments using either a ball-on-disc setup or a hip simulator to determine friction and wear of DLC-coated hip joint balls sliding against UHMWPE or of metal/metal joints with one or both sides coated with DLC. Depending on the test setup and especially the liquid lubricant used, different and partially contradicting results can be found in the literature.

DLC films can be deposited with ultra smooth surface finish. Figure 8.15 represents the AFM topography of a DLC film obtained by pulsed direct current sputtering. The roughness parameter for this surface is computed to be 0.6 nm. High amorphicity of the deposited films can also be proven by high resolution transmission electron microscopic investigations. Figure 8.16 shows high resolution transmission electron microscopic (HRTEM) image of DLC film obtained by pulsed direct current sputtering. The non-crystalline nature of deposited films is evident. However, small atomic aggregates in the range of 2–3 nm are found in the DLC matrix.

In order to describe the structure of the diamond-like carbon films with Raman spectroscopy, it is necessary to discuss the intensity ratio I_D/I_G, the full width at half maximum of the G-band (FWHM (G)) and the position of the G-band. First, it has to be considered that the intensity ratio I_D/I_G is a measure of the size of the sp^2 phase organised in rings [120]. If the intensity ratio I_D/I_G becomes lower or zero, the sp^2

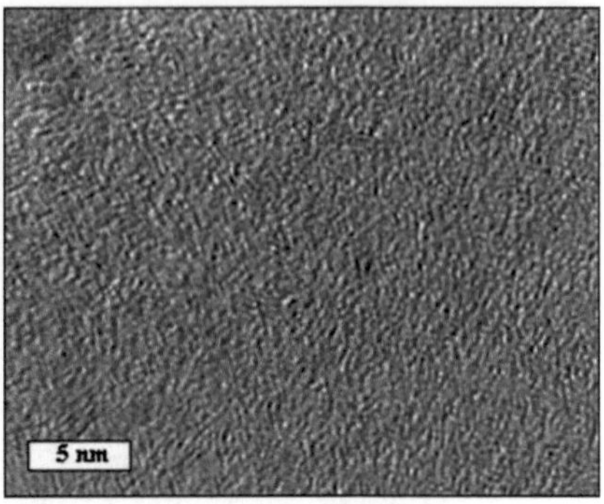

Fig. 8.16 TEM image of ultra smooth carbon film obtained by direct current sputtering

phase is organised rather in chains, whereas a higher intensity ratio I_D/I_G is an indication of an increase of the sp^2 phase in aromatic rings. Thus, if no D-band is visible, no aromatic carbon ring exists in the investigated material [121].

For a-C:H films, where the sp^2 phases and the sp^3 phases are linked together, the visible Raman parameters (532 nm) can be used to reveal information about the sp^3-hybridisated carbon fraction of the films. A lower intensity ratio I_D/I_G is connected with higher overall sp^3 content [122]. Since carbon is sp^3 bonded in a-C:H films also to hydrogen, an increase in sp^3 content does not always mean an increase in density, hardness and other mechanical properties [120, 121]. For a-C:H with hydrogen content over 25 at.% the overall sp^3 content can increase, but not the C–C sp^3 content.

To investigate the structure of the deposited films in more detail it is necessary to focus also on the FWHM (G). The FWHM (G) is a key parameter of monitoring structural disorder in DLC films [117]. This structural disorder arises from the bond angle and bond length distortions in DLC. The FWHM (G) is small when sp^2 clusters are more defect-free and ordered and a higher FWHM (G) is thus indicative for an increase in disorder [120]. The effect originates in the higher bond length and higher bond angle in a more disordered material [120]. The increase in disorder is linked to an increase in C–C sp^3 content, density and hardness [120].

An I_D/I_G ratio of 0.61 ± 0.03 was calculated with peak amplitudes of the bands shown in Fig. 8.17. Intensities from D and G-bands in spectra taken from samples deposited at higher C_2H_2/Ar-ratios of 0.11–0.25 at similar process parameters show relatively constant I_D/I_G ratios of 0.62 ± 0.04 and 0.64 ± 0.01, respectively. Therefore the size of graphitic clusters is not much influenced in this process window. Sputtering at the highest used C_2H_2/Ar-ratio of 0.43 resulted in a film structure with an I_D/I_G ratio of 0.55 ± 0.05. From these data, the trend of a decreased clustering of sp^2 phases in aromatic rings in the film deposited with the highest C_2H_2/Ar-ratio can be deduced [120, 121, 123, 124].

The position of the G-band moved down from $1{,}550 \pm 4\ cm^{-1}$ (C_2H_2/Ar-ratio of 0) to $1{,}546 \pm 1\ cm^{-1}$ (C_2H_2/Ar-ratio of 0.43). The lower I_D/I_G ratio for this film is evident for a higher overall sp^3 binding content [120, 121, 123, 124]. Here it has to be considered, that in a-C:H the carbon atom can form bonds to carbon as

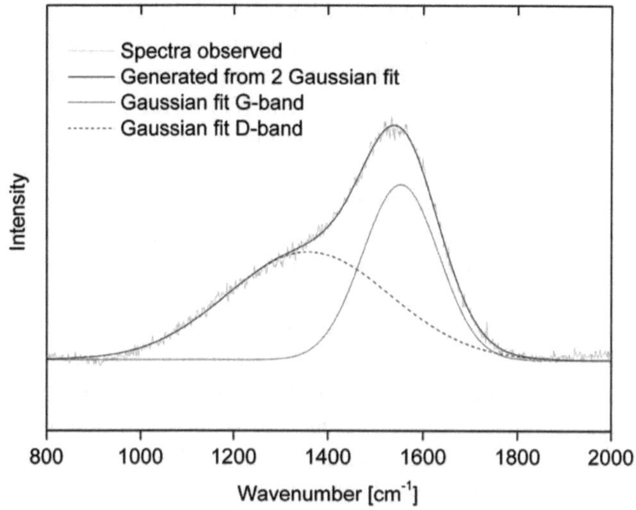

Fig. 8.17 A typical Raman spectra obtained from ultra smooth carbon film obtained by direct current sputtering

well as to hydrogen. Therefore it is difficult to identify variances in C–C and C–H bonds in a-C:H films by using the I_D/I_G ratio only. In order to investigate the trends in the C–C sp^3 binding content, the FWHM (G) is used to reveal information about the structural variance induced by changes in the C_2H_2/Ar-rate. It is found, that to the increase of the C_2H_2/Ar-rate shows a decreased structural disorder verified by the decreased FWHM (G) from 191 ± 3 cm^{-1} to 173 ± 1 cm^{-1} (Fig. 8.17). The decrease in disorder can here be linked to a decrease in C–C sp^3 content in the films [120]. Additionally, the film deposited with the highest used C_2H_2/Ar-ratio of 0.43 has the highest C–H sp^3 binding content in the investigated film series.

Evolution of the background slopes arising in the spectra showed unaffected backgrounds with no increasing photoluminescence caused by elevated hydrogen contents in DLC films when using a relatively high excitation wavelength of 532 nm in Raman spectroscopy. The Raman spectra indicate a hydrogen content of roughly $\leq$25 at.%, which could be estimated from the ratio of the fitted background slope (m) per intensity of the G-band (I_G) [121] of ~0 μm.

Biotribology of DLC films is studied extensively. Hydrogen-free DLC, also named ta-C, coated metal hip joint balls tested in 1 wt.% NaCl water by ball-on-disc and in a hip joint simulator showed a reduced wear of the UHMWPE cup by a factor of 10–100, compared to the uncoated samples [125, 126]. Additionally, an increase in the corrosion resistance of the metallic material was obtained by the coating [126]. A decrease of a factor of five in wear of the UHMWPE was obtained by coating the cobalt chromium counterface with DLC when tested in a knee wear simulator using distilled water as a lubricant [127]. Different coatings have been tested with a ball-on-disc, also using distilled water as a lubricant, and a large decrease in UHMWPE wear was obtained with all the coatings. However, the thermally oxidised Ti–6Al–4V surface still performed about eight times better than the DLC coating [128]. Analogously, DLC-coated stainless steel femoral heads have been tested against UHMWPE cups in a hip joint simulator using distilled water as a lubricant. A decrease of the UHMWPE wear by a factor of six was obtained due to the DLC coating. The same low wear of the UHMWPE was also obtained when using a zirconia femoral head under the same test conditions [129]. Sheeja et al. [130] prepared multilayer ta-C films by the filtered cathodic arc method from pure C targets. The wear tests of the CoCrMo/UHMWPE and DLC/UHMWPE sliding pairs have been made in water and simulated body fluid on a ball-on-disc apparatus and, in contradiction to the results presented above, no significant difference in wear could be measured between the coated and the uncoated samples [131]. Saikko et al. [132] compared the wear of UHMWPE cups operated against CoCr, alumina and DLC-coated CoCr hip joint balls in a biaxial hip wear simulator in the presence of diluted calf serum. For all three combinations tested, they obtained wear rates of the UHMWPE cups between 48 and 57 mg per one million cycles. There was no significant difference in the wear due to the DLC coating and all wear values obtained were in the range known from clinical observations with CoCr and alumina hip joint balls. Similar results have been obtained also by Affatato et al. [133]. Femoral heads made from 316L stainless steel, alumina, CoCrMo and DLC-coated TiAlV have been tested in a hip joint simulator using bovine calf serum as a lubricant. They obtained wear rates of the UHMWPE cups between 25 and 37 mg per million cycles for all four material combinations tested. From the results shown above, it can be seen that apparently contradicting results on the wear of DLC-coated joints sliding against UHMWPE are obtained. There are several issues presented below which may explain these differences found in the literature. The experimental setup and especially the liquid lubricant used in a tribological test are found to have a crucial influence on the friction and wear values obtained as well as on the type of wear particles produced [134–136]. It was suggested that when bovine serum or synovial fluid was used as a lubricant, the different proteins, especially phospholipids, adsorbed on the surfaces, strongly influences the tribological behaviour in the joints [136] and that also when the protein concentration is too low. The results may show a non-clinically relevant wear morphology [135]. Further, the surface texture has a decisive influence on the wear behaviour of a joint. Even single scratch, which may not be detected by an average surface roughness measurement, is capable of increasing the wear rate of UHMWPE by a factor of 30–70 [137]. Depending on the coating and the tribological conditions, DLC is able to form a transfer layer on the counterpart even in distilled water [138]. However, no transfer layer is formed in biological media against UHMWPE and the UHMWPE still shows wear. To summarise, it should be stated that wear tests on load-bearing implants having a polymer as a counterpart should be made in an appropriate tribological setup such as an implant joint simulator. As a lubricant, a supply of a solution containing an adequate distribution of proteins has to be maintained to compensate for the proteins decomposed in the test due to high pressures between contact spots of the bearing. Additionally, the surface texture of the areas involved in the tribological process must be characterised carefully.

Platon et al. [139] compared the friction and wear on different material couples used for hip prostheses. Contact pressure was found to be the main parameter which governs wear of materials especially for UHMWPE. The contact pressure in ball-on-disc tests can be controlled in the whole range by the value of ball radius. Results obtained with several different tested materials (stainless steel/UHMWPE, stainless steel + DLC coating/UHMWPE, stainless steel + DLC coating/stainless steel + DLC coating, titanium alloy + DLC coating/UHMWPE, titanium alloy + DLC coating/titanium alloy + DLC coating, zirconium dioxide/UHMWPE, alumina/UHMWPE, alumina/alumina) have shown the superiority of DLC coatings. Dry friction of DLC coatings, in comparison with UHMWPE, leads to a very low wear rate, approximately equal to an alumina couple. Thus, the use of metallic heads with DLC coating in hip prostheses can avoid ruptures observed with ceramic materials at the head–neck joint.

In a study due to Shi et al. [140], the tribological performances of several candidate implant materials, including the diamond-like carbon (DLC) thin film coating on stainless steel were investigated. A pin-on-flat contact configuration in reciprocating sliding was used for preliminary materials evaluation and friction and wear testing. Test pairs were lubricated with bovine blood serum. The DLC coating sliding against uncoated stainless steel showed the lowest friction coefficient and very little, if any, wear as shown in Fig. 8.18. Wear mechanisms in tests of ceramics and steel pairs were primarily abrasive. Similarly Kim et al. [141] reported improved performance of DLC-coated Ti-based dental abutment screw.

In the case of a metal/metal joint with both sides coated with DLC, very low wear rates have been reported. Lappalainen et al. [142], Tiainen [125] investigated hip joints with both sides having a 100-μm thick layer of hydrogen-free DLC made by filtered pulsed plasma arc discharge (85 % sp^3 bonding). Using a hip joint simulator and aqueous NaCl as a lubricant, a reduction of the wear rates, by a factor of about 10,000, corresponding to wear rates of 10^{-3}–10^{-4} mm^3 $year^{-1}$, has been reported [125]. The long duration wear tests of 15 million cycles (corresponding to about 15 years of use) performed by Lappalainen et al. with a hip joint simulator and using bovine serum as a lubricant, which was replaced regularly to compensate for depleted proteins, showed extremely low wear below 10^{-4} mm^3 $year^{-1}$ [142]. Shi et al. [140] tested steel, ceramic and a 2-μm thick DLC (a-C:H)-coated steel ball all sliding against a flat steel plate and using bovine serum as lubricant. They found a severe reduction of the ball wear by about a factor of 100 as well as reduced wear rate in the stainless steel plate. The situation of DLC sliding against DLC may be different than DLC sliding against UHMWPE. The test performed in aqueous NaCl did lead to very low wear values, comparable to those obtained in bovine serum, indicating that a similar wear mechanism takes place in both cases. The build up of a transfer layer may not be a key requirement for low wear (analogous to DLC/sapphire in atmosphere [143]) or may not be severely altered by the presence of proteins.

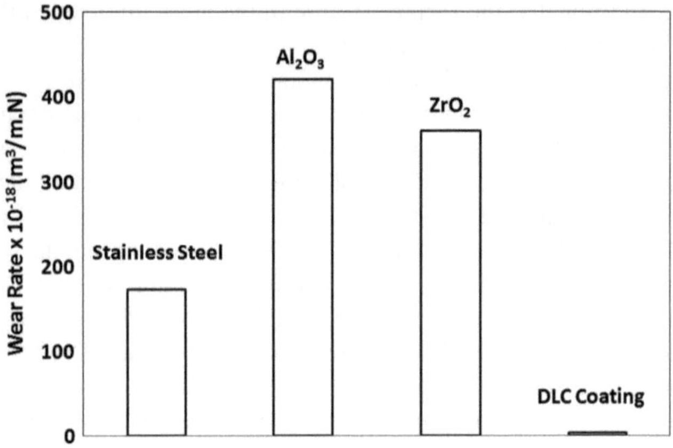

Fig. 8.18 Bar diagram showing the wear rate of Al_2O_3, ZrO_2 and DLC coating sliding against uncoated stainless steel [140]

French company 'Matériels Implants du Limousin SA (M.I.L. SA)' commercially offered DLC-coated titanium shoulder joint balls and ankle-joints with both parts (the tibial and the talar component) made from a nitrided AISI Z5 CNMD 21 steel and coated with DLC. In 2001, the company 'Implant Design AG' sold knee joints under the trade name 'Diamond Rota Gliding' with the sliding area of the femur component coated with DL N (diamond-like nanocomposite, a SiO_x containing DLC described in [144]) which was sliding against UHMWPE. Recently Teager et al. [145] published data about the clinical failure rate of a 8-year follow-up on 101 patients with implanted DLC-coated femoral balls articulating against polyethylene. The DLC-coated femoral heads with the trade name Adamante had been obtained from Biomecanique, France. They consisted of a 2–3-μm thick DLC coating on a Ti–6Al–4V alloy ball, made by ion-beam deposition. Whereas the DLC-coated implants showed no sign of problems within the first 1.5 years. Subsequently, more and more DLC implants showed aseptic loosening requiring revision (replacement) of the implant. The currently accepted explanation for aseptic loosening is that wear debris generated from the prosthesis initiates a macrophage-mediated inflammatory response, leading to osteoclast cells activation resulting in bone resorption. Within 8.5 years, 45 % of the originally implanted DLC-coated joints had to be replaced. The DLC coating on the retrieved joint heads showed numerous mostly round shaped pits. When a coating is deposited onto a metallic substrate, usually about 1 nm thick reaction layer is formed at the interface and this layer is responsible for good adhesion. Depending on the precleaning and the conditions at the very beginning of the DLC deposition process, the interface reaction layer consists usually of a metal-carbide or metalhydro-oxy-carbide. The long-term chemical stability of this reaction layer has to be guaranteed also under in vivo conditions. That biological fluids, particularly phosphate buffered saline solution (PBS), can penetrate the coating through pinholes and slowly corrode the interface between DLC and a-Si:H/DLC is described in [146]. An example of a

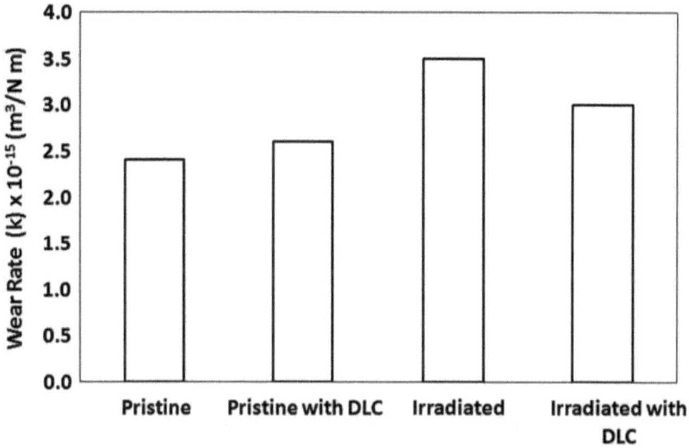

Fig. 8.19 The bar diagram showing the improved wear rate of DLC-coated UHMWPE than uncoated UHMWPE [148]

chemical instable interface leading to ongoing delamination even during storage in ambient atmosphere is given in [147]. Interface delamination due to interface corrosion involves also residual stress and electrochemical aspects and is not fully understood today.

Puértolas et al. [148] deposited hydrogenated diamond-like carbon (DLCH) thin films on medical grade ultra high molecular weight polyethylene (UHMWPE) by radio frequency plasma-enhanced chemical vapour deposition. The substrates were discs made of UHMWPEs typically used for soft components in artificial joints, namely virgin GUR 1050 and highly crosslinked (gamma irradiated in air to 100 kGy) UHMWPEs. Tribological properties under bovine serum lubrication at body temperature were assessed on coated and uncoated polyethylenes by means ball-on-disc tests. Morphological features of the worn surfaces were obtained by confocal microscopy and scanning electron microscopy. The wear factors k of the coated and uncoated surfaces were determined according to the following equation.

$$k = \frac{2\pi r A_m}{Ls} \tag{8.1}$$

where r is the wear track radius, A_m is the average worn area, L is the applied load and s is the sliding distance. The result of their observation is given in Fig. 8.19 in the form of bar diagram. This study confirms an increase in wear resistance for coated materials after 4,400 m of sliding test compared to uncoated polyethylene. These results point out that UHMWPE coated with DLCH films could be a potential method to reduce backside wear in total hip and knee arthroplasties.

Varieties of metal containing nanocomposite DLCs such as Ti alloyed, Cr alloyed and Si alloyed are developed [149–152] and these films have promising future in biomedical application. In general, these nanocomposite films exhibit better response to cell seeding than amorphous carbon films. The seeding behaviour

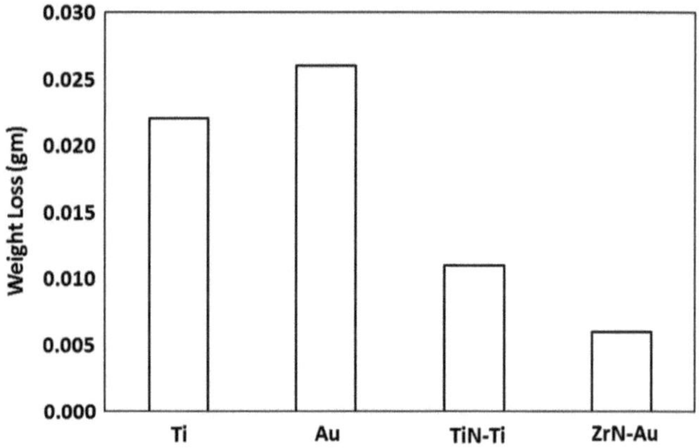

Fig. 8.20 Bar diagram showing the weight loss of TiN and ZrN coating ion plated on abutment screw made of Ti and gold respectively [153]

of endothelial cell on diamond-like carbon surface is governed by contact angle, surface energy, resistivity, work function, contact potential difference and atomic dopant concentration. Consequently these coatings have potential applications not only in prosthetic joints but also in cardiovascular devices such as heart valves, etc.

Another important biotolerant coating is TiN. Choe et al. [153] found improved wear performance of TiN ion-plated Ti dental screw and ZrN ion-plated dental gold screw as shown in Fig. 8.20. Similar observation for abutment screw was also made by Jung et al. [154]. Gispert et al. [155] attempted to improve the tribological behaviour of the prosthetic pair TiN-coated stainless steel/ultra high molecular weight polyethylene (UHMWPE) by chlorine implantation of the TiN surface. Friction and wear were determined using a pin-on-disc apparatus and the wear mechanisms were investigated through scanning electron microscopy (SEM) and atomic force microscopy (AFM). Chlorine implantation led to a significant polymeric wear reduction when the lubricant was Hanks' balanced salt solution (HBSS). If bovine serum albumin (BSA) was added to HBSS, a strong decrease of both friction and polymeric wear was observed for implanted and non-implanted TiN coatings. The former case was explained by the formation of a titanium oxide layer on the TiN surface, while the latter derived from albumin adsorption. Wear reduction may be attributed to the substitution of the hard TiN counterface by a titanium oxide layer with lower hardness which is a less wear aggressive for the polymer surface. The behaviour of the friction coefficient was attributed to the slow precipitation of calcium phosphate. When albumin was added to HBSS, friction and wear decreased significantly due to protein adsorption but still, the Cl-implanted TiN coating led to the best tribological results.

Wang et al. [156] studied the tribological properties of TiN and DLC coatings in a simulated body fluid (SBF) environment. The ball-on-plate impact tests were conducted on the coatings under a combined force of a 700 N static load and a 700 N dynamic impact load for 10,000 impacting cycles. The results indicated that the TiN and DLC coatings exhibited an excellent wear resistance and chemical stability during the sliding tests against a high density polyethylene (HDPE) biomaterial. Compared to the DLC coating, the TiN coating has a better compatibility with the HDPE. However, the impact tests showed that the fatigue cracks and the coating chipping occurred on the TiN coating but not on the DLC coating. In a separate study, the performance of three titanium nitride coatings: TiN, TiNbN and TiCN for biomedical applications were assessed in terms of their surface properties, cytotoxicity and tribological performances by Serro et al. [157]. The tribological behaviour of the coatings rubbing against UHMWPE in lubricated conditions was investigated using a pin-on-disc apparatus. Albumin adsorption on the three coatings was studied with a quartz crystal microbalance with dissipation (QCM-D) and AFM scratching. Cytotoxicity was determined both in direct or indirect contact of the cells with the coating materials. The results demonstrate that the three coatings have similar surface properties and are not cytotoxic. TiNbN showed the best tribological performance in the presence of albumin, although albumin adsorption was slightly higher on TiN. Friction and wear of polyethylene counterfaces against the three coatings were found to depend on the lubricant in HBSS. TiN was the best solution while TiNbN seemed to be a better choice when albumin was added to HBSS.

A modified pin-on-disc machine was used by Hoseini et al. [158] for the tribological investigation of ultrahigh molecular weight polyethylene (UHMWPE) sliding on stainless steel or stainless steel coated with diamond-like carbon, titanium nitride or 'Micronite'. 'Micronite' is a new type of coating applied by a physical vapour deposition technique combined with a very low friction coating material giving improved tribological properties. The tribological parameters used were chosen to mimic the conditions prevailing in the human body. The wear debris and the counter-surfaces were analysed. The surface analysis showed that the coating changed the roughness of the counter-surfaces. The diamond-like carbon and 'Micronite' coatings had a much higher surface roughness than the titanium nitride coating. The results shown in Fig. 8.21 indicated that the enhanced tribological behaviour of the 'Micronite'/UHMWPE sliding pair justified their usage as a material combination in artificial joints. Figure 8.21 essentially indicates friction coefficient and specific wear of UHMWPE against various coatings.

Tribological screening tests (simple, reciprocating ball-on-flat tests) were performed with the objective to identify an appropriate coating for the articulating surfaces of artificial hip joints whose acetabular cups and femoral stems are made from Ti–6Al–4V alloy by Osterle et al. [159]. Their results are presented in Fig. 8.22. Standard coatings like TiN or CrN performed better than more complicated multi-layer systems. These coatings, however, did not perform as good as different types of amorphous carbon coatings, generally referred to as diamond-like carbon or DLC coatings. Among the DLC coatings, hydrogenated amorphous carbon (a-C:H) displayed the best properties, especially if the hydrogen content was increased by reducing the bias voltage during PA-CVD-deposition. The optimised a-C:H coating revealed the most promising wear behaviour under the applied testing conditions [149]. Aluminised and chromised high carbon steel also exhibited promising performances in physiological solution [160]. Regarding the

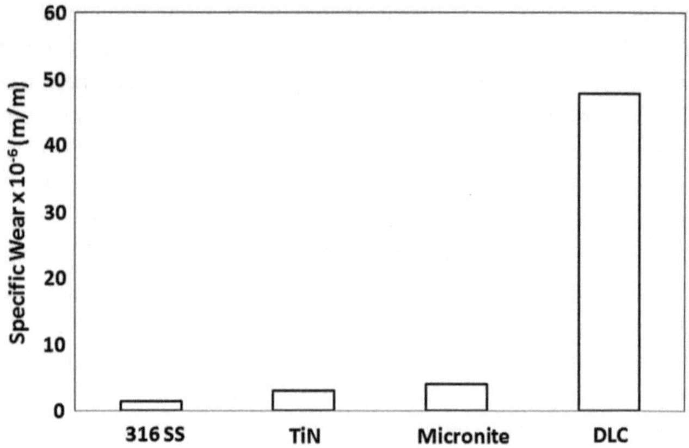

Fig. 8.21 Bar diagram indicating the enhanced tribological behaviour of the Micronite/UHMWPE sliding pair used as a material combination in artificial joints [158]

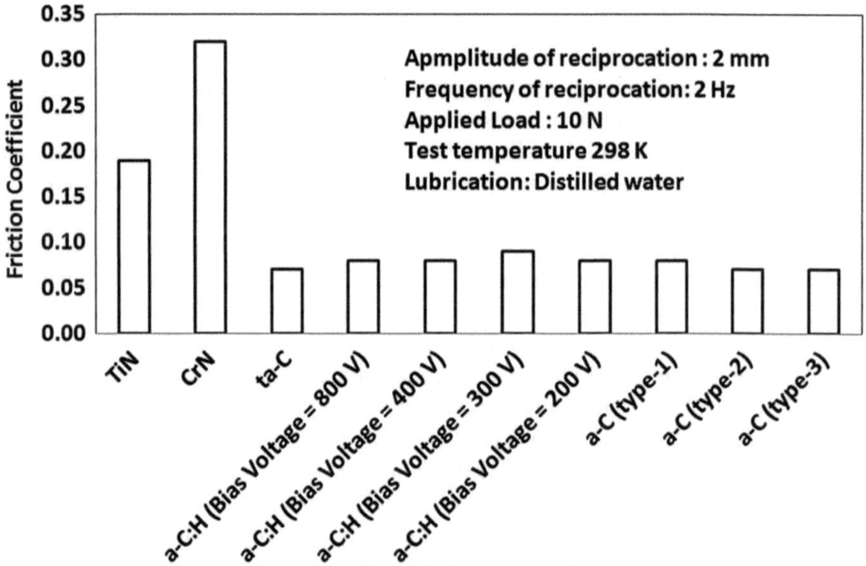

Fig. 8.22 Bar diagram showing the friction behaviour of various coatings primarily DLC films [159]

materials examined, correlation of wear with mechanical properties obtained by nano-indentation revealed that high hardness was not an adequate criterion for selecting appropriate coatings. A high ratio of hardness and elastic modulus (H/E) proved to be more important [161]. Microstructural and micro-analytical investigations revealed transformation of TiN and CrN to TiO_2 and Cr_2O_3, respectively, and amorphous carbon was, at least partly, transformed to graphite. Furthermore, incorporation of Al_2O_3 from the ball was observed at a very fine scale. The wear debris of favourable coatings always formed agglomerates of nanoscale particles. It was shown that commercial nano-particles of Al_2O_3, Cr_2O_3 and carbon black are comparable to particles generated by the tribological tests. However, it is uncertain whether they are comparable to those formed during simulator studies or in vivo.

Unsworth [162] defined a parameter similar to friction coefficient and known as friction factor given as

$$f = \frac{T}{rL} \tag{8.2}$$

Where T is the torque, r is the radius of the femoral head and L is the applied load. With the help of a hip simulator, the friction factor and film thickness was calculated [126] for metal/metal, metal/ceramic and metal/polymer hip joint lubricated with bovine serum or aqueous solution of carboxyl methyl cellulose. Calculated film thickness varied from 0.05 to 0.09 μm.

8.4 Summary and Outlook

Thus a variety of coatings and surface modification techniques have been identified as potential methods for improved performances of large variety of tribological issues of human body. Even modification of amorphous diamond-like carbon coatings with addition of metals such as Si, Cr results in formation of nanocomposite coatings with better biocompatibility. A successful exploitation of these devices is the main aim of scientific research and clinical application. Success of such application depends on surgery process, size and design of implants, their anatomical location, biological environment and age and sex of the patients. Laser-assisted direct metal deposition technique also offers huge potential as future manufacturing technologies for orthopaedic implants. Since this method is based on the concept of three-dimensional stereolithography and is capable of restructuring a 3D shape using computer in a short time, this technology can be adopted to manufacture custom design implant. DLC and TiN coatings can be considered to be potential areas of application in artificial heart valves, bearings of heart pump, left ventricular assist device, catheters, etc.

Biometric approach appears to offer new horizon in clinical research for tribology-related application. This method allows desired coating with bone-like homogenous composition. Nanoapetite coatings can also be deposited by this method. The next stage of development in orthopaedic and dental field is biomimetic bone implant. Biomimetic implants are not available commercially at this moment as the possibility of undesirable bone tissue reaction is still not eliminated. The combination of biomimetic nanoapatite and bone implants may result in a nanophase osteogenesis on orthopaedic and dental implant. Various nanomimetic materials investigated so far include ceramics, metals, polymers, composites and carbon nanofibre and nanotubes. The hybrid coatings with nanophase biomaterials provide a new method to design implant surfaces for next generation prostheses. The unique nanoporous microstructural properties of nanomaterials allow strong selective adsorption of molecular architectures making hybrid coating a potential carrier for therapeutic agents [163, 164]. It is possible to enhance the regeneration and healing of bone and soft tissues by nanoscale modification of bone implant surfaces by incorporating biological agents into nanostructure materials.

The usage of traditional coarse-grained plasma-sprayed HA coatings on joint replacements and dental implants will be replaced by nanostructured CaP coatings. Nano-porous HA-coated drug-eluting stents will partly replace the current polymer-coated drug-eluting stents. Owing to high biocompatibility of HA nanocrystals for microvascular endothelium [165], such stents will be less thrombogenic and will probably require only short-term anticoagulant therapy [166]. New future directions are aimed at creating therapeutic coatings that have a dual beneficial effect namely osteoconductive properties combined with the ability to deliver therapeutic agents, proteins and growth factors directly into the coating. These new coatings may offer the ability to stimulate bone growth, combat infection, and ultimately, increase implant lifetime [167]. A different route in the quest for new promising biomaterials is based on ceramic–polymer composites that are being designed to mimic the mechanical and biological performance of natural bone. Such novel materials have been designed ‘intelligent’ and are defined as human-friendly materials that can change their characteristics in response to surrounding conditions, for example, varying stress fields [168]. Tribological studies of all these new coatings in physiological solution will certainly give new dimension on joint replacement or other implants of human body.

The combination of bioactivity and biodegradability is probably the most relevant characteristics of the next generation of the biomaterials. These properties should merge with the ability to signal and stimulate specific cellular activity and behaviour. Biodegradable composite scaffolds, unresorbable porous metallic and polymer foams can be considered as perspective materials. Their surface bioactivation can be achieved by functionalizing surfaces with different biomolecules by applying a variety of methods where both chemical bonding and physical adsorption take place. Some more sophisticated ‘bottom-up’ and ‘top-down’ techniques should be developed to engineer surfaces with high specificity levels. The task of tailoring the biomaterials’ surfaces for at least some of the different purposes of implant integration and tissue regeneration seems a feasible challenge in the future, probably at midterm by the synergistic interdisciplinary work of materials science, engineering, biology, chemistry, physics and medicine.

References

1. Gawaz M, Neumann FJ, Ott I, May A, Schomig A (1996) Circulation 94:279
2. Lahann J, Klee D, Thelen H, Bienert H, Vorwerk D, Hocker H (1999) J Mater Sci Mater Med 10:443

3. Bittl J (1996) N Engl J Med 335:1290
4. Haycox CL, Ratner BD (1993) J Biomed Mater Res 27:1181
5. Yang Y, Franzen SF, Olin CL (1996) J Heart Valve Dis 5:532
6. Dowson D (1995) Wear 190:171
7. Podsiadlo P, Stachowiak GW (1999) Wear 230:184
8. Campbel AA (2003) Mater Today 11:26
9. Barril S, Mischler S, Landolt D (2005) Wear 259:282
10. Windler M, Klabunde R, Brunette DM et al (eds) (2001) Titanium in medicine. Springer, Berlin, p 703
11. Hallab NJ, Jacobs JJ (2003) Corros Rev 21:183
12. Hiromoto S, Mischler S (2006) Wear 261:1002
13. Dobies DR, Cohoon A (2006) J Invasive Cardiol 18(5):E146–E148
14. Daculsi G (1998) Biomaterials 19:1473
15. Ducheyne P, Qiu Q (1999) Biomaterials 20:2287
16. Busur D, Ruskin J, Higginbottom F, Hartdwick R, Dahlin CC, Schenk R (1995) Intl J Oral Maxillofac Implants 10:666
17. Ong JL, Carnes DL, Bessho K (2004) Biomaterials 25:4601
18. Loty C, Sautier JM, Boulekbache H, Kokubo T, Kim HM, Forest N (2000) J Biomed Mater Res 49:423
19. Kokubo T, Kim HM, Kawashita M (2003) Biomaterials 24:2161
20. Li P, Kangasniemi I, De Groot K (1993) Bioceramics 6:41
21. Piconi C, Maccauro G (1999) Biomaterials 20:20
22. Adolfsson E, Hermanson L (1999) Biomaterials 20:1263
23. Kola PV, Daniels S, Cameron DC, Hashmi MSJ (1996) J Mater Process Technol 56(1–4):422
24. Knotek O, Loffler F, Weitkamp K (1992) Surf Coat Technol 55(1–3):536
25. Dion I, Roques X, More N (1993) Biomaterials 14:712
26. Van Raay JJAM, Rozing PM, Van Blitterswijk CA, Van Haastert RM, Koerten HK (1995) J Mater Sci Med 6:80
27. Huang N, Yang P, Leng YX, Chen JY, Sun H, Wang J, Wang GJ, Ding PD, Xi TF, Leng Y (2003) Biomaterials 24:2177
28. Li P, Ohtsuki C, Kokubo T, Nakanishi K, Soga N, de Groot K (1994) J Biomed Mater Res 28:7
29. Hench LL, Spinter RJ, Allen WC, Greenlee TK Jr (1971) J Biomed Mater Res 2(1):117
30. De Groot K (1991) Centennial Mem Issue 99(10):943
31. Bolz A, Schaldach M (1990) Artif Organs 14:260–269
32. Shtansky DV, Gloushankovab NA, Sheveiko AN, Kharitonova MA, Moizhess TG, Levashov EA, Rossi F (2005) Biomaterials 26:2909
33. Shtansky DV, Gloushankova NA, Bashkova IA, Petrzhik MI, Sheveiko AN, Kiryukhantsev-Korneev FV, Reshetov IV, Grigoryan AS, Levashov EA (2006) Surf Coat Technol 201:4111
34. Shtansky DV, Levashov EA, Glushankova NA, D'yakonova NB, Kulinich SA, Petrzhik MI, Kiryukhantsev-Korneev FV, Rossi F (2004) Surf Coat Technol 182:101
35. Shtansky DV, Gloushankova NA, Bashkova IA, Kharitonova MA, Moizhess TG, Sheveiko AN, Kiryukhantsev-Korneev FV, Osaka A, Mavrin BN, Levashov EA (2008) Surf Coat Technol 202:3615–3624
36. Shtansky DV, Gloushankova NA, Sheveiko AN, Kiryukhantsev-Korneev PV, Bashkova IA, Mavrin BN, Ignatov SG, Filippovich SY, Rojas C (2010) Surf Coat Technol 205:728–739
37. Leyland A, Matthews A (2000) Wear 246:1
38. Shtansky DV, Grigoryan AS, Toporkova AK, Sheveiko AN, Kiruykhantsev-Korneev PV (2011) Surf Coat Technol 206:1188–1195
39. Shtansky DV, Gloushankova NA, Bashkova IA, Kharitonova MA, Moizhess TG, Sheveiko AN, Kiryukhantsev-Korneev FV, Petrzhik MI, Levashov EA (2006) Biomaterials 27:3519
40. Park JH, Lee YK, Kim KM, Kim KN (2005) Surf Coat Technol 195:252
41. Kokubo T (1985) J Mater Sci 20:2001

42. Holmes RE (1986) J Bone Joint Surg Am 68:904
43. Bucholz RW et al (1989) Clin Orthop 240:53
44. Osborne JF, Newesley H (1980) In: Heimke G (ed) Dental implants. Carl Hansen, Munich
45. Orly I et al (1989) Calcif Tissue Int 45:45
46. Ducheyne P (1987) J Biomed Mater Res 21(A2):219
47. Fujiu T, Ogion M (1984) J Biomed Mater Res 18(7):845
48. Daculsi G et al (1989) J Biomed Mater Res 23:849
49. Tracy BM, Doremus RH (1984) J Biomed Mater Res 18:719
50. Hench LL (1998) J Am Ceram Soc 81:1705
51. Wen J et al (2000) Biomaterials 21:1339
52. Dong ZL et al (2003) Biomaterials 24:97
53. Zhang C et al (2001) Biomaterials 22:1357
54. Thomas MB et al (1980) J Mater Sci 15:891
55. de With G et al (1981) Mater Sci 16:1592
56. Buser D et al (1991) J Biomed Mater Res 25:889
57. Jansen JA et al (1991) J Biomed Mater Res 25:973
58. Soballe K et al (1990) Acta Orthop Scand 61(4):299
59. Choi WJ et al (1998) J Am Ceram Soc 81:1743
60. Kong YM et al (1999) J Am Ceram Soc 82:2963
61. Lemons JE (1988) Clin Orthop 235:220
62. Kyeck S, Remer P (1999) Mater Sci Forum 308–311:308
63. Ong JL et al (1991) J Am Ceram Soc 74:2301
64. Dasarathy H et al (1996) J Biomed Mater Res 31:81
65. Yamashita K et al (1996) J Am Ceram Soc 79:3313
66. Hero H et al (1994) J Biomed Mater Res 28:343
67. de Bruijn JD et al (1992) J Biomed Mater Res 26:1365
68. Hench LL, Paschall HA (1973) J Biomed Mater Res Symp 7(3):25
69. Ban S et al (1994) J Biomed Mater Res 28:65
70. Guipont V, Espanol M, Borit F, Llorca-Isern N, Jeandin M, Khor KA, Cheang P (2002) Mater Sci Eng A 325:9
71. Chang E, Chang WJ, Wang C, Yang CY (1997) J Mater Sci Mater Med 8:193
72. Gaona M, Fernandez J, Guilemany JM (2006) In: Proceedings of international thermal spray conference, 15–17 May, Seattle, WA
73. Ning CY, Wang YJ, Lu WW, Qiu QX, Lam RWM, Chen XF, Chiu KY, Ye JD, Wu G, Wu ZH, Chow SP (2006) J Mater Sci Mater Med 17:875
74. Lamy D et al (1996) J Mater Res 11:680
75. Liu X, Ding C (2002) Biomaterials 23:4065
76. Fu Y, Batchelor AW, Khor KA (1999) Wear 230:98
77. Diomidis N, Mischler S, More NS, Roy M (2012) Acta Biomater 8:852
78. More N, Diomidis N, Paul SN, Roy M, Mischler S (2011) Mater Sci Eng C 31:400
79. Sahu S, Palaniappa M, Paul SN, Roy M (2010) Mater Lett 64:12
80. Keicher DM, Smugeresky JE (1997) J Met 49(5):51
81. Griffith M, Schlienger ME, Harwell LD, Oliver MS, Baldwin MD, Ensz MT, Esien M, Brooks J, Robino CE, Smugeresky JE, Hofmeister W, Wert MJ, Nelson DV (1999) Mater Des 20(2–3):107
82. Long M, Rack HJ (2005) Mater Sci Eng C 25(3):382
83. Long M, Rack HJ (2001) Wear 249:158
84. Samuel S, Nag S, Scharf T, Banerjee R (2007) Mater Sci Eng C 28(3):414
85. Erdemir A, Fenske GR, Krauss AR, Gruen DM, McCauley T, Csencsits RT (1999) Surf Coat Technol 120–121:565
86. Roy M, Steinmuller-Nethl D, Tomala A, Tomastik C, Koch T, Pauschitz A (2011) Diam Relat Mater 20:573
87. Papo MJ, Catledge SA, Vohra YK (2004) J Mater Sci Mater Med 15:773

88. Fries MD, Vohra YK (2002) J Phys D Appl Phys 35:L105
89. Met C, Vandenbulcke L, Sainte Catherine MC (2003) Wear 255:1022
90. Amaral M, Abreu CS, Oliveira FJ, Gomes JR, Silva RF (2007) Diam Relat Mater 16:790
91. Amaral M, Abreu CS, Oliveira FJ, Gomes JR, Silva RF (2008) Diam Relat Mater 17:848
92. Shenhar A, Gotman I, Gutmanas EY, Ducheyne P (1999) Mater Sci Eng A268:40
93. Rie K-T, Lampe TH (1985) Mater Sci Eng 69:473
94. Manso M et al (2002) Biomaterials 23:349
95. Long M, Rack HJ (1998) Biomaterials 19:1621
96. Pouilleau J, Devillers D, Garrido F, Durand-Vidal S, Mahe E (1997) Mater Sci Eng B 47:235
97. Lin C-M, Yen S-K (2004) J Electrochem Soc 151(12):D127–D133
98. Bendavid A, Martin PJ, Takikawa Thin H (2000) Solid Films 360:241
99. Amor SB, Baud G, Besse JP, Jacquet M (1997) Mater Sci Eng B 47:110
100. Mandl S, Thorwarth G, Schreck M, Stritzker B, Rauschenbach B (2000) Surf Coat Technol 125:84
101. Liqiang J, Xiaojum S, Weimin C, Zili X, Yaoguo D, Honggang F (2003) J Phys Chem Solids 64:615
102. Natarajan C, Nogami G (1996) J Electrochem Soc 143:1547
103. Li P, Kangasniemi I, De Groo K (1993) Bioceramics 6:41
104. Huang N, Yang P, Chen X (1998) Biomaterials 19:771
105. Liu X, Chu PK, Ding C (2004) Mater Sci Eng R47:49
106. Carinci F, Pezzetti F, Volinia S, Francioso F, Arcelli D, Farina E (2004) Biomaterials 25 (2):215
107. Ferraris M, Verné E, Appendino P, Moisescu C, Krajewski A, Ravaglioli A, Piancastelli A (2000) Biomaterials 21:765
108. Piconi C, Maccauro G (1999) Biomaterials 20(1):1–25
109. Akagawa Y, Ichikawa Y, Nikai H, Tsuru H (1993) J Prosthet Dent 69:599–604
110. Rosengren A, Pavlovic E, Oscarsson S, Krajewski A, Ravaglioli A, Pincastelli A (2002) Biomaterials 23(4):1237
111. Hulbert SF, Morrison SJ, Klavitter JJ, Biomed J (1972) Mater Res 6:347–374
112. Covacci V, Bruzzese N, Maccauro G, Andreassi C, Ricci GA, Piconi C (1999) Biomaterials 20(4):371–376
113. Kim BK, Bae HE, Shim JS, Lee KW (2005) J Prosthet Dent 4:357
114. Akagawa Y, Hosokawa R, Sato Y, Kameyama K (1998) J Prosthet Dent 80:551–558
115. Lappalainen R, Anttila A, Heinonen H (1998) Clin Orthop Relat Res 352:118–127
116. Kokubo T (1990) J Non Cryst Solids 120:138
117. Cui FZ, Li DJ (2000) Surf Coat Technol 131:481–487
118. Quiros C, Nunez R, Priet P et al (1999) Vacuum 52:199
119. Mcaughlin JA, Maguire PD (2008) Diam Relat Mater 17:873
120. Casiraghi C, Ferrari AC, Robertson J (2005) Phys Rev B 72:085401
121. Ferrari AC, Robertson J (2004) Philos Trans R Soc Lond A 362:2477
122. Ronkainen H, Koskinen J, Anttila A, Holmburg K, Hirvinen JP (1992) Diam Relat Mater 1:639
123. Ferrari AC, Robertson J (2000) Phys Rev B 61:14095
124. Ferrari AC, Robertson J (2001) Phys Rev B 64:075414
125. Tiainen VM (2001) Diam Relat Mater 10:153
126. Lappalainen R, Heinonen H, Anttila A, Santavirta S (1998) Diam Relat Mater 7:482
127. Onate JI, Comin M, Braceras I, Garcia A, Viviente JL, Brizuela M et al (2001) Surf Coat Technol 142–144:1056
128. Dong H, Shi W, Bell T (1999) Wear 225–229:146
129. Sheeja D, Tay BK, Shi X, Lau SP, Daniel C, Krishnan SM (2001) Diam Relat Mater 10:1043
130. Dowling DP, Kola PV, Donelly K, Kelly TC, Brumitt K, Lloyd L et al (1997) Diam Relat Mater 6:390
131. Sheeja D, Tay BK, Lau SP, Nung LN (2001) Surf Coat Technol 146–147:410

132. Saikko V, Ahlroos T, Calonius O, Keraen J (2001) J Biomater 22:1507
133. Affatato S, Frigo M, Toni A (2000) J Biomed Mater Res 53:221
134. Ahlroos T, Saikko V (1997) Wear 211:113
135. Liao YS, McNulty D, Hanes M (2003) Wear 255:1051
136. Saikko V, Ahlroos T (1997) Wear 207:86
137. Fisher J, Firkins P, Reeves EA, Hailey JL, Isaac GH (1995) Proc Inst Mech Eng H J Eng Med 209:263
138. Ronkainen H, Varjus S, Holmberg K (2001) Wear 249:267
139. Platon F, Fournier P, Rouxel S (2001) Wear 250:227
140. Shi B, Ajayi OO, Fenske G, Erdemir A, Liang H (2003) Wear 255:1015
141. Kim SK, Lee JB, Koak JY, Heo SJ, Lee KR, Cho LR, Lee SS (2005) J Oral Rehabil 32(5):346
142. Lappalainen R, Selenius M, Anttila A, Konttinen YT, Santavirta SS (2003) J Biomed Mater Res B Appl Biomater 66B:410
143. Ronkainen H, Likonen J, Koskinen J, Varjus S (1996) Surf Coat Technol 79:87
144. Neerinck D, Persoone P, Sercu M, Goel A, Venkatraman C, Kester D et al (1998) Thin Solid Films 317:402
145. Taeger G, Podleska LE, Schmidt B, Ziegler M, Nast-Kolb D (2003) Mat-wiss u Werkstofftech 34:1094
146. Chandra L, Allen R, Butter M, Rushton N, Lettington AH, Clyne TW (1995) J Mater Sci Mater Med 6:581
147. Muller U, Hauert R, Oral B, Tobler M (1995) Surf Coat Technol 71:233
148. Puértolas JA, Martínez-Nogués V, Martínez-Morlanes MJ, Mariscal MD, Medel FJ, López-Santos C, Yuberoc F (2010) Wear 269:458
149. Kvasnica S, Schalko J, Benardi J, Eisenmenger-Sittner C, Pauschitz A, Roy M (2006) Diam Relat Mater 15:1743
150. Ali N, Kousar Y, Okpalugo TI, Singh V, Pease M, Ogwu AA, Gracio J, Titus E, Meletis EI, Jackson MJ (2006) Thin Solid Films 515:59
151. Grischke M, Bewilogua K, Trojan K, Demigen H (1995) Surf Coat Technol 74–75:739
152. Memming R (1986) Thin Solid Films 143:279
153. Choe HC, Chung CH, Brantley W (2007) Key Eng Mater 345–346:1201
154. Jung SW, Son MK, Chung CH, Kim HJ (2009) J Adv Prosthodont 1(2):102
155. Gispert MP, Serro AP, Colaco R, Botelho do Rego AM, Alves E, da Silva RC, Brogueira P, Pires E, Saramago B (2007) Wear 262:1337
156. Wang L, Su JF, Nie X (2010) Surf Coat Technol 205:1599
157. Serro AP, Completo C, Colaço R, dos Santos F, Lobato da Silva C, Cabral JMS, Araújo H, Pires E, Saramago B (2009) Surf Coat Technol 203:3701
158. Hoseini M, Jedenmalm A, Boldizar A (2008) Wear 264:958
159. Osterle W, Klaffke D, Griepentrog M, Gross U, Kranz I, Knabe C (2008) Wear 264:505
160. Gulhane UD, Roy M, Sapate SG, Mishra SB, Mishra PK (2009) In: Proceedings of ASME/STLE International Joint Tribology Conference, IJTC 2009, October 19–21, Memphis, TN
161. Roy M, Koch T, Pauschitz A (2010) Adv Surf Sci 256:6850
162. Unsworth A (1991) J Eng Med 205:163
163. Zhu X, Eibl O, Scheider L, Geis-Gerstofer J (2006) J Biomed Mater Res 79A:114
164. Liao SS, Cui FZ, Zhang W, Feng QL (2004) J Biomed Mater Res B Appl Biomatter 69(2):158
165. Pezzatini S, Solito R, Morbidelli L et al (2006) J Biomed Mater Res 76A:656
166. Rajtar A, Kaluza GL, Yang Q et al (2006) EuroIntervention 2:113
167. Campbell AA (2003) Mater Today 11:26
168. Balani K, Anderson R, Lahaa T et al (2007) Biomaterials 28:618

Index

M. Roy (ed.), *Surface Engineering for Enhanced Performance against Wear*, DOI 10.1007/978-3-7091-0101-8, © Springer-Verlag Wien 2013

O

P

U

Y

Z

MIX
Papier aus verantwortungsvollen Quellen
Paper from responsible sources
FSC® C105338

If you have any concerns about our products,
you can contact us on
ProductSafety@springernature.com

In case Publisher is established outside the EU,
the EU authorized representative is:
Springer Nature Customer Service Center GmbH
Europaplatz 3, 69115 Heidelberg, Germany

Printed by Libri Plureos GmbH
in Hamburg, Germany